Science Policy under Thatcher

Science Policy under Thatcher

Jon Agar

UCL PRESS

First published in 2019 by
UCL Press
University College London
Gower Street
London WC1E 6BT

Available to download free: www.uclpress.co.uk

A CIP catalogue record for this book is available from The British Library.

ISBN: 978-1-78735-343-5 (Hbk)
ISBN: 978-1-78735-342-8 (Pbk)
ISBN: 978-1-78735-341-1 (PDF)
ISBN: 978-1-78735-344-2 (epub)
ISBN: 978-1-78735-345-9 (mobi)
ISBN: 978-1-78735-346-6 (html)
DOI: https://doi.org/10.14324/111.9781787353411

For Kathryn, Hal and Max, and my parents Ann and Nigel Agar.

Contents

List of tables

Acknowledgements

I have benefited enormously from discussions and comments from many people. I would like to thank Duncan Wilson, Michael Huber, Stuart Butler, Jacob Ward, Emmeline Ledgerwood; Stephen Davies, Mat Paskins, John Langrish, Jack Stilgoe, Steve Miller, Tiago Mata, Dimitriy Myelnikov, Brian Balmer, Melanie Smallman, Simon Werrett, Jon Turney, Dominic Berry, Simone Turchetti, Carsten Timmermann, Ian Miles, Claire Craig, Kieron Flanagan, David Aubin, Graeme Reid, Joe Cain, Frank James, Norma Morris and Kathryn Packer. I would also like to thank Efstathios Arapostathis, Stephen Pollock, Tatjana Buklijas, Marina J. S. Schütz, Solveig Crompton, Peter Reed and Bex Coates for their interest.

Thanks to the audiences who have engaged talks on my research for this book, including Histoire des sciences, histoire de l'innovation, Paris, 2016; the Royal Society Wilkins-Bernal-Medawar Prize Lecture 2017; BSHS Annual Conference, York, 2017; Fred Jevons Science Policy Lecture, Manchester Institute of Innovation Research 2017; Thatcherism Now conference, Liverpool, 2018; and the Nucleus archive centre, Wick, 2018. I am especially grateful to Peter Collins and UCL Press's anonymous reader for critical comments on the whole manuscript. Thanks also to Catherine Bradley for editing the text, and Sarah M. Hall for proofreading it.

I would like to thank all the staff and students of the Department of Science and Technology Studies, UCL for their support.

I am very grateful to Rachel Hall and colleagues at the Campaign for Science and Engineering (CaSE) for welcoming me into their archives. As ever, I have found the National Archives at Kew to be a friendly, calm and efficient place to study.

Finally, thanks to Chris Penfold, Alison Fox, Pauline Hubner and all at UCL Press.

1 Introduction

It was the winter of late 1985, halfway through the decade of British politics dominated by a science-trained Prime Minister, and Britain's scientists were in open revolt. In October a small group of academics had met and shared stories of low morale, declining science funding and rumours of a new brain drain of disaffected talent. Quietly, they had begun canvassing fellow scientists, seeking modest sums to support the placing of a half-page advertisement in *The Times*, aiming to draw attention to their collective plight. 'Within a few days of the first letters going out subscriptions began to flow in, and by Christmas £12,000 had been donated by over 1500 supporters from across the scientific spectrum … The initiators, having tentatively probed for signs of support, were overwhelmed by a gusher'.[1]

The advert, which duly appeared on page 5 of *The Times* on the morning of Monday 13 January 1986, announced the existence and aims of a new, urgent, campaigning organisation, Save British Science:

> Basic science has given us radio and television, plastics, computers, penicillin, X-rays, transistors and microchips, lasers, nuclear power, body-scanners, the genetic code, … All modern technology is based on discoveries made by scientists seeking an understanding of how the world works, what it is made of and what forces shape its behaviour. …
>
> Today's basic research enlarges our conceptions of the world and our place in it and underlies tomorrow's technologies, the basis of future prosperity and employment.
>
> Yet British science is in crisis: opportunities are missed, scientists emigrate, whole areas of research are in jeopardy. The Government's support for research is declining, falling further behind that of our main industrial competitors in Europe whose policy is to increase investment in scientific research.

> There is no excuse: rescue requires a rise in expenditure of only about one percent of the Government's annual revenue from North Sea oil. We can and must afford basic research, Britain's investment for the future.[2]

Supporters were urged to contact the Save British Science campaign, via a PO Box in Oxford, and to 'ask your member of Parliament to help save British science before it is too late'.

Several observations can be made about the abrupt revolt of the scientists expressed by Save British Science. First, the mood was one of widespread despair tending to existential crisis, of historically low morale and of suffering in silence.[3] 'The state of morale among British scientists is at its lowest ebb in living memory,' Tony Watts, a biochemist who had been lured back from West Germany to a lectureship at Oxford, told a BBC reporter. 'In other countries, scientists feel they're doing a job that someone wants them to do. Here, we're fighting for survival.'[4] The 'senseless mutilation of the best elements of the research base' was 'totally demoralizing and is making us the laughing stock of our scientific and industrial competitors', according to Colin Blakemore, the neurobiologist and epitome of fast-rising, British scientific talent. In the same letter to his MP, he added that for 'the first time in my life, I am now thinking seriously of leaving this country'.[5] 'The real significance of the extraordinary explosion of support for the advertisement,' argued John Mulvey, the Oxford scientist who co-founded Save British Science, 'is the measure it gives of the intensity of frustration, the depth of the concern so widely felt about the damage being done to scientific research – and the teaching of science – in this country'.

Second, the feeling was that the old political settlement of science was failing. Funding through the research councils had been kept level, but the funding going to universities to maintain these institutions in a condition able to conduct research had been cut. Even the leading figure of the research councils, Sir David Phillips, an able friend of science, had told scientists that the best they could hope for was a steady state: 'What scientists are rebelling against is the realisation that to meet the challenges of the future, we have to ask how we can raise the money by stopping something'.[6] While a few leaders among the science community had spoken out – for example, Sir Hans Kornberg, as President of the British Association for the Advancement of Science had used his presidential address of 1985 to complain of the cuts – the rank and file of Save British Science strongly felt that their needs were not being defended in political debate. This silence was seen as having deep, cultural roots:

scientists, to preserve their autonomy and independence, should refrain from political intervention; politicians should likewise desire an autonomous and independent science and preserve it by ensuring it was financially secure. Nevertheless, as Save British Science told a Parliamentary select committee in 1986: 'Ministers have stated that they were surprised by the "silence" of the scientific community. We fear that … while the damage to science research in the UK has become alarming to those in a position to understand it, the Government has been misled by the traditional hesitation to speak out'.[7]

Third, attention should be paid to the relationship assumed by these academic scientists between 'basic' science and industry. The technologies of the modern world, and the economic prosperity they generated, were founded on prior basic science. Cut basic science, therefore, and future prosperity would be undermined. The 'great damage that is being done … to the research base of this country by the Government's funding policy', wrote Blakemore, was 'a national disaster that has unthinkable implications … for Britain's industrial performance in the coming decades'.[8] On one hand, this outburst reflected an understandable self-interest of an academic scientist, who was more likely to be conducting basic research. But, on the other, there was also what science policy commentators label the 'linear model' at work here: basic science leads to applied science which leads to technological innovation and wealth generation.

The linear model was not new in the 1980s and its naivety as a model of innovation was well known.[9] Yet it survived then, as it does now, because it does political work. 'Basic' science was a clever and effective conceptual invention. As the 'base' it was the foundation for later applications. But since those applications might be unpredictable and far in the future, 'basic' science should be supported and protected from demands for immediate practical application and relevance. Base talk was therefore also a way of delineating and protecting the independence and autonomy of science. However, the lack of a sophisticated way of analysing the contribution of science to technological change and industrial performance meant that science policy in the 1980s was open to a far more ideological attack – one in which the supposed national characteristics of strength in basic research and weakness in application could be mobilised with effect. The relationship between science – 'basic', 'applied', 'strategic', 'near-market' or 'curiosity-driven' – and innovation and industry became the focus, as I will show, of the central debate in science policy in the 1980s.

Finally, who was responsible for this 'crisis'?[10] Save British Science's view, in internal deliberations on the targets of their complaints, was clear: 'We feel that the arguments should be directed principally at the

Government'.[11] The cuts in university funding were a consequence of the much wider programme of reduction of public spending, a central policy aim, alongside privatisation, of Margaret Thatcher's radical Conservative administration. This reduction was itself part of a Thatcherite package that included anti-Keynesian economic theory and a set of political values that privileged the individual over the collective. At first glance science policy might seem a minor, subsidiary, technical part of this programme. But since it contained within it claims about the ultimate source of wealth, the importance of science policy was, I argue, far greater than its relative invisibility might suggest.

An unusually long editorial was published on the subject of 'research and prosperity' in *The Times* on 19 February 1987. It drew attention to a House of Lords select committee report that had made the 'case that the future of Britain demands an attention to scientific enquiry that is manifestly absent in Britain today'.[12] After a familiar set of arguments – Britain's political class knew more of the classical world than it did of Silicon Valley, Britain failed to turn ideas into products, Britain was falling behind international competitors as measured by the funding of research and development, talented scientists were moving abroad – the editors made clear where responsibility for this lay:

> To see the demands of the future, and, by seeing, meet them, is a prime duty of government. Britain has to change the way it sees science. Its government has to help create that new vision. It has to act upon it without rancour over past failures and without dogmas of past success in other fields. The Prime Minister has a unique opportunity to set the agenda for the future as she has changed the agenda of the past. She has a unique responsibility too.[13]

Margaret Thatcher was seen as having such a 'unique responsibility' for two reasons. First, as Prime Minister, she possessed the straightforward constitutional power to lead on such an important issue, and as a Conservative leader her conceptions of science policy must be seen in the light of the history of Conservative thinking on science. Science, when remembered as a concern of party politics, is usually – and, if taken on its own, misleadingly – associated with the left. The rhetorical power of Harold Wilson's 'white heat of technology' speech at the Labour Party conference of 1963, which he later claimed had 'replaced the cloth cap with the white laboratory coat as the symbol of British labour',[14] had a major impact; its association has retained a powerful grip on popular historiography of science and British politics.[15]

Wilson had used science successfully to rebrand Labour in a modern image.[16] His speech cemented an association that can be traced from radical, left-wing scientists in the 1930s[17] through the mobilisation of scientists during the Second World War as part of a planned war economy and that continued in the 1950s as influential socialist scientists, such as the physicist Patrick Blackett, advised Labour behind the scenes. Yet the rebranding served to mask considerable continuity. Administrations under Clement Attlee, Winston Churchill and Anthony Eden had invested in Cold War science,[18] not least the nuclear projects. Harold Macmillan's Conservative government had been the first to appoint a Minister for Science, in 1959, while Macmillan and Alex Douglas-Home, whatever their fusty image might suggest, 'encouraged new, large scale projects of scientific and technological modernity': the Post Office Tower, Concorde and the modernisation of British Rail were all begun by Conservatives before Wilson's 'white heat'.[19] Indeed, to restrict historical attention only to left-wing technocracy would, according to David Edgerton, be 'to miss most of the history of British technocracy, and most of the history of British science and technology policy'.[20]

The second reason Margaret Thatcher might have been said to have had a 'unique responsibility' to lead the harnessing of science policy to a 'new vision' of the future of the nation was that she had been a scientist herself. She had been trained in Chemistry at Oxford in the 1940s and had worked as an industrial chemist at the companies British Xylonite and Lyons & Co. until the early 1950s. She actively maintained her interest in science as Prime Minister, and it provided a point of contrast with the officials and ministers around her. As her last chief scientist recalled:

> Few senior civil servants understand science, think science, or promote science; they would prefer their ministers not to be distracted by it. Prime Minister Margaret Thatcher was exceptional. She wanted to understand and engage in science. She was interested in science as a subject, listened to scientific reasoning, was happy to talk about science and enjoyed it.[21]

Furthermore, this contrasting interest in science was seen by commentators as implying the possession of different analytical skills. 'Mrs Thatcher is unusual in being a Prime Minister with a science degree,' observed a *Financial Times* journalist, noting that 'it would be surprising if that did not influence her thinking'.[22] Her supervisor and mentor at Oxford, the outstanding x-ray crystallographer Dorothy Hodgkin, thought her training equipped her to 'to see what the scientists are doing', while one of

Thatcher's biographers concluded that it gave her a 'blueprint for the practical mind' – a 'rare capacity to understand the scientific mind at all'.[23] Many similar quotations could be given. They all assume that her scientific training and working experience made a difference to how she thought or what she valued.

In a mid-1980s opinion poll, commissioned by the BBC, 67 per cent thought that 'politicians don't know enough about science to judge its importance'.[24] This view was not new, and can be seen as the product of a century-long lobbying campaign. The historian Frank Turner notes that 'Scientists, like other groups of intellectuals who during the 1860s had hoped to participate broadly in public life, found themselves able to exert relatively little direct power or influence in the civic arena'. Therefore after the 1870s the group Turner calls 'public scientists' became increasingly critical of 'politicians and complacent manufacturers', now seen as enemies of the 'progress and application of scientific knowledge'.[25] The public scientists now attacked the political system, in which party politics rather than science guided policy. They promoted science education, as means of instilling the desired virtues of truthfulness and endurance in citizens (and eventually politicians), and eugenics, as a means by which science could deliver 'direct civic benefits to the nation-state'.[26] An alliance between public scientists, pre-eminently Norman Lockyer, the editor of *Nature*, and sympathetic social imperialist politicians, was institutionalised in the British Science Guild in 1904. The Guild lobbied hard for science. In particular, it viewed science as a solution to political problems:

> [The purpose of the British Science Guild] is to stimulate, not so much the acquisition of scientific knowledge, as the appreciation of its value, and the advantage of employing the methods of scientific inquiry, the study of cause and effect, in affairs of every kind. Such methods are not less applicable to the *problems* which confront the statesman, the official, the merchant, the manufacturer, the soldier, and the schoolmaster, than those of the chemist or the biologist; and the value of a scientific education lies in the cultivation which it gives of the power to grasp and apply the principles of investigation employed in the laboratory to the *problems* which modern life presents in peace or war.[27]

'Edwardian public science centred on [the] technocratic BSG/Nature axis,' argues the historian Andrew Hull. He agrees with Turner, noting that such an axis 'continually pressed for executive influence over government policy for scientists, arguing both that scientific method was transferable to social problems and that science was the key component in national power in a modern state faced with constant economic competition which might at any time become war'.[28] From the turn of the nineteenth century, through the First World War and beyond, the public scientists continued to argue publicly that 'politicians were ignorant of scientific matters', and that the Government neglected to support or use science properly.[29]

Fast forward to the 2010s and we find a very similar set of complaints. For example, Mark Henderson, Head of Communications at the Wellcome Trust, asks why is it that politicians do not grasp that increased public funding for science would more than repay its costs in economic benefits? Why do politicians misunderstand, misuse or disrespect empirical evidence? Why can they not learn from scientific values and methods? 'The answer lies chiefly in the wider failings of the political classes' understanding and experience of science,' Henderson states, adding 'only one of the 650 MPs in the UK's House of Commons was a scientist in his previous career … There is a lack of familiarity with the practice of science, of what it needs to succeed, which blinds politicians to the consequences that their funding decisions will have'.[30]

There has been, therefore, a long-standing view that politics would be better if more politicians were scientists. So what happened when the leading politician of her generation, a powerful Prime Minister who served 11 years, with commanding authority, was indeed a scientist? One immediate point to make is that Margaret Thatcher's understanding of science was not as deep, nor her embrace of scientific rationality as warm, as at least two of her Conservative predecessors. Arthur J. Balfour, Prime Minister between 1902 and 1905, was a philosopher of science before he was a politician. Thatcher's predecessor as leader of the Conservative Party, Edward Heath, acted on his conviction that technocratic rationality could be built into government through innovations such as the Central Policy Review Staff, the 'think-tank' that was home to the chief scientific adviser. William Waldegrave, future science minister, recalled Heath should correctly be labelled 'Technocratic, I mean, exactly. He greatly respected the French Civil Service, the French technocratic elite, he greatly respected them, we saw it even in those days. I went with him to China in '74 and there were Chinese technocrats even then. Ted's view of how the world should be run really was by rational, non-ideological people'.[31]

The contrast with Margaret Thatcher could not be greater. Edgerton and Hughes have argued that 'what is distinctive about Mrs Thatcher is not that she is a scientist, but rather that she is the first anti-technocratic Prime Minister Britain has had in the [twentieth] century'.[32] By 'anti-technocratic' they meant that Thatcher rejected the views that science and technological change were determinants of economic growth and development, and that an interventionist state was necessary to deliver them. Thatcher's overall political aim was instead to free private enterprise, reduce state intervention and cut public expenditure, 'and to shape what is left to serve industry directly'.

Often the antithesis of the technocrat – but also the populist – is taken to be the 'conviction politician'. Thatcher's status as the pre-eminent conviction politician was the keynote of obituarists on her death in 2013.[33] But another antonym to the technocrat is the narrative politician – the politician whose tools of persuasion are the stories that can be told. Moral stories, as we shall see, were remarkably powerful in how Thatcher intervened in science policy. The story she heard on visiting Cambridge in 1980 about the apparent failure to patent monoclonal antibodies (as discussed in Chapter 3) quickly became the font of her furious analysis of what she saw as wrong in the relationship of science, industry and nation. In the critical years of the mid-1980s, just as Save British Science was telling its own public story of plummeting morale and brain drains, the heart of government science policy-making was the location for a private battle between two different visions of how to reconnect science to industry for national prosperity. One side – loosely the research councils, the Department of Trade and Industry and the chief scientific adviser – saw the route to persuasion in figures and data, and lost; the other – located in the Number 10 Policy Unit – deployed stories, anecdotes with lessons from historical and contemporary science, and won. I trace this central debate in Chapter 3. However, the contrast between the technocratic versus the story-led form of policy-making should not be turned too high.

In this book I will ask whether it mattered that Thatcher had been a scientist. In some ways it clearly did. Science was, albeit intermittently, an important part of her public image. In 1951 she had dressed in a white coat and been photographed in front of scientific apparatus, instruments of her then working life, for electioneering publicity purposes. In the 1970s, when she was a minister responsible for science and the new, untested Party leader, journalists recalled her science training. In 1979, within days of taking office, Thatcher reserved the political right to speak on science matters. At Number 10 Downing Street she installed a bust of Michael Faraday,[34] icon of applied science, and a portrait of Isaac Newton – a 'great star for her', recalled a Chief Scientific Adviser; the works symbolised both unparalleled scientific achievement and her Lincolnshire origins.[35] In the late 1980s, at the peak of her prestige, Thatcher made science integral to her leadership in speeches to both national and international audiences. Of particular note were her 1988 speech to the Royal Society, in which she confirmed a new science and innovation policy and highlighted global environmental challenges, and her 1989 speech to the General Assembly of the United Nations, emphasising the need for action on the ozone hole and climate change.

Perhaps most significantly, a reason it mattered that one of the pre-eminent world leaders of the twentieth century, one who was at the peak of her powers in the 1980s, possessed not only scientific training but also a working knowledge of applied science, was that so many issues of the decade were saturated with science. The 1980s was the decade of acid rain, AIDS, the imminent threat of nuclear apocalypse, the discovery of the ozone hole, the first release of genetically modified organisms into the environment, the first cases of 'Mad Cow' disease, personal computers and Chernobyl. Any British prime minister would have had to confront such issues, and formulate and decide policies, in these areas where knowledge of science was critical. But we can add to this picture a prime minister who pursued a radical conservative agenda of cutting public expenditure and privatisation, who challenged union power and who fought and won a military and naval campaign halfway round the world in the South Atlantic. How did science and science policy fit with these ambitions?

What I do in this book is use a wealth of new primary sources to explore science policy under Margaret Thatcher. By 'science policy' I mean not only 'science-for-policy', which is how science informed decisions on a range of issues, but also 'policy-for-science': how decisions were taken about the ways in which science should funded, managed and deployed.[36] The new primary sources are predominantly the documents relating to science policy produced by central government – especially within Number 10 Downing Street, but also encompassing a diverse range of departments and ministries – that have been released at the National Archives. For the past few years the National Archives has been on overdrive as the United Kingdom moved from a 30-year-rule to a 20-year-rule guiding the release of public records. This shift in the archival horizon has meant that nearly all governmental records of the 1980s have been released (although, of course, a significant minority are still retained for a variety of reasons, not least national security) and historians can already examine records of John Major's premiership. With these new sources the time was right to reassess the government's actions in the 1980s.

One of the opportunities invited by the availability of these documents is to contrast what we, as analysts of science policy, knew (or thought we knew) from the 1980s public record with what we can now trace actually happened, in detail, by whom, when and why, from the previously closed record. There are surprises – for example, the extent to which the consequences for science of the cuts in public spending were not anticipated – and, I think more significant, the identification of precisely who persuaded Margaret Thatcher to reshape science policy in her final years as Prime Minister. I have talked to journalists who covered science policy in the 1980s and to academic science policy experts whose careers spanned the 1980s to the present, and neither knew the identity of this agent.[37] Historical documents have the considerable advantage of revealing individuals and their roles within institutions in ways not as easily revealed through investigations at the time. Nevertheless, it is also true that historical documents, not least those written by senior civil servants, celebrated masters of word-craft, have to be read carefully, critically and in context. Even then the documents do not capture the informal, off-the-record discussions nor the conversations in the stairwell. Some of this culture can be, and has been, captured through oral history, although such methods also have their blind spots. Ultimately primary source documents are the securest foundation for historical interpretation.

In Chapter 2 I sketch the landscape of science policy-making in the United Kingdom, providing a 'who's who' of the people, roles and bodies that together shaped the direction of travel. I trace Margaret Thatcher's life from chemistry graduate to industrial scientist and then through her political career as a Member of Parliament, Minister and Prime Minister. I survey the government departments of Whitehall, some of which were important because of the funds they channelled towards science, others because of their powerful influence on how decisions were taken or how money could be spent. I introduce the committees and other structures through which advice on science was given. In particular, I show how the role of an individual, the chief scientific adviser, has changed over time. Chief scientific advisers are crucial figures in the chapters that follow. However, some of the most important decisions taken under Thatcher on science policy were influenced more by political than scientific advice (although the boundary was never sharp). The Number 10 Policy Unit is a particularly important source of this political wisdom. Finally I also introduce other bodies, including Parliamentary ones as well as those of civil society – notably the Royal Society and campaigning groups such as Save British Science – which sought to intervene to change policy.

In Chapter 3 I present my main findings on the causes, topics and protagonists of the central debates on science and innovation. In particular, I show that gathering tensions around a set of complex and interconnecting issues – the effects of cuts in public expenditure on university science, appropriate policies for government support of research-intensive industry, entrepreneurialism in academia and industry, the preponderance of defence research and the UK's involvement in CERN – were indeed resolved (if not solved) by a decisive 1987 reversal in science and industrial policy. In this sense, as I discuss in detail in Chapter 8, all of the commentators – Edgerton and Hughes,[38] Wilkie,[39] Williams,[40] Christie,[41] Wilks and Cini,[42] von Tunzelmann[43] – are right to say something happened in science policy, that there had been a 'radical change' and a 'great debate'. However, their details and explanations go awry. Cuts were part of the mix, but by no means the main factor (contra Williams and Christie). The new policy sought to end an industrial strategy of government funding of 'near-market' research, but simultaneously (contra Wilkie) celebrated, rather than curtailed, 'curiosity-driven research'. Some businesses – notably big pharmaceutical concerns – would thrive in the post-near-market-funding world; others, notably GEC, would not, and this was anticipated by advisers.[44] Thatcherite policy was not (contra Edgerton and Hughes) a paradoxical yet coherent mix of the centralising and the liberating; rather these were expressions of different and conflicting protagonists within the science policy system. Specifically, I demonstrate that the Number 10 Policy Unit (and the adviser George Guise in particular) and Thatcher came to have very different views from those expressed by the institutional experts on science policy. I also show that the reversal happened when her Chief Scientific Adviser was sidelined, and also, although the argument has to be counterfactual, after Michael Heseltine, a potential counterweight in favour of an industrial strategy, had resigned.

While it is clear from the primary source documents that this reversal of science policy and industrial strategy was indeed the 'great debate', an observer at the end of the 1980s would probably have listed the challenge of the devastating and novel disease of AIDS and the controversy over embryos as the two biggest science-related public issues. In a Coda to Chapter 3 I trace how central government, especially Margaret Thatcher, responded to these challenges. I will show that the ways in which they were deliberated and handled were exceptional rather than typical of science policy matters under Thatcher.

The next three chapters examine aspects of Thatcher's nuclear policies during the 1980s, a decade that, surprisingly, was the last of the Cold War. Chapter 4, as its title 'Power/leaks' suggests, has a divided structure.

In the first half I follow the Prime Minister's visits to UK nuclear sites, including the fast-breeder reactors at Dounreay in North Scotland and the site in Cumbria, first called Windscale and later Sellafield, which contained the first nuclear power station, Calder Hall. This was built in the 1950s, and by the 1980s had become a vast complex mostly devoted to reprocessing and storage. Thatcher was avowedly pro-nuclear, and these visits reveal a determination to present the UK nuclear project as necessary, modern and safe.

The key words here are 'power' and 'containment'. Both political image and the technical achievement of making, using and reprocessing nuclear fuel needed to be carefully packaged and managed to be powerful: the lustre of one sought to bolster the fortunes of the other. However, as I show in the second half of the chapter, both political image and technological system, could be – and were – challenged and undermined, not least by losses of control, or 'leaks'. High politics and nuclear power are both tightly coupled systems, in which the aim of complete containment was unattainable. I examine the Thatcher government's response to incidents such as the pollution of beaches near Sellafield in 1983 and the Chernobyl accident of 1986, as well as issues such as leukaemia clusters, the challenges to the search for an underground nuclear waste storage site and the Sizewell B inquiry into the next generation of nuclear power stations.

Chapter 5 investigates how Thatcher's nuclear plans intersected with the flagship policy aim of privatisation. I made a conscious decision to focus research fairly narrowly for this chapter. Privatisation, in general, was an immense and complex undertaking, affecting swathes of the UK industrial, public and financial sectors. Like all large, modern, technical enterprises, the nationalised industries needed and used science in many ways, and it was not possible for me to research all of these topics. The history of science stories of the privatisation of industries, such as aerospace and telecommunications, will be told by others. Chapter 5, therefore, selects an extreme but important sector, one in which the role of the state was extraordinarily dominant, and for which postwar investment in science was paramount: the nuclear.

I start with the story of the privatisation of the state-owned Radiochemical Centre as it became the private company Amersham International. While small, this project was one of the very first, if not *the* first, privatisation under Thatcher, and was therefore an experiment of sorts. Next I survey the faltering steps towards privatisation of the bulk of the civil nuclear project: the United Kingdom Atomic Energy Authority (UKAEA), British Nuclear Fuels Ltd and the nuclear power stations.

Thatcher and her advisers were convinced that only full exposure to the market would make nuclear power economic. Yet the irony was that only in preparation for privatisation were the costs of nuclear power made public – after which, having seen the numbers, few private investors were tempted. I also suggest there may be a connection between the rejection of government funding of near-market research (in other words the argument of Chapter 3) and the unwillingness of Thatcher's government to continue to support industrial nuclear projects.

Chapter 6 looks at the military side of the nuclear, but is also the result of a decision to focus my research. There are many developments of the UK's military nuclear project in the 1980s that will need to await the greater disclosure of primary sources before their history can be adequately told. One example would be the decision to purchase Trident and its implications for nuclear weapons laboratories, submarine bases and other sites of technical expertise. Instead, I examine the response in the UK to President Ronald Reagan's vision of missile defence, the Strategic Defense Initiative (SDI). In her autobiography Thatcher recalled that 'This was one of those areas in which only a firm grasp of the scientific concepts involved allows the right decisions to be made'. But the project was also deeply criticised by scientific experts and advisers, not only in academia, but even, for example, from within the UK Ministry of Defence.

The chapter therefore raises and answers questions about the roles of technical knowledge and advice in policy-making. But interwoven with scepticism about whether SDI would ever work were concerns about the UK losing out in a new technological race, as well as, frankly, an element of avarice. 'Isn't there high grade employment here for a lot of British brains?' one senior diplomat asked. I trace the dramatic proposal from Michael Heseltine, the defence minister and Thatcher's political rival, for American funding for missile defence research in the United Kingdom – a proposal which, if it had been granted, would have been equivalent in spending power to a major new research council. Heseltine is a fascinating although intermittent figure throughout this book, both as a minister for environment and for defence. His proposals tended to be bold and interventionist, pointing towards an integrated industrial strategy and science policy eventually quite at odds with Thatcher's direction of travel.

Chapter 7 explores five case studies of science, policy and the environment. Acid rain was a major issue andone in which scientific evidence was central, although uncertainty within such evidence could be, and was, used both to support strong remedial action and to delay it. I will show how different interests at play encouraged different interpretations and deployments of research findings. Acid rain was also intrinsically an international issue. An important turning point occurred when Gro Harlem Brundtland, the Norwegian Prime Minister, reached out to Thatcher (and, in doing so, emphasised their unusual hybrid status as leading politicians with scientific training) in 1986.

The second case study concerns Antarctic science, before and after the Falklands War. The diminishing of support for South Atlantic science was one of the signals read by Argentina as evidence of a declining UK commitment to defence of the Falkland Islands. The strategic importance of Antarctic science, post-Falklands, was such that Thatcher overrode the priorities set by the research councils. Research conducted by the British Antarctic Survey, published in 1985, revealed the ozone hole. International agreements to reduce emissions of the chemicals causing the ozone hole, notably the Montreal Protocol of 1987, became folded into the late, green pivot of Thatcher's administration.

The third case study, anthropogenic climate change, also reflects this turn. Both the ozone hole and climate change featured in Thatcher's famous science speech, made to the Royal Society in 1988. Thatcher, almost certainly a climate sceptic earlier in her career, became briefly an evangelist for international action in the late 1980s, before relapsing into scepticism in retirement. There is a causal chain here: from the increased receptivity to Antarctic science post-Falklands, to Thatcher's well-received championing of action on the ozone hole, to the search for another international, science-based, green cause, to her alighting on climate change – which also aligned to her new, post-1987 emphasis of restricting government's role to supporting non-industrial science. Thatcher's green turn of the late 1980s was partly political judgement as green politics spiked in importance (the Green Party received 15 per cent of the popular vote in the 1989 European elections, by far its best-ever showing). It was also partly a way of intervening on the international stage which made the most advantageous use of her attributes (including her scientific training) – perhaps partly enabled by her grasp of the science, but also, crucially, a product of the conversations she had with advisers.

Some environmental issues involving science occupied Thatcher's attention. However, it is important to note that others did not. The fourth and fifth case studies, on biodiversity conservation and the release of genetically modified organisms into the environment respectively, are two examples of issues where, despite their importance, policy was not driven by prime ministerial concern. The case of Antarctic research after the Falklands conflict – in which Thatcher insisted that research council funds be redirected, but refused to consider an increase in the overall research council budget to accommodate it – also shows that the overriding concern for reducing public expenditure, in other words aspects of the debates of Chapter 3, was relatively more important than other political targets.

In the final chapter I briefly summarise my main findings before reviewing some of the changes and continuities in policies for and affecting science after Thatcher fell from power in 1990. The new Prime Minister, John Major, had, as was entirely typical of the British political class, no substantial training in the sciences. However, when attention did turn to reforming science policy, notably as expressed in the 1993 white paper, *Realising our Potential*, I will show that it was constrained by, and largely followed, rather than broke from, Thatcher's science policy.

Notes

1. John H. Mulvey, 'Can we save British science?', *Laboratory Practice* (February 1986), p.5.
2. Advertisement, 'Save British Science', *The Times* (13 January 1986), p.5.
3. 'Listen to the cries of anguish from scientists in research council labs and universities and it sounds as if their part of Britain's research system is coming apart at the seams'. Martin Ince (and Jon Turney), *The Politics of British Science* Brighton: Wheatsheaf, 1986, p.37.
4. Watts, quoted in Stuart Simon, 'Starving today's research means destroying tomorrow's industries', *The Listener* (29 January 1987).
5. Blakemore to George Walden, 13 March 1987. CaSE archives.
6. Phillips, quoted in Simon.
7. 'British science in decline. Evidence prepared by Save British Science for the House of Lords Select Committee on Science and Technology', 3 June 1986. Campaign for Science and Engineering (CaSE) archives.
8. Blakemore to George Walden, 13 March 1987. CaSE archives.
9. Karl Grandin, Nina Wormbs and Sven Widmalm, *The Science-Industry Nexus: History, Policy, Implications*. New York: Science History Publications, 2004. Benoît Godin, 'The linear model of innovation: the historical construction of an analytical framework', *Science, Technology, & Human Values* 31 (2006): 639–67.
10. 'Crisis' talk was widespread, not least from the leaders of Save British Science. For example: Denis Noble, 'Britain's culture in crisis', *The Independent* (13 January 1987). Noble, 'Funding for basic research: retrievable crisis or inevitable decline', speech to AGM of the Fine Chemicals Group of the Society for Chemical Industry, 24 April 1986.
11. Blakemore to Saunders, 10 June 1988. CaSE archives.
12. Editorial, 'Her unique responsibility', *The Times* (19 February 1987), p.13.
13. Editorial, 'Her unique responsibility', *The Times* (19 February 1987), p.13. For the attribution of personal responsibility, see also Eugene Garfield, editorial, 'Shame on you, Mrs Thatcher', *The Scientist* (9 March 1987), p.9. Garfield draws on statistics from D. C. Smith, P. M. D. Collins, D. M. Hicks and S. Wyatt, 'National performance in basic research', *Nature* (23 October 1986), pp.681–4.
14. Harold Wilson, *The Relevance of British Socialism*. London: Weidenfeld and Nicolson, 1964, p.54.
15. Richard Coopey, 'The white heat of scientific revolution', *Contemporary British History* 5 (1991): 115–27.
16. Guy Ortolano, *The Two Cultures Controversy: Science, Literature and Cultural Politics in Postwar Britain*. Cambridge: Cambridge University Press, 2009, p.182.
17. Gary Werskey, *The Visible College*. London : Allen Lane, 1978.
18. David Edgerton, 'Liberal militarism and the British state', *New Left Review* 185 (1991): 138–69.

19. Stuart Butler, 'National prestige and in(ter)dependence: British space research policy 1959–73', PhD thesis, University of Manchester, 2016, p.35.
20. David Edgerton, 'C.P. Snow as anti-historian of British science: revisiting the technocratic moment, 1959–1964', *History of Science* 43 (2005): 187–208, 201.
21. 'Sir William Stewart (FRS 1977; Chief Scientific Adviser, 1990–5): Edited transcript of a discussion between Bill Stewart and Peter Collins at the Health Protection Agency London office on 8 January 2009', Royal Society Library.
22. Malcolm Rutherford, 'Who is Robin Nicholson?', *Financial Times* (5 March 1982).
23. Hugo Young, *One of Us: a Biography of Margaret Thatcher*. London: Macmillan, 1989, p.15.
24. Quoted in Denis Noble, 'Science, culture and wealth. Lloyd Roberts Memorial Lecture. 8 December 1987', copy in CaSE archives.
25. Three specific factors, suggests Turner, account for this shift: the identification of Germany as an industrial and military threat, the political success of anti-vivisectionism and the perceived 'unresponsiveness of the political system'.
26. Frank M. Turner, 'Public science in Britain, 1880–1919', *Isis* 71 (1980), pp.589–608, pp.596–7.
27. The announcement of the British Science Guild in *Nature* 72 (1904) 72: 586. My emphasis. Quoted in Turner, *op. cit.*, pp. 601–2.
28. Andrew Hull, 'War of words: the public science of the British scientific community and the origins of the Department of Scientific and Industrial Research, 1914–16', *British Journal for the History of Science* 42 (1999): 461–81, 466–7.
29. Turner notes the new Committee on the Neglect of Science criticising the government at length in *The Times* on 2 February 1916, for example.
30. Mark Henderson, *The Geek Manifesto: Why Science Matters*. London: Bantam Press, 2012, p.147.
31. 'Lord Waldegrave (Chancellor of the Duchy of Lancaster [ie Science Minister], 1992–4, and numerous other science-related Government posts). Edited transcript of a conversation between William Waldegrave and Peter Collins at the Royal Society on 13 July and 14 September 2011', Royal Society Library.
32. David Edgerton and Kirsty Hughes, 'The poverty of science: a critical analysis of scientific and industrial policy under Mrs Thatcher', *Public Administration* 67 (1989): 419–33.
33. See, as one example, Mark Easton, 'Margaret Thatcher: a tribute to conviction politics', 17 April 2013. https://www.bbc.co.uk/news/uk-22183714. See also Eric J. Evans, *Thatcher and Thatcherism*. London: Routledge, 1997, p.107; and the collection of MPs' and Lords' tributes in John Blundell (ed.). *Remembering Margaret Thatcher: Commemorations, Tributes and Assessments*. New York: Algora, 2013, p.88, p.113.
34. Charles Moore, *Margaret Thatcher. The Authorized Biography. Volume Two: Everything She Wants*. London: Allen Lane, 2015, p.6.
35. 'Sir Robin Nicholson (FRS 1978). Edited transcript of a conversation between Robin Nicholson and Peter Collins at the Royal Society on 6 May and 18 May 2009', Royal Society Library. 'She had something of Fleming's as well, and possibly Darwin, and she had a number of artefacts', Nicholson recalled, and added that their function was diplomatic: 'if I was there obviously the discussion would usually be on some science and it would be a President or Prime Minister from some other country and she would say, well of course Britain has been very great in science for a very long time'.
36. Harvey Brooks, 'The scientific advisor', in R. Gilpin and C. Wright, eds, *Scientists and National Policy-making*. New York: Columbia University Press, 1964, pp.73–96.
37. One comment I received I think demonstrates the difficulty of reporting science policy in the 1980s: 'I think the main thing I note … is the poor access one had, despite fairly assiduous cultivation of sources, to what was actually going on. All the spending papers and so on were Officially Secret, so reporting was an odd game, labour-intensive but in a slightly pointless-seeming way. I would talk regularly to heads of RCs [Research Councils], members of ABRC, chair of UGC, chief scientists etc, but they would of course only say what they wanted you to know'. Jon Turney, personal communication.
38. David Edgerton and Kirsty Hughes, 'The poverty of science: a critical analysis of scientific and industrial policy under Mrs Thatcher', *Public Administration* 67 (1989): 419–33.
39. Tom Wilkie, 'The Thatcher effect in science', in Dennis Kavanagh and Anthony Seldon, eds, *The Thatcher Effect*. Oxford: Clarendon Press, 1989. Wilkie, *British Science and Politics since 1945*. Oxford: Blackwell, 1991.

40. Roger Williams, 'UK science and technology: policy, controversy and advice', *The Political Quarterly* 59 (1988): 132–44.
41. Ian Christie, 'Research and development policy: the great debate', *Policy Studies* 8 (1988): 11–22.
42. Stephen Wilks and Michelle Cini, 'The redirection of science and technology policy under the Thatcher governments', *Policy Money & Management* 11 (1991): 49–56.
43. Nick von Tunzelmann, 'Technology in post-war Britain', in R. Floud and P. Johnson, eds, *The Cambridge Economic History of Modern Britain, Volume III: Structural Change and Growth, 1939–2000*. Cambridge: Cambridge University Press, 2004, pp.299–31.
44. PREM 19/2835. Guise to Turnbull, 27 April 1989.

2
Who made science policy?

Thirteen days after the 1979 general election the Cabinet Secretary, John Hunt, asked the new Prime Minister how she wanted to organise science matters. Since Margaret Thatcher had not appointed a Minister for Science, he asked whether he was right to assume that the 'arrangements of recent years' (under both Conservative and Labour governments) would continue, namely:

> (i) DES [Department for Education and Science] is responsible for the Research Councils and for the Science Budget.
> (ii) Each major Department is responsible for determining and financing its own R and D programme broadly on the 'customer/contractor' principle.
> (iii) There is no 'scientific overlord' who tries to run science and R and D from the centre, but a Cabinet Minister is given a general co-ordinating role and acts as the Government's spokesman on general scientific affairs.[1]

Thatcher confirmed that Neil Macfarlane as minister at the Department for Education and Science would '"do" the science side of things of the Ministry', but added that she herself would take a role: 'I will answer questions on science if need be'.[2]

Hunt's summary of science in government hid an enormous amount of complexity. The Junior Minister at the Department for Education and Science would in practice play an almost negligible part in policy-making, while the Prime Minister did indeed pay close attention to science matters and would often go much further than merely answering questions. In between were advisers, official committees, civil servants, ministers and Parliamentary bodies, as well as a host of interested parties external to government but which also argued, lobbied and produced evidence that shaped debate. The purpose of this chapter is to provide a guide to the people, roles, bodies and mechanisms that together produced and administered decisions on science and on science-laden issues.

The Prime Minister

As Prime Minister, Margaret Thatcher's roles in science policy-making were various and powerful: as the head of the Cabinet that took ultimate responsibility for political decisions, including those on science and on issues which depended on science; as the senior authority in constant communication with ministers on affairs of state; as someone who might choose to lead on particular matters (such as missile defence, discussed in Chapter 6); as a chair, if she so wished, of the key committees in which issues were discussed; and as a recipient for the special channels of advice. Such roles were those taken by prime ministers before and after her. Thatcher, Prime Minister, had two extra unique dimensions: she had said she would 'answer questions if need be', and she had a training in, and working industrial experience of, science.

The young Margaret Roberts chose to narrow her sixth-form studies at Kesteven and Grantham Girls' School to chemistry, biology and physics. Her biographers record that the choice of chemistry was one that 'suited her practical bent of mind' and a subject that had 'good employment prospects'; she was also 'inspired by the excellent teaching of the chemistry mistress at Kesteven, Miss Kay'.[3] Roberts applied to the University of Oxford to read chemistry, arriving at Somerville College in 1943, when the campus was eerily empty during the Second World War. One early biographer suggested, with little evidence, that 'she seems to have decided rather coolly and calculatingly that for a girl Chemistry was the best examination bet', while there 'was also the attraction of invading and succeeding in what was considered a man's domain'.[4] (As Prime Minister, however, Thatcher would only rarely accept invitations that sought to identify her as a high-achieving *female* politician who would support the recruitment of women into science and engineering.)[5]

Somerville was a women's college, with a progressive left reputation, led by haematologist and radiologist Dame Janet Vaughan. Roberts' fourth-year dissertation was an investigation of antibiotic gramicidin S, guided by a refugee German postgraduate Gerhardt Schmidt and Dorothy Hodgkin, one of the leading x-ray crystallographers of the mid-twentieth century – a woman who would not only win a Nobel Prize, but also be a figure of life-long respect for the young chemist-politician.[6] Neither senior woman rated Margaret Roberts particularly highly at her subject. She was a 'perfectly good second-class chemist', according to Vaughan. 'I came to rate her as good,' recalled Hodgkin, 'One could always rely on her producing a sensible, well-read essay.'[7]

While already active in university Conservative circles and considering training as a lawyer as a step towards politics as a career, Roberts chose instead to work for British Xylonite (BX) Plastics after graduation in 1947. She would work in the Essex firm's new Research and Development Section as an industrial chemist for two years before moving to another chemistry job, this time in food science at J. Lyons & Company, at the cakes and teashop business's headquarters in Hammersmith, London. One reason she wanted the move was that it was nearer to a constituency in north Kent, where she had been selected to fight a then unwinnable seat for the Conservatives. As part of the publicity for her 1951 Dartford election campaign, Margaret Roberts was photographed wearing the white coat of a laboratory chemist.

In December 1951 Margaret married Denis Thatcher, who worked in his family's paint and preservatives firm. There was a chemical affinity – the couple had met at a paint trade fair in Dartford. In the same year Margaret left Lyons and trained as a tax lawyer. Denis took a back seat as Margaret's political career took off, beginning with her success at the polls to become Member of Parliament for Finchley in 1959. His business life very rarely intersected with her political work, an almost vanishing intersection when it came to science policy. However, one exception was policy on agricultural pesticides. In 1980 Thatcher's private secretary decided she should be informed of an otherwise routine regulatory decision of the Advisory Committee on Pesticide's that 2,4,5-T herbicides – one-half of the constituent chemicals of Agent Orange – could safely be continued to be used. 'You may get some reaction to this because of DT's firm,' he noted.[8] As regulation of 2,4,5-T, following changing scientific advice, became tighter in the 1980s – as evidence accumulated that the chemical was carcinogenic – so Thatcher was kept informed.[9] She expressed her displeasure at the regulatory bureaucracy, although, as we shall see, this was not atypical of her.

In 1970 Edward Heath appointed Margaret Thatcher as his Secretary of State for Education and Science. She was therefore the minister responsible for matters of school and university education and relevant aspects of civil science policy. Perhaps the most consequential decision on science policy during her ministerial career was on the research and development funded by government departments. Heath had established a Central Policy Review Staff, a 'think-tank' charged with providing original and radical examinations and recommendations. Staffed by a mixture of Whitehall hotshots and talented outsiders, the CPRS was led by Lord Rothschild. Victor Rothschild had been trained in biophysics at Cambridge in the mid-1930s before embarking on a postwar career that encompassed both government work (he was chair of the Agricultural Research Council in the 1950s) and industry (as director of Shell UK's research programmes in the 1960s). In 1971 Rothschild proposed a new way of framing, understanding and managing the research and development of civil government departments. This framing was the 'customer-contractor' principle: the department (the customer) says what it wants, science (in the form of research institutes, research council-funded scientists or others) contracts to provide it, and the customer pays.[10] The proposals, published in the 'Rothschild' report in November 1971,[11] were immediately controversial and opposed by the established institutions of UK science, including the Royal Society.[12]

The crucial meeting to consider the Rothschild reforms took place on 20 April 1971 at 10 Downing Street. Present were Edward Heath, Rothschild, the two most senior civil servants (Sir William Armstrong, head of the home civil service, and Robert Armstrong, Heath's Principal Private Secretary) and the minister responsible, Margaret Thatcher. What is most intriguing about the meeting is that the minutes show Thatcher opening with a strong defence of the status quo, as she had been briefed by her department and a line supported by the Royal Society. Then, after presumably intense argument, recorded by Robert Armstrong in his artfully abbreviated summary as 'discussion [in which it was] recognised that this would be fundamentally different from the present system', Thatcher emerged convinced that the 'fundamental change', the marketised framing of government research, should be adopted. Episodes such as this one, crunch situations of political choice where market ideas were embraced, are more likely stages in the extraordinary journey of Thatcher, previously quite an ordinary Conservative minister, to Thatcherism than the standard historiography which sees her being persuaded by weakly institutionalised Hayekian supply-side economics ideas.[13]

The 1971 reassessment of science policy stands out as a concrete and early moment when Thatcher chose the market as an alternative to established models of resource allocation.[14] Having worked as an industrial chemist, and therefore witnessed an engaged, problem-solving side to science very different to the vision of autonomy championed by the Royal Society, Thatcher nevertheless also saw science as the best of the public economy. She viewed science as a source of wealth, and therefore a justified expenditure from the public purse. Yet this elevation made science even more of a test case for her developing views on economic liberalism. If markets could work for science policy, they could work anywhere.[15]

Rothschild was patchily implemented: completely in many departments, but tried and rejected in medical research, for example,[16] in the 1970s. Contrary to the view that it was a minor reform (in that it only, initially, applied to the research internally contracted by government departments, and that it was incomplete), Rothschild was of lasting significance. Once departments were buying research services from an internal market, it was a natural extension to ask those internal suppliers to compete against external suppliers, or indeed become privatised suppliers themselves. The significance was partly discursive (framing how research could be talked about) and suggested a direction of travel.

In 1974 the Heath government fell, and a year later Margaret Thatcher became Leader in opposition of the Conservative Party. Developing a set of radical policies, science would never be the overriding concern of Thatcher when she became Prime Minister in 1979. However, as I show in the following chapters, science would nevertheless be intertwined with some of the major issues she confronted and policies she pursued. The question of whether her scientific background mattered to her prime ministerial career is therefore still an intriguing one. In addition to the possibility that practical science policy decision-making in the 1970s was a step towards Thatcher becoming Thatcherite, other commentators have noted either the application of a scientific frame of mind (of which I am sceptical) or the relevance of her scientific knowledge. Hugo Young drew on Dorothy Hodgkin's judgement:

> What, according to the Nobel Laureate, does the study of chemistry do to a person's mind? 'I think it should interest you in problems of finding out as much as you can about the way we work, the way matter is put together. And it should give you an interest in using the results.'
>
> This blueprint for the practical mind, a marriage between speculative and empirical habits, is one which as a politican Mrs Thatcher consistently made much of. She retained a genuine interest in science, which Dorothy Hodgkin concedes. It equipped her, says the professor, to take serious decisions on scientific matters and 'to see what scientists are doing'. In the politician, her lack of any outstanding scientific talent was less significant than her rare capacity to understand the scientific mind at all.[17]

'Margaret Thatcher much prided herself on being the only scientist in her government,' noted Sir Crispin Tickell, whose advice on climate change to the Prime Minister is discussed in Chapter 7. 'Anything that related to science she took a particular interest in, and almost felt that she owned it. Some of her views were radical and didn't always fit the other views she heard from others.'[18] Jonathon Porritt, the green advocate who was granted a meeting with Thatcher in 1989, believed that 'The ozone layer got through to the Prime Minister because she got high-level briefings from NASA, and her chemist's training enabled her to take it seriously'.[19] Thatcher requested and received unleavened and technical scientific advice from her advisers relating to policy issues that were underpinned by science, for example, lists of equations describing the chemical reactions of the stratosphere as part of acid rain discussions (see Chapter 7) or the qualities of α-, β-, and γ-radiation as part of nuclear policy-making (see Chapter 4).[20] The information was provided with the justified expectation that she would understand and use the knowledge. Her blue pen underlining passages demonstrates she read them. Her chief scientific advisers would occasionally address her with the phrasing 'as a scientist … ' if they wanted to appeal to this instinct. The Norwegian Prime Minister Gro Harlem Brundtland addressed her as a fellow scientist (with effect, as I show in Chapter 7). I will analyse plenty of examples of Thatcher's special interest in science-related policies in the chapters that follow.

However, her relationship with scientific institutions could also be problematic. As citizens and voters, British scientists were just as likely to have strong views, positive and negative, about Thatcher's politics as anyone else. They also worked in organisations – universities, research institutes – that were deeply affected by her policies, not least the commencement of public sector cuts. Thatcher's communications through the scientific press, such as an interview with the editor of *Nature*, John Maddox, were typically forthright and divisive.[21] When Thatcher's name was put forward for election as a Fellow of the Royal Society in 1983, not in itself unusual for a prime minister, but one with much added significance given her politics and background, 'her candidature split the Fellowship, and the normally placid election meeting attracted an unprecedented turnout, marshalled by the key protagonists on each side,' records the Society's historian.[22] She was narrowly elected as an FRS, a prestigious title in science, but the furore meant that the admission ceremony had to be delayed. 'I need hardly say that this incident does not detract from the honour which I feel in having been elected to the Fellowship,' Thatcher wrote to the Royal Society President.[23] In 1985 she was denied an Oxford honorary degree, 'the opposition led by scientists'.[24]

In this context, Thatcher adopted a noticeably distinctive way of doing the business of communicating science policy and gathering scientists' views: she hosted seminars and receptions in which communication could happen directly. Examples include:

Reception for inventors and innovators 26 January 1981
Luncheon with scientists 19 February 1982
Seminar on science, technology and industry 12 September 1983
Recent developments in scientific research 8 July 1984
Meeting with industrialists 21 May 1985
Seminar on engineering and technology February 1986
(Proposed) Seminar on Priorities for Science and Technology Spring 1987
Seminar with young scientists 13 September 1989
Seminar with young scientists 28 September 1990

Invitees to the 1981 reception for inventors and innovators included P. C. Dowles ('inventor of new type of hotel room safe'), P. Gotley ('invented a new form of gas detection equipment, employing microprocessor'), Maurice A. Hiles (inventor of '"Sorbothane", a polymer that simulates the energy-absorbing properties of human flesh'), F. B. Mercer (invented a 'new way of making nets', Netlon).[25] These people were among those who had received government assistance from an organisation called the National Research Development Corporation (NRDC, discussed below). They therefore represented the older, collective support of intellectual property exploitation. But alongside them were others, including E. Biss (an inventor of a new 'note weighing machine, controlled by microprocessor'), L. Brownlow (managing director of Rodime Ltd, a new company, established 1980, manufacturing computer peripherals), T. J. Parker (inventor of 'novel containers for pharmaceuticals') and Howard Calvert (a young inventor of a 'portable gymnasium', who had run a 'family software firm while studying for A levels').[26] Here was a new class of inventors, unencumbered by public sector support. Filling the room with these 'Inventors' were 'Entrepreneurs' (it is interesting that inventors were not themselves categorised as entrepreneurs), venture capitalists, financiers and bankers, as well as 'Industrialists' the representatives of bigger engineering businesses.[27]

Thatcher addressed the reception, opening by saying that the 'principal aim of the Government's economic policies has been to stimulate individual initiative by encouraging the formation of new businesses and enabling their owners to retain more of the wealth that they have created'.[28] It was to smaller firms that would be looked to 'to take up much of the labour now being shed because of the rundown of our older industries'. She asked what were the 'barriers to the exploitation of new ideas', and began a discussion that covered such issues as the relations between individual inventors and banks, the mismatch in expectations between the inventor ('absorbed in his own vision of his concept in operation') and the financier (who was looking for management skills first before technical viability), as well as the possibilities of awards and tax incentives.[29] While the record of the meeting suggests that the tone of the discussion, perhaps shaped by the realism of the established interests present, was fairly sober – 'Inventions did not necessarily lead to innovation: innovation did not necessarily lead to business success. The individual invention could only play a small part in meeting national requirements for new products' – this was in contrast to Thatcher's opening words, the paean to the individual inventor.

One can fairly say there were two groupings present. On one side those championing the individual (Thatcher and the lone inventors) and on the other the organisations, including those charged with holding the individuals back. The fact that this gala was held in Downing Street was a statement that the lone inventors should be valued. They were a type to champion.[30] I will discuss other cases of these events – especially the most consequential, the Seminar on Science, Technology and Industry at Lancaster House in September 1983 – in later chapters.

In 1988 Thatcher made her most famous speech on science, hosted by the Royal Society. As I show in Chapters 3 and 7, while the speech has mostly been remembered as the moment a leader of a Western industrial nation made a call to arms for action on anthropogenic climate change, it also highlighted government support for curiosity-driven science. This positive framing of pure science in largely academic settings was, I show, the obverse to the cutting of government funding for near-market research. In many ways it was a clever, rhetorical move – who could be against curiosity? – but it was also the direct product of an ideological desire to remove the state from industrially relevant applied research.

Studying the science policy papers that were included in the boxes of files that were prepared for the Prime Minister, and read by her, usually overnight – the PREM series files now at the National Archives – it is hard not to be impressed by her sheer capacity to read, absorb information and arguments, and to think politically. And I have only focused on the files dealing with science-related issues – such papers were a fraction of her daily work. Yet in the last two years of her premiership a new note can be detected. She became more impatient of long documents: 'this is yet another lengthy woolly minute which avoids the real issues', she wrote on Geoffrey Howe's plans for international scientific collaboration, while a review of agricultural R&D was dismissed as 'guffy'.[31] What had been forthright now reads as tetchiness. Thatcher, more than before, resorted to clichés. For example, she railed at the regulation of agricultural pesticides (including 2,4,5-T) in 1990. 'It sounds like bureaucracy gone mad – and all new accommodation to go mad in,' she wrote of expanded pesticide evaluation units'[32]; it was a 'bureaucrats' paradise'[33] – a phrase that she began to turn to often. In November 1990, following Howe's resignation speech, Conservative MPs, enough of whom now thought she was no longer an electoral asset, brought her prime ministership to an end.

Departments

Central government is divided into departments, each of which has one or several politician ministers at its head and is staffed by civil servants.[34] Whitehall was therefore formed of a patchwork of departments, some powerful, some weaker, some with wide-ranging concerns, some managing specific interests – all of which could, and did, conflict when it came to the use and direction of science. 'In short', a review had concluded just prior to the 1979 election, 'the Government does not have a single science policy: it has a whole range of policies – relating eg to defence, industry, agriculture and the environment – and the role of scientific R&D is to contribute to the achievement of those policies'.[35]

The senior ministers of departments are members of Cabinet, which is chaired by the Prime Minister and which possesses executive power, collective decision-making and shared responsibility. The Cabinet Office supports the work of Cabinet, in particular through providing a secretariat for the Cabinet committees, which address specific topics. Cabinet committees are typically either ministerial (ie the members are politicians) or official (the members are civil servants). Important Cabinet committee relating to science in the 1980s included E(ST) – E for economics, ST for science and technology – and STO – O for official.[36] Cabinet committees do not have budgets, but they could both advise and make decisions. They were therefore crucial sites for the formation, negotiation and settlement of science policy. Nevertheless, key decisions might be made outside of committees, although some formal approval at full Cabinet level would be necessary.

In addition to providing the secretariat for the Cabinet committees, the Cabinet Office was also home to the Central Policy Review Staff

(CPRS) and the Central Statistical Office. The Cabinet Office and system date from the early twentieth century, while the CPRS was set up by Heath. A second body intimately close to the centre of government, the Number 10 Policy Unit, is younger still. Essentially, whereas the CPRS offered critical, far-reaching reviews and comments across government, the staff of the Number 10 Policy Unit, usually recruited from outside, brought these functions closer to the Prime Minister. The CPRS and the Number 10 Policy Unit were both sources of influential thinking on science policy, although in the Thatcher years the latter thrived and the former was abolished.

In 1979, out of about £3 billion total UK R&D spending (3 per cent of GDP), the UK government planned to spend £2,141 million on research and development.[37] Roughly half of this – £1,160 million – was budgeted through the Ministry of Defence.[38] By 1981/82 the expected expenditure on research and development by the Ministry was £1,680 million, £260 million of which was classed as research.[39] The leading ministers during the Thatcher administration were Francis Pym (1979–81), John Nott (1981–3), Michael Heseltine (1983–6), George Younger (1986–9) and Tom King (1989–92). The Ministry employed a chief scientific adviser, who, during the same period was Ronald Mason (1977–83) and Richard Norman (1983–8), both chemists, followed by the geophysicist Ron Oxburgh (1988–93). The Ministry had a system of advisory committees, including an internal, priority-setting Defence Research Committee and a network of subject-specific sub-committees under a Defence Scientific Advisory Council, designed to infuse academic expertise into the advisory process. The Ministry ran a number of defence science laboratories, including major ones for nuclear weapons at Aldermaston, radar research at Malvern, aeronautics at Farnborough – the Royal Aircraft Establishment was the 'largest research establishment in Western Europe' at the beginning of the 1980s – and naval research and development on the south coast.[40] It was also responsible for the Meteorological Office, which had mostly civil significance. In the late 1980s and 1990s the defence research laboratories faced transformation as targets for privatisation contingent with the end of the Cold War.

While the criticism is sometimes made that the UK did not have a unified science policy since defence research and development was considered separate from civil (and that historians of science policy have only focused on the civil side),[41] this charge is in fact largely an artefact created by what was publicly visible. For example, the large and important 1984 annual review of R&D surveyed both defence and civil science, but when published (as had happened in 1983) only the civil information was contained.[42] The impression from the outside was consequently that only civil science was being reviewed. From inside the highest levels of central government, however, both sets of data were placed side by side, and, as I show in Chapter 3, a ministerial argument over respective balance of civil and defence R&D took place.

The interconnected military and civil nuclear projects were of central importance to postwar British science. Both had begun in the late 1940s as the UK was excluded from postwar nuclear collaboration with the United States, and an overview of their history is given in Chapter 4. By 1979 the UK Atomic Energy Authority was formally under the Department of Energy (the responsibility for UKAEA's laboratory for nuclear weapons, Aldermaston, having been passed to the Ministry of Defence). The Department of Energy in that year had a budget of £155 million for research and development, of which the vast bulk was devoted to UKAEA's civil nuclear work. Chapter 4 discusses Thatcher's symbolic visits to UKAEA and other nuclear establishments, including those of British Nuclear Fuels Ltd, while Chapter 5 examines the moves to privatise the radioisotopes factory at Amersham as well as UKAEA itself. Scientists managed nuclear institutions, and the Department of Energy possessed a chief scientific adviser and expert advisory committees. The leading minister, the Secretary of State for Energy, during the Thatcher administration was David Howell (1979–81), Nigel Lawson (1981–3), Peter Walker (1983–7), Cecil Parkinson (1987–9) and John Wakeham (1989–92).

The Department of Trade and Industry (the two had been split in 1974, but merged again in 1983) possessed a sizeable research and development budget (£132 million in 1979), Requirement Boards to discuss science policy,[43] and a Chief Engineer and Scientist. It was responsible for one of the major laboratories (the centre of precise measurement and the maintenance of standards, the National Physical Laboratory at Teddington, London), but its importance to the themes of this book went much further, since the Department was responsible for policies in support of industry, technology and the innovation process more generally. Peter Hennessy, a long-standing analyst of Whitehall, notes that it has housed a long-running conflict between *dirigiste* (interventionist) and *laissez faire* traditions.[44] Under the DTI's wing was the National Research Development Corporation (NRDC), which collectively held and exploited patents derived from publicly funded research. As I discuss, the NRDC was a target for privatisation and dismantlement by Thatcherites. During the Thatcher years there was a frequent turn over of ministers, including Keith Joseph ('who arrived … in 1979 armed with copies of Adam

Smith which he urged his officials to read'), Patrick Jenkin (1981–3), Cecil Parkinson (1983, almost immediately to resign because of scandal), Norman Tebbit (1983–5), Leon Brittan (1985–6), Lord Young (1987–9) and Nicholas Ridley (1989–90). Nevertheless, it was in the Department of Trade and Industry that strongest support for using R&D as a means of promoting industrial competitiveness – that is to say an integral part of an industrial strategy – was expressed. For example, Norman Tebbit writing to Geoffrey Howe in 1984, in the context of a discussion of European collaboration, could say:

> My primary concern is to allocate our limited R&D resources to the more immediate challenge of ensuring that we keep [our] industrial capability in place. This means domestically funded programmes directed at national objectives.[45]

Such thinking justified extensive programmes of support for what was later described as 'near-market' science. As I show in Chapter 3, the DTI therefore was a source of opposition to the radical new science policy of cutting near-market research in the late 1980s.

The two departments related to land use and natural resources, the Department of the Environment (combined initially with the Department of Transport) and the Ministry of Agriculture, Fisheries and Food (MAFF), had comparable, middling-sized research and development budgets, organised by the Rothschild principles, in 1979 of £59 million and £50 million respectively. Each had a chief scientist (known in the Environment as the director of research) and a network of research laboratories. Environment was also responsible for the Nature Conservancy Council, which offered advice on nature conservation and supported research.

The fact that responsibilities and budget for research and development were placed in other ministries meant the Department of Education and Science did not have the full influence on science policy that its name might initially suggest. However, it channelled considerable public funds through a 'dual-support system' to what came to be called in the 1980s the 'science base': £190 million direct to universities and colleges via the University Grants Committee (UGC) in 1979, and twice that figure through the research councils. Research councils are semi-autonomous bodies that distribute research funding, not all of which went to universities, guided by expert advice and peer review. In 1979, there were five research councils, and the funding broke down as follows:[46]

Science Research Council (SRC)(later Science and Engineering Research Council, SERC)	£177 million
Medical Research Council (MRC)	£68 million
Agricultural Research Council (ARC)(later Agriculture and Food Research Council, AFRC)	£58 million
Natural Environment Research Council (NERC)	£55 million
Social Science Research Council (SSRC)	£18 million

The belief that the research councils have been and should be autonomous, and in particular should not be directed by ministerial (or prime ministerial) wishes, is referred to as the 'Haldane principle'.[47] (In fact, under Section 2(1) of the 1965 Science and Technology Act, the responsible Secretary of State could give 'directions'.)[48] Whether or not it was ever articulated by Lord Haldane is secondary to the effective belief that the principle exists. However, as I show in Chapter 3, in which the Chief Scientific Adviser sought legal advice on the extent that research council work could indeed be directed, or in Chapter 7 where, in the case of Antarctic science, the effective existence of the principle was demonstrated by its breach, the autonomy of the research councils was put under pressure in the 1980s.

The final stop on our tour of Whitehall, the Treasury, is perhaps the most powerful of all, although it commissioned a negligible amount of research directly and itself had no chief scientific adviser. (Indeed, as was noted within the department, 'the Treasury does not have knowledge in-house to enable it to put forward a scientific view, and prefers it that way'.)[49] The Treasury controlled the purse strings, and each department would have to negotiate with it over the amount and the rules of distribution of public expenditure. The Treasury also influenced taxation policies, some of which were, in effect, research and development policies. For example, in 1984 a 'scientific research allowance' (SRA), essentially a 100 per cent tax credit on capital expenditure incurred by a trader on scientific research (undertaken either by the company or on the company's behalf), amounted to support of R&D to the tune of £100 million a year. It was in the Treasury that such schemes were examined, often sceptically. For example, one official asked 'to what extent do we want to single out research as a priority area? Is not the main priority support for development?'; he also questioned whether other schemes, such as the industrial strategy in support of advanced information technology – known as Alvey (see Chapter 3) – were 'cost effective'.[50] SRAs were the subject of departmental horse-trading: the DTI were keen on them, while the Treasury looked askance.[51] Finally, the Treasury built and ran the models on which the economy's performance was predicted and understood, and against which policies were evaluated. The Treasury model did 'not include any variable solely and specifically related to scientific and technological change'.[52]

Advice and advisors

Above I quoted Sir Crispin Tickell recalling that 'Thatcher's view on science and science policy were 'radical and didn't always fit the other views she heard from others'; he also remarked that the 'main advice she got was, of course, from the civil service machine'.[53] I have noted bodies that possessed departmental scientific advisers, a presence that was a consequence of the Rothschild reforms. These departmental advisers were sometimes part-time or appointed from within (rather than outsiders to Whitehall, and therefore lacking academic or industrial experience).[54] Further advice was provided in a collective form, through committees that pooled ranges of expertise, including from academia and industry.

Royal Commission on Environmental Pollution

Some of these bodies sat aside of central government, such as the Royal Commission on Environmental Pollution, which produced a series of reports on environmental matters beginning in 1971.[55] Royal commissions are ad hoc advisory and investigatory bodies that, while appointed by government, have a useful quasi-independent standing. Thatcher's government received seven reports from the RCEP, as follows.

7th Report	Agriculture and pollution	Cmnd 7644, September 1979
8th Report	Oil pollution of the sea	Cmnd 8358, October 1981
9th Report	Lead in the environment	Cmnd 8852, April 1983
10th Report	Tackling pollution – experience and prospects	Cmnd 9149, February 1984
11th Report	Managing waste: the duty of care	Cmnd 9675, December 1985
12th Report	Best practicable environmental option	Cm 310, February 1988
13th Report	The release of genetically engineered organisms into the environment	Cm 720, July 1989

Other committees, which I now turn to, were more centrally located.

ACARD and ACOST

The views of industry were represented by the Advisory Council for Applied Research and Development (ACARD), first appointed in 1976.[56] When John Hunt had asked Thatcher in May 1979 how she wanted to organise science matters, he asked whether ACARD should continue:

> After a shaky start ACARD now seems to be doing a useful job. Its first two reports, on microelectronics and on the encouragement of innovation, were well received. It has four more now in progress on the employment implications of technological change; joining and assembly techniques; computer-aided design and manufacture; and the implications for the private sector of the public sector's R and D capacity: and they are also doing a joint study with the Royal Society and the ABRC studying biotechnology. The 'co-ordinating' Minister chairs ACARD (the scientific community have welcomed this as a visible token that Government takes the applied end of civil science seriously): but there are two working Deputy Chairmen (Dr Alfred Spinks and Sir James Menter): all the other members are outsiders but Sir Kenneth Berrill and some Departmental Chief Scientists attend as assessors; and some support ... is provided by the Chief Scientist in the CPRS (John Ashworth) and the Cabinet Secretariat.[57]

In 1979 the companies that were represented on ACARD by individuals were ICI (Spinks was former director of research), the Scottish Offshore Partnership, GEC, Ricardo Consulting Engineers Ltd, Delta Materials Research Ltd, Vickers and British Petroleum. In addition there were four academics and two trade unionists. 'The point is not how many interesting reports ACARD produces but whether those reports have practical value,' noted Thatcher; sceptical of the contribution of ACARD, she added: 'I doubt it'.[58] Her reviewer of quangos, Leo Pliatzky, suggested giving ACARD two more years, after which the case for continuing it would be reconsidered. Thatcher agreed.[59] Major reports produced by ACARD during the Thatcher years included:

Joining and Assembly: the Impact of Robots and Automation	October 1979, published 1979
Computer Aided Design and Manufacture	October 1979, published 1980
Technological Change: Threats and Opportunities for the United Kingdom	November 1979, published 1980
Biotechnology (the 'Spinks Report')	Joint with ABRC and Royal Society, published 1980
Information Technology	August 1980, published 1980
Exploiting Invention	December 1980
The Food Industry and Technology	July 1982, published 1982
Improving Research Links between Higher Education and Industry	Joint with ABRC, June 1983
First Joint Report by the Chairmen of the ACARD and ABRC	Joint with ABRC. Published July 1983
New Opportunities in Manufacturing	Published in October 1983
Exploitable Areas of Science	September 1985. Published in May 1986
Software: a Vital Key to UK Competitiveness	March 1986. Published in June 1986
Medical Equipment	Published July 1986

In 1980 the membership of ACARD was freshened up, with the lacklustre Menter allowed to retire and the vice-chancellor of Cranfield Institute of Technology, Sir Henry Chilver, stepping up from deputy to chair. New industrial representatives came from companies such as Unilever and Dunlop and, in order to link academic to applied science, the chair of the Advisory Board for the Research Councils (ABRC) was invited on to the Council. Thatcher complained that 'there is no one from the information technology area' (an interesting, early use of the term).[60] By 1983 ACARD was a rather unwieldy committee of 16 members (including the chair, Chilver), six 'assessors' and a secretariat.[61] In November 1985 Sir Francis Tombs, who had a background in the electricity supply business and was chairman of Rolls-Royce, replaced Chilver as ACARD chair.

ACARD's reports tended to be detailed and lengthy. Nevertheless, the submission of a new report, and other occasions too, permitted the chair of ACARD to communicate direct with the Prime Minister and offer advice. Sometimes the advice was influential, especially when it aligned with senior politicians' analyses. Such was the case in ACARD's criticisms of the National Research Development Corporation (see Chapter 3).

At other times the advice fell on infertile soil, as when ACARD urged, in the same report that criticised the NRDC, that the BBC and ITV be encouraged to package entrepreneurial tips into its programming 'rather as farming advice used to be given in every episode of "The Archers"'.[62] Thatcher, reading the report, picked out the NRDC criticisms as the ones for further action.[63]

ACARD clashed with the government on science funding. In a 1979 report, read by Thatcher, ACARD highlighted the fact that 'in recent years R and D expenditure by United Kingdom industry has declined disturbingly compared with that of other countries'.[64] It contested the public statement made by the government in 1985 that 'the UK's R&D expenditure as a proportion of GDP was sufficient'.[65] ACARD was also a key proponent of continuing the policy of the UK pursuing an active industrial strategy in which public bodies and funding would be coordinated to support promising sectors. This message, for example, came through loud and clear in the first recommendation of one of the first ACARD reports to be received by Thatcher, *Technological Change*, in November 1979:

> The development of technology strategies for different industrial sectors should form an integral part of the [the work of the National Economic Development Council] NEDC ... The R and D programmes of the Department of Industry's Requirement Boards and of Research Associations, and the relevant parts of Research Council programmes, should be aligned with these strategies.[66]

Likewise in a 1983 report on links between higher education and industry, ACARD, jointly with the ABRC, called for ways to be found actively to channel funding to 'areas of research which are both academically worthwhile and have industrial relevance'; it also called for an 'industrial seedcorn fund' to support 'research that will complement effective, industrially-financed applied research'.[67] In 1985 – in its most lengthy report, produced over two years with funding from ICI and the help of academics at the Science Policy Research Unit at Sussex University – ACARD argued that:

> There is a thesis, widely accepted in the United Kingdom, that basic research cannot be organised to deliver economic return. The thesis is not generally accepted in other countries. They believe that science is now so important to a country's future that some attempt

must be made to structure support, and achieve more effective exploitation of science. ...

[S]ome mechanism is needed in the best interests of the country to prioritise and guide a fairly high proportion of that part of the national scientific resource paid for by the taxpayer, and to stimulate its effective exploitation to the benefit of the United Kingdom.[68]

In Chapter 3 I will show that in the later 1980s, in order to solve the second issue of underinvestment by UK companies in research and development, Thatcher chose to cut precisely the type of support of near-market research that ACARD had lobbied for. Thatcher's government therefore would sharply diverge from ACARD's view of how to support industry through applied research.

In February 1987 the Chief Scientific Adviser, John Fairclough, told Thatcher that 'We need a renaissance of our industrial prowess through the contribution from research and development to again become an effective competitor to Japan, Germany and the United States'. He therefore proposed moves to streamline and strengthen decision-making machinery, notably new Cabinet committees, including E(ST) at the centre of government. To offer this structure good advice, Fairclough recommended that ACARD:

> be extended to cover basic and strategic science in addition to its role in applied research and development. This would create an advisory body that would look across the whole subject and so balance our priorities between fields of endeavour which currently underlay the current division between basic and applied science.[69]

In April 1987 Fairclough had a name for the extended ACARD: the Advisory Council on Science and Technology (ACOST).[70] E(ST), advised by ACOST, would set 'national priorities for science and technology' and have 'responsibility from science right through to the exploitation stage'. ACOST first met later in the year, and continued to be chaired by Tombs. But by then, as I show in the next chapter, science policy had radically shifted, and the centralised, informed industrial strategy on which E(ST) and ACOST had been called into being had been ended. ACOST was now regarded as a quango, second-guessing the market, and a symbol of a 'hankering after some bureaucratic direction of research'.[71] When ACOST suggested, as ACARD had done before it, that the government should increase science funding, it too was told in no uncertain terms that such advice was neither welcome nor expected.[72] Even though Thatcher would occasionally chair ACOST (for example, on 11 May 1988[73] and 14 March 1990)[74], the ground had been cut from beneath ACOST's feet. In July 1990, when Tombs finished his role of chair of ACOST, he used his final meeting with Thatcher, with Guise present, to complain about the ending of governmental support for applied R&D.[75]

Advisory Board for the Research Councils

The second important central committee, the Advisory Board for the Research Councils (ABRC), gathered together academic scientists, chief scientists from the departments and the Central Policy Review Staff (in effect, chief scientific adviser to the government), the chair of ACARD and the heads of the research councils. It was chaired until January 1983 by the physicist and vice-chancellor of Bristol University Sir Alex Merrison, and thereafter by the Professor of Molecular Biophysics at the University of Oxford, Sir David Phillips – a figure 'very powerful through the '80s and very good', according to a former chief scientific adviser.[76]

The ABRC had been established in 1972 with terms of reference to advise the Secretary of State for Education and Science on 'his responsibilities for civil science' and the 'allocation of the Science Budget', and to 'promote close liaison between the Councils and the users of their research'. In 1982 it had published its advice for the first time, revealing a view that a supposed 'swing away from "big science" ... had gone as far as it could if Britain was to maintain a stake in high energy physics and astronomy'.[77] Like ACARD, the ABRC published reports, based on the findings of working groups, an example being *A Study of Commissioned Research* led by Sir Ronald Mason, published in November 1983; it was critical of the practice, if not the 'logic', of the Rothschild customer/contractor approach, and recommended that the ABRC be strengthened 'by giving it enhanced authority and responsibilities'.[78]

The chairs of the ABRC and ACARD began issuing joint periodic reports on the state of science and technology in the United Kingdom from 1983. The ABRC report that gathered the most public comment and debate was *A Strategy for the Science Base*, published in 1987; this was assumed to be influential, or even a statement of UK science policy.[79] However, as I show in the next chapter, government science policy moved in a quite different direction. Following the establishment of ACOST, the ABRC was eventually, in 1990, given a slimmed-down role and needed fewer members.[80]

Chief Scientific Adviser

ACARD, ABRC and later ACOST were the leading bodies for collective provision of advice. However, for much of the postwar period, a single individual has played an important, central role in the provision and organisation of scientific advice. The role has changed over the years, shaped by circumstance, personality and policy. Chief scientific advisers will make many appearances in the following chapters, so it is valuable to trace how the role developed in some detail.

Individual scientists, such as Henry Tizard and Frederick Lindemann (Lord Cherwell), had advised the centre of government in the 1940s, but in the 1950s and early 1960s a more distributed model of the provision of scientific advice had prevailed. The return to the individual role, and indeed the official inauguration of the title 'Chief Scientific Adviser' to the government, occurred in 1964. Solly Zuckerman was a South African-born zoologist who had conducted the gory but necessary work of investigating the effects of explosives on bodies; he also carried out statistical assessments of bombing operations during the Second World War.[81] He was a trusted, independent insider. Zuckerman had already served in many capacities in government before he was appointed Chief Scientific Adviser to the Ministry of Defence in late 1959.[82] 'No one ever more completely stormed every bastion of the British establishment,' observed Roy Jenkins.[83] Interestingly Zuckerman insisted on a change of name from 'Chief Scientist' ('inappropriate', he thought, 'for someone who knew as little as I did about the "hardware" side of things') to Chief Scientific Adviser.[84]

Zuckerman repeatedly stressed the requirement of an adviser to challenge received opinions and entrenched interests. His views could be 'heterodox', rejecting battlefield nuclear weapons, for example, against the view of chiefs of staff. In 1964 Harold Wilson wanted to make Zuckerman a minister of state, leading on disarmament issues. Zuckerman declined. But his role as CSA for MoD was also untenable, perhaps because the Minister of Defence Denis Healey and Zuckerman never quite saw eye to eye. The role of Government Chief Scientific Adviser (GCSA) was therefore created for him. He also, and he never tired of telling people of the fact, was made Head of the Scientific Civil Service, a managerial responsibility for 10,000 people – larger than the body of 3,000 administrative civil servants.

As GCSA, Zuckerman advised on large defence projects – including controversial cancellations, for example the RAF's 'pet project', the supersonic, low-flying, multipurpose TSR-2 aircraft.[85] He was instrumental in recruiting the cosmologist Hermann Bondi to review the UK's space policy in the late 1960s.[86] Environmental issues were increasingly prominent, and Zuckerman's advice included responding to immediate crises (such as the giant Torrey Canyon oil spill off Cornwall in 1967)[87] as well as instituting longer term bureaucratic mechanisms for collating information and reviewing courses of action, notably the Royal Commission for Environmental Pollution in 1969–70.[88] Other issues included London flood planning (leading to the Thames Barrier) and the panic over the migration of scientific talent labelled the first 'brain drain'. He also attempted to review R&D spending across departments, through a new Central Advisory Council for Science and Technology, set up in 1967, which brought him into conflict with ministers.[89]

Zuckerman retired in 1971, but he continued to chip in his views about science and government right up through the 1980s (indeed he retained rooms in the Cabinet Office, and advised on the badger/bovine TB issue during the Thatcher administration). His style was to be the trusted consultant, the challenger of received views, and he relied on good, wide, informal networking. Zuckerman's list of attributes of an ideal GCSA can be extracted from his comments in speeches and private correspondence. An ideal GCSA would:

1. offer up sensible, reasoned, informed advice;
2. be independent of vested interests;
3. keep in touch (inwards, with the civil service, and outwards, with the scientific community);
4. answer requests for information (CSAs play this role in departments);
5. anticipate information that will be needed, and therefore commission research if necessary;
6. sometimes manage staff;
7. should not be excluded from key discussions;
8. be personally trusted by the Prime Minister; and
9. be personally trusted by the Cabinet Secretary.[90]

The technocratic Heath government brought in the era of the Central Policy Review Staff (CPRS), the think-tank assigned the general task of wide and deep critical review. It was led by a scientist, Victor Rothschild. Therefore it was a moot point whether there should be another chief scientific adviser to the government after Zuckerman. The Treasury was against; so was Burke Trend, the Cabinet Secretary, who smoothly said Zuckerman was 'sui generis'; Zuckerman insisted, arguing that since 'Permanent

Secretaries do not reproduce themselves from the same mould' neither should the next GCSA be expected to have the same background and experience as the last.[91] In the event Alan Cottrell, a science adviser to the Ministry of Defence and 'the outstanding physical metallurgist of the twentieth century' was appointed – albeit, as Zuckerman noted, at a rank 'one pip lower than mine'.[92] Cottrell also threatened to resign if, instead of being retained as an independent GCSA, he had been placed under Rothschild.[93] Nevertheless during the 1970s it was the CPRS – a team of talents – rather than the individual GCSA that mobilised specialist expertise for the guidance of government.

When Cottrell became full GCSA in 1971, the division of labour was, in effect, split three ways. The CPRS led on any issue, including science-based issues, that took its interest and Zuckerman, although retired, retained a role advising on nuclear weapons matters – leaving Cottrell with a rump including civil nuclear policy, space policy, the environment, communications and Europe. The 'scientific role' of the CPRS was 'that of asking the fundamental or innovatory question and of undertaking certain studies or projects which are best conducted or led from the centre', while the GCSA's role was primarily one of coordination.[94] Nevertheless, Cottrell did make substantial contributions at the interface of science and government. He was instrumental in the decision to deepen the UK Government's horizon scanning work, exemplified by the establishment of the Cabinet World Trends committee, part of a complex response to the well-publicised computer simulations of the Club of Rome.[95] He was most effective, however, on an issue that directly demanded his expertise. The choice between types of reactor for civil nuclear power was a fraught one, with the candidates for the second, post-Magnox generation including British Advanced Gas-cooled Reactors (AGRs), American Pressurised Water Reactors (PWR), a Steam-Generating Heavy Water Reactor (SGHWR, based on a Canadian system) and fast reactors that transmuted uranium to plutonium. Cottrell weighed in against the Central Electricity Generating Board's choice, the PWR, warning that its large reactor pressure vessels were liable to sudden, brittle fracture.[96] The result was considerable, further investigation.

A second major issue of the Cottrell years was the content and consequences of the Rothschild report of 1971, a green paper that introduced the 'customer-contractor' principle to guide the relationship between government departments. While Cottrell seems to have suggested this principle to Rothschild, the 'issue … caused quite a furore, and much of Cottrell's time in late 1971 and early 1972 was spent "clearing up the mess"'.[97] One strategy to smooth ruffled feathers was to insist, in a draft

White Paper,[98],that the principle marked 'not an arms-length contractual arrangement, but a partnership between the Research Councils and the executive Departments, held together financially'; in addition, departments would be urged to appoint chief scientists 'with responsibilities to make the partnerships work effectively'.[99]

Alan Cottrell resigned in April 1974 to become Master of Jesus College, Cambridge, and he, although less often than Zuckerman, occasionally contributed to the national debate about science policy thereafter.[100] His responsibilities were taken over by Robert Press. Press had worked as Assistant Chief Scientific Adviser, Nuclear, in the Ministry of Defence from 1964 until 1967 when he became Chief Scientific Officer in the Cabinet Office Science and Technology Group, also specialising in nuclear matters. His formal position, after Cottrell left, was Deputy Secretary, Science and Technology, within the Cabinet Office. Press advised on nuclear security, the disposal of radioactive waste and the supply of uranium for the nuclear projects.[101]

The diffusion of advisers encouraged by the Rothschild reforms, and the continuing greater influence of the CPRS, when combined with the temporary and lowly status of Press, means that he can barely be considered to be the successor to Cottrell at all. Zuckerman had a jaundiced view of Press, describing him as 'really a note taker ... kept on to deal with nuclear weapons matters' and in doing so 'merely became the mouthpiece of the Aldermaston interests'.[102] Unlike both his predecessor and successor, Press had neither academic professorial rank nor was made a Fellow of the Royal Society.

Beginning in 1974 there was considerable debate about what to do after Cottrell. This intensified when Press too retired in 1976. In 1974, and again in 1976, Sir John Hunt, Cabinet Secretary, had framed the debate by offering a three-way choice.[103] The first was to 'go for a replacement at the same level' as Cottrell. But 'following the appointment of more Chief Scientists in Departments and the increasing emphasis on the customer/contractor relationship', he felt there 'was no need for a full-time [GCSA] job in the Cabinet Office'. Hunt claimed that, in 1974, Sir William Armstrong (head of the Home Civil Service), Zuckerman, Rothschild, Bondi and Cottrell himself had 'all agreed with me on this'. The second option was not to appoint a GCSA, but instead disperse the staff, 'putting their "advice" responsibilities with the CPRS and integrating their "secretarial" responsibilities with the rest of the Cabinet Secretariat'. 'However,' Hunt noted, presciently, 'it seemed difficult presentationally to disband Cottrell's unit.' The third choice was a 'short-term expedient': to appoint someone at lower level and not with the title of

GCSA, 'who might be more successful in co-ordinating the scientific machine than Cottrell had been'; this optionhad the 'additional advantage in that it could be adapted in light of experience'. In 1974 this third option had been followed with the appointment of Press. Now, in 1976, Hunt urged Prime Minister Harold Wilson to go the whole hog and end the GCSA role permanently:

> I am sure that we were right to give up the CSA post: and all those whom I have consulted agree that of we were to revert to it on Dr Press' retirement we should only be looking for trouble. We would not want a second-rate CSA, and if we get someone first-rate he would either become frustrated or get in the hair of the Departmental Chief Scientists like Hermann Bondi or Walter Marshall. In other words the days when a Solly Zuckerman could virtually direct all our scientific activities from the centre have gone for good.

Indeed, Wilson agreed with his private secretary's view – and this speaks to the relative insignificance of the GCSA post at this time – that the post could be usefully sacrificed to counter impressions of empire building around the Prime Minister.[104] In Wilson's and Hunt's eyes, the CPRS was enough.

But word leaked out. There was a concerted campaign to reverse the decision from MPs on the Science and Technology Select Committee, the Labour MP Tam Dalyell and the President of the Royal Society.[105] 'Clearly reflecting the views of senior fellows', Alan Hodgkin raised the issue in his retirement speech as President, invoking an imaginary rabies epidemic to show what would be missed without a GCSA.[106] Stung, Wilson agreed an avowedly 'cosmetic change': a new appointment, John Ashworth, could be called 'Chief Scientist, CPRS'.[107] The job was not considered, at least by the Cabinet Secretary, as equal to a 'full' GCSA.

John Michael Ashworth had studied chemistry and biochemistry at Oxford before completing, in 1965, a PhD at Leicester University.[108] He briefly joined the 'brain drain' of scientific talent to the United States before returning to the United Kingdom to take up academic posts – first at Leicester in 1967 and then, from 1973, at the University of Essex. He took a secondment in 1976 from being Head of the Department of Biology to take up the new appointment. As Chief Scientist, CPRS, Ashworth played much of the GCSA role, but he was also only second-in-command in the organisation, deputy to Sir Kenneth Berrill, an economist who had taken over from Lord Rothschild as head of CPRS.

Nevertheless, in practice, Ashworth grasped the 'opportunity for a different kind of influence' to that operated by the Chief Scientist, CPRS being the centre of a Venn diagram of overlapping bodies.[109] First, at CPRS he carefully avoided 'being seen to set up a sub-unit within the CPRS'. Ashworth, as an integrated member of the think-tank therefore contributed to the general reports on issues submitted to Cabinet. Second, the CPRS 'got more than another scientifically qualified member with a fancy title' because of his membership of two committees, one attending to internal issues and the other to external matters. The Official Committee on Science and Technology (STP) committee, a committee of chief scientists (including Ashworth) and permanent secretaries, chaired by the Secretary to the Cabinet, was tasked with 'co-ordination'. 'To help them and to provide a needed interface between Government and organisations outside Government' ACARD had been established in 1976, serviced by the Cabinet Office. Third, a link to the research councils under the Department of Education and Science was made through the invitation to the Chief Scientist, CPRS, to attend the ABRC. Fourth and last, the Chief Scientist, CPRS, also took on international representative roles, such as being the UK chair to the body advising the Council and Commission of the EEC. In summary, these intersecting roles meant that the Chief Scientist, CPRS, possessed influential links to the worlds of industrial (ACARD), academic (ABRC), governmental (STP) and international science.

The issues Ashworth influenced were therefore numerous. They included industrial policy (especially microelectronics, the subject of the first ACARD report[110] and a controversial topic stoked by a very effective BBC *Horizon* programme, 'When the chips are down', in 1978), a review of the Scientific Civil Service,[111] UK–USSR scientific relations, the definition of 'genetic engineering' in a context complicated by the threat of closure to the Microbiological Research Establishment, and anthropogenic climate change.[112] Climate change (discussed in Chapter 7) was raised in Ashworth's first meeting with Thatcher, which did not take place, the biographers tell us, until 1980. Thatcher stopped him mid-flow and asked Ashworth 'incredulously, "Are you telling me I should worry about the weather?"'.[113]

Ashworth continued until September 1981, when he took up the offer of the Vice-Chancellorship of Salford University. Again there was a transition point, when what the chief scientist should be was open to renegotiation and change. Robert Armstrong raised the issue with Margaret Thatcher in April 1981. 'Dr Ashworth has done the job admirably, with energy and drive as well as good sense,' he wrote, adding that it was 'no reflection on him when I say that I think that we are not quite strong enough on the co-ordination of scientific policy and the provision of scientific advice at the centre; indeed, it is partly his particular personal qualities that have masked what is, I believe, a deficiency in organisational terms'.[114] But Armstrong did not think it was possible 'to go back to having a Chief Scientific Adviser to the Government'; science-based decision-making was now too widely distributed 'down the line in Departments' and, furthermore, 'we can no longer hope to find, or would want to have, a "political" scientist like Lord Cherwell or Lord Zuckerman'. But the 'centre, and particularly the Prime Minister' still needed 'access to scientific advice over the whole range when necessary'. Armstrong therefore proposed two new developments: first, identifying 'a small number of scientists of the highest eminence' who, by 'informal arrangement' could be tapped by the Prime Minister, and second, keeping a Chief Scientist, CPRS, but upgrading the post to Deputy Secretary level.

Part of the context for this discussion was increased lobbying for upgrading the status of scientific advisors to government – the latest instance in the century-long 'neglect of science' complaint. The key linkage here was between the Royal Society and the House of Lords Select Committee on Science and Technology, with Lord (Alexander) Todd, recently retired Royal Society president, as the linchpin. The Royal Society submitted evidence to the Select Committee's inquiry that called for enhanced status of chief scientific advisers.[115] Armstrong read it as a desire 'to go back to the good old days of a Minister for Science and a Council of Scientific Advisers'.[116]

Thatcher's first response was vehement. Should the provision of scientific advice to the centre be strengthened? 'No,' she wrote, 'the advice available through the ABRC should be available to me and it is much more <u>varied</u> than that of any <u>one</u> scientific adviser.'[117] In discussion she relented, slightly, saying that she had no objection to the appointment of a new Chief Scientist, CPRS, 'provided there was an off setting reduction of a scientific post in the same grade elsewhere in government'.[118] Armstrong and the director of the CPRS, Robin Ibbs, began the search. When the Cambridge professor of engineering Michael Ashby declined for 'personal reasons', the outstanding candidate became Robin Nicholson.

Suggested by the new Royal Society president Andrew Huxley as his 'first choice', Nicholson was 47, a Cambridge-trained metallurgist who had worked both in academia (he had been professor of metallurgy at Manchester) and industry. Indeed, as managing director of Inco Europe Ltd and co-chairman of Biogen, he possessed, wrote Armstrong to Thatcher, 'from our point of view, the great advantage of having seven years in industry, and having a strong interest and experience of the application of science in industry'. He had, declared Armstrong, a 'lively mind and attractive personality'.[119] Thatcher had indicated that she wanted a scientist with experience of industry and also the mark of scientific status that came with being a Fellow of the Royal Society.[120]

Yet even then Thatcher was not happy. She complained she was not convinced that the CPRS needed a chief scientist, and wondered if Nicholson might work better at Number 10. The deeper issue here was the CPRS itself.[121] Thatcher rapidly developed an extreme antipathy to the role and products of the think-tank. So while Nicholson was first indeed appointed as Chief Scientist, CPRS, she was ready to accept the urging of the Select Committee on Science and Technology when it called, in December 1981, for the development of the post of Chief Scientist, CPRS 'into a post of Government Chief Scientist'.[122] Nicholson was also given 'the right of direct access to the Prime Minister'.[123] With the bloody demise of CPRS in 1983 Nicholson became Chief Scientific Adviser, Cabinet Office, a title Thatcher had rejected emphatically one year previously.[124]

Nevertheless, it was from this unpromising beginning that the current GCSA role became entrenched. At his first meeting, Thatcher asked Nicholson to advise her on the feasibility of a project she had just agreed with President Mitterrand: a fixed Channel link. 'My jaw dropped,' he later recalled, 'I knew nothing about bridges and I knew nothing about tunnels.'[125] Yet he immediately drew on networks of contacts to gather informed advice that satisfied the Prime Minister – a good illustration of how a single individual chief scientific adviser, who could only ever be an expert on a fraction of the topics demanded, could perform a general function. However, that the role became entrenched at all was largely a result of the strong relationship between Nicholson and Thatcher, one based on shared ideological convictions. (Perhaps this is what Armstrong had signalled when he had written to Thatcher of Nicholson's 'attractive personality'.) When the Select Committee on Science and Technology, for example, produced a report on engineering R&D in 1983 calling for a 'national strategy for technology and manufactured products', Nicholson led the charge that it 'was not work of the highest quality', and ran 'completely counter to the Government's view that market considerations should guide the support of technological developments'.[126] Likewise his view about academic tenure was that it was anti-market. 'One of the problems for universities … has been their inflexibility in retraining/hiring/losing the appropriate staff,' Nicholson wrote to Keith Joseph. He added: 'the tenure system means that universities can simply not respond in the way a business can to changes in demand for teaching and research', a system that 'greatly hinders any rational response to the market'.[127] Nicholson shared with Thatcher a faith in market solutions. In a discussion of defence R&D, Nicholson argued that the present situation of government-funded research and industrial development 'under contract and at no risk to the company' was the cause of the low level of civil spin-off. He continued:

> Frankly I believe that a significant improvement will only occur if defence procurement changes radically from the present system where Government pays first for the research, then for the development and finally for the equipment, to a more normal commercial arrangement where Government buys defence goods at a price which allows the manufacturer to carry out and pay for his own R&D.[128]

This policy of privatisation of defence research would indeed be followed, albeit at a later date.

Nicholson reserved particular venom for other organisations. The British Technology Group, which had been formed in 1981 out of the old Attlee-era National Research Development Corporation, was described as having 'all the sloth and leaden-footedness characteristic of a state-owned monopoly' whose 'eventual reward for success must be privatisation'.[129] The European high energy physics facility, CERN, was another subject where Nicholson and Thatcher shared views. Thatcher, in a meeting with Keith Joseph, Secretary of State for Education and Science, had described CERN as 'extravagant' and its UK funding open to review.[130] Nicholson's advice again chimed with his Prime Minister's view:

> Withdrawal from CERN must be contemplated as one option … I personally doubt it will come to that. More likely will be recommendations to improve the cost-effectiveness of CERN (you've seen the gold plating yourself) and, crucially, to slow down the pace and hence the rate of spend on this area of research. There is no reason why the tax-payers of Europe and the USA should have to fund a private race between two scientific cliques carried out at a pace determined largely by their own curiosity and arrogance.[131]

This discussion was occurring during the time between 1983, when CERN discovered the W boson particles, and 1984, when CERN scientists won the Nobel Prize for Physics for their discovery.

In summary, when Nicholson became, quite contingently, Chief Scientist, Cabinet Office on the demise of the CPRS, he was able to develop a free-standing role that re-established the full GCSA model. Like Zuckerman, he expressed his views with admirable pungency and force. Like Ashworth, he was effective by having influence on many issues. However, the added difference, indicated by his being granted the right of direct access to the Prime Minister, was a personal rapport with the most influential figure of all.

In 1984 Nicholson's attention was drawn to another major state-funded institution concerned with nuclear research: the UK Atomic Energy Authority. In March the Secretary of State for Energy, Peter Walker, started a comprehensive review of the activities and role of UKAEA, and asked for Nicholson's participation.[132] The state provided half of UKAEA's funding, with the rest coming from the electricity utilities and British Nuclear Fuels Ltd. Thatcher's Private Secretary (Economics) suggested that a review was certainly 'overdue'. He asked: 'Has the nation got value for money? Or is it a producer dominated organisation? Does it need to be financed so much ... by the taxpayer or could its customers ... contribute more? Has one branch of science absorbed too much of our research effort?'[133] When it emerged in August, the report concluded that continued government support of nuclear R&D was justified.[134] But it is clear that Nicholson wanted to go further. In his view, UKAEA's 30-year history had seen the 'UK slide from first rank to second rank position in civil atomic energy', in which the customer (CEGB) had been forced to buy the wrong reactors, export had been negligible, taxpayers' money had been wasted and 'by virtue of the interest of its research and its employment conditions, [UKAEA had] creamed off a significant slice of the UK's R&D talent'.[135] While this talent had produced a 'substantial technological asset', the 'contribution of this technological asset to the wealth-creating sector of the economy had been negligible'.

In the words of Nicholas Owen, of Thatcher's own think-tank, the Number 10 Policy Unit:

> Robin Nicholson has reminded us that some of the best scientific brains of a generation have been squandered on misdirected work on civil nuclear power. Since our scientists are among the ablest in the world, the finger points to political misjudgements and intervention over a long period. Can we do better over the next 30 years?
>
> Our priority should be to develop a framework for nuclear research which allows the maximum role for the market and little at all for political interference. The best solution would be to privatise the AEA.[136]

By April 1985, in fact, so involved did he feel about the issue that Nicholson entered into discussions about becoming the new chair of UKAEA. 'We need somebody who can combine a reputation for scientific excellence with determination to improve the Authority's commercial performance,' wrote Peter Walker.[137] On being called in to a Friday meeting with the Prime Minister, Thatcher had told Nicholson that 'Ministers would very much like Sir Robin to take over the Chairmanship'. Nicholson replied that he was 'interested but that he was by nature "a private sector man"'; his decision would depend 'on what he was being asked to do'.[138]

Disagreements subsequently emerged about what UKAEA might be expected to do in the future: Nicholson wanted UKAEA a free hand to enter non-nuclear manufacture, to better exploit its assets, while Thatcher thought this would be 'inconsistent' with her government's 'policy on the role of public sector bodies and would be bound to create difficulties with private sector companies'. Nicholson let it be known he was being headhunted for a post at the glass company Pilkington, perhaps a ploy to put gentle pressure on Thatcher.[139] But Thatcher in June 1985 decided not to appoint Nicholson as chair of UKAEA.[140] Left in the breach, Nicholson soon ended his career (which was renewable on a two-yearly period) as Chief Scientific Adviser.

The new Chief Scientific Adviser was John Fairclough. He was described as 'the first industrialist' to take up the position (although, as we have seen, Nicholson had considerable business experience).[141] He was also the first, indeed the only, GCSA not to be a Fellow of the Royal Society (excluding Press). Born in Yorkshire, Fairclough had joined the computer department of Ferranti in 1954, where he worked on the design of the Ferranti Pegasus mainframe computer. He joined IBM in 1957, where he was instrumental in the design of the influential, inter-compatible fleet of computers, the System/360.[142] By the time of his appointment as GCSA Fairclough had risen to be Director of Manufacturing and Development and Chairman of Laboratories for IBM (UK). The choice of a computer expert as GCSA in the 1980s is no surprise. Information technology (IT) had become one of the central targets for policy as well as a prominent area of technological change, associated with the spread of small office and home computers, many of which were brought to market by UK entrepreneurs. But the choice also reflected Thatcher's preference for industrial innovators over pure research scientists. Fairclough found the job 'a very different sort of experience from the work and environment I was used to in IBM ... a mixture of daunting freedom on the one hand ... and a lack of specific budgetary responsibility on the other – I have no responsibility for any individual science and technology programme, no laboratories in which to test out my advice'.[143]

Fairclough's career as GCSA has been associated, for example by Edgerton and Hughes, with the distinction drawn between 'near-market' and 'far-market' research, in which the former should be organised and funded by industry, while the government retained responsibilities for public funding the latter.[144] Indeed, government funding would be 'confined to areas where the market would "fail to operate to produce maximum benefits to the economy as a whole"'. Such a decision led, according to one group of academic analysts, 'to the withdrawal of most government support for civil near market and single company R&D, the main exception being in aerospace'.[145] However, I show in the next chapter that the drive for this new science policy did not come from Fairclough, but rather from the Number 10 Policy Unit.

The Number 10 Policy Unit has been created to serve Harold Wilson in 1974. Like the CPRS, it contained a small staff of bright insiders and outside talent. Unlike the CPRS, it served the Prime Minister, not the Cabinet as a whole. When the CPRS was demolished, the Number 10 Policy Unit became ever more influential; it was led by Ferdinand Mount (1982–3), John Redwood (1983–4) and Brian Griffiths (from 1985) not least as a formidable and caustic source of Thatcherite advice. In 1986 the Unit had eight members, each with an area of specialisation. One newcomer was George Guise, described by his fellow Unit member David Willetts as 'on secondment from Consolidated Gold Fields: specialising in industry and research and development'.[146] As I show in the next chapter, it was Guise who persuaded Thatcher to follow the new science policy.[147] Guise's rhetorical method was surprisingly anecdotal, but then again, as Willetts noted, the Policy Unit was not 'objective' in a straightforward sense.[148]

I show that Fairclough was essentially sidelined, and a new science policy introduced, during a period of months when Thatcher hardly spoke to her Chief Scientific Adviser. The new science policy was able to be driven through, however, by a centralisation of science policy-making that was promoted by Fairclough – notably a strengthened Cabinet committee system – and which was described as being inspired by his experience of IBM's directed management of research.[149] Ironically this system enabled the unpicking of the UK's industrial strategy for research.

The following year, in 1989, Fairclough 'also restated the customer–contractor principle with the aim of encouraging development of an internal market in which public sector research providers would compete for public funds for R&D'.[150] Privatisation was seen as taking the 'customer/contractor principle to its logical conclusion'.[151] It was in this context that he oversaw the setting up of government research establishments as Next Steps Agencies. The approach was encapsulated in so-called 'Fairclough Guidelines'.[152] In September 1990 Fairclough left his GCSA position and became chair of NM Rothschild and Sons' venture capital arm.

The new GCSA was Professor William Stewart. A Scot, raised on the island of Islay, Stewart was an academic 'biologist with environmental interests'; he had built a strong life sciences department at Dundee University, become an FRS at the age of 42 and in 1987 been headhunted to lead the Agriculture and Food Research Council.[153] It is clear that Thatcher had Stewart in mind for the GCSA as early as 1989 as, when his name was mentioned in connection with a more minor appointment, she ruled it out, noting 'we have other plans for him'.[154] His environmental credentials, exemplified by a 1983 Royal Society report on nitrogen in the environment and work for the Royal Commission on Environmental Pollution, appealed to Thatcher during her late Prime Ministerial green pivot.

At her first meeting with Stewart she told her GCSA that her main concerns were that the apparatus for scientific advice and for research funding were too bureaucratic, that Big Science was too greedy of resources, and of uncertainties and gaps in the scientific evidence for global warming.[155] Stewart replied that he 'shared many of these concerns'. (Interestingly, in an oral history interview Stewart later recalled that when 'I turned up as the new CSA her first two sentences to me were: "Good morning Professor Stewart" and "Sort out intellectual property".')[156] However, two months later Stewart was officially advising a new Prime Minister, although John Major, with less interest in science than Thatcher, did not meet his GCSA until 1991.[157]

Parties and Parliament

Despite its significance, science was never a major component of party politics in the 1980s. The 1979 election manifestos made barely any reference to science or technology;[158] the Conservative manifesto of 1983 promised to 'accelerate the transfer of technology from university laboratories to the market place' by encouraging Science Parks as well as funding for 'new blood' in higher education, while Labour promised a reversal of Tory cuts and vague support for new science-based industries (the new Liberal–SDP Alliance merely promised raising school standards).[159] The 1987 manifestos reflected, without fanfare, the great shift in science policy, with a statement that the 'task of government is to support basic research and to contribute where business cannot realistically be expected to carry all the risks'. Labour proposed a new Ministry of Science and Technology to coordinate government activities and budgets, while the Liberal–SDP Alliance promised to reform A-Levels to heal the 'arts-science divide', suggesting that their science policy thinking was stuck in the times of C. P. Snow.[160] Nevertheless, science-based issues (such as climate change) were sometimes viewed through the prism of party presentational tactics. More importantly the debate around science and industrial strategy was, as I show in Chapter 3, deeply freighted by ideology. Otherwise, in a parliamentary context, science featured in certain prominent debates and in the multi-party work of select committees and individual Members of Parliament.

Debates in the Houses of Parliament could influence government thinking. One example was the 10 February 1984 debate in the House of Lords on science funding, in which concerns about the effects of cuts were raised and debated (see Chapter 3).[161] Lord Flowers had concluded, in a speech that was brought to Thatcher's attention by William Waldegrave:

> It is often said that we produce more Nobel prizes per head of population than any other country, so that there can be little the matter with the level of support that we give to science. I am sad to have to say that I doubt whether that is any longer the case. Our research facilities in general no longer bear fair comparison with those of similar countries such as Germany, France and the United States. It seems to me that our research is no longer held in quite the same high esteem internationally as it once was … . Much of the blame must fall on the decline of the dual support system and the resulting stifling of initiative.[162]

Such debates, therefore, were avenues by which concerns over the effects of cuts in public expenditure on science were articulated.

Rarely, parliamentary debate could be a prime mover on a whole area of science policy. An outstanding example of this proved to be the embryology debate around the Warnock report, discussed in the coda to Chapter 3. However, such an example was very much the exception.

Parliamentary select committees were also an important cog in the science policy machine, examining witnesses and generating reports to which the government had to respond. Particularly important was the House of Lords Select Committee on Science and Technology (established after the House of Commons equivalent was wound up in a reorganisation of Parliamentary structures), while the Select Committee on the Environment appears in Chapter 7. The annual reviews of R&D and the periodic joint reports on the state of UK science and technology by the chairs of ACARD and ABRC were both instituted in 1983 as part of the government's response to a House of Lords Select Committee on Science and Technology's report.[163] Likewise, the establishment of a Chief Scientific Adviser, Cabinet Office in 1983, when the CPRS ended, was also cast as a response to this Committee's report, although a parallel call for a Minister of Science was rebuffed.[164]

Thatcher's decision to reserve the right to respond to science questions created awkwardness when it came to the work of select committees. Since select committees called witnesses to scrutinise the work of government, a science select committee could reasonably ask to hear from the minister responsible for science. For example, the House of Lords Select Committee on Science and Technology pointed out, in 1986, that since 'you [Thatcher] have accepted responsibility for co-ordinating questions on science and technology, they feel it would be discourteous not to give you the opportunity to come to give evidence'.[165] In the event, citing busyness and on the advice of her Cabinet Secretary and her Chief Scientific Adviser, Thatcher sent her secretaries of state.

Finally, individual Members of Parliament could take an influential interest in science matters. Jeremy Bray, for example, was a mathematical economist who paid attention to technical matters, such as computer modelling of the economy and the environment in the 1970s, and was Neil Kinnock's Opposition spokesperson on science and technology after 1983. However, the outstanding case was the Labour MP for West Lothian (1962–83) and Linlithgow (1983–2005), Tam Dalyell. Dalyell was independent-minded, and fiercely criticised both Conservative and Labour governments on a range of issues, including the sinking of the *Belgrano* in the Falklands War and the 'dossier' on weapons of mass destruction integral to the second Iraq conflict. While his education was in history and economics, he watched, critiqued and published on science policy, and contributed a column to *New Scientist*.[166] Dalyell was a skilled parliamentarian, and treated as a gift Thatcher's statement that she might be responsible for science issues. Thatcher would receive letters from Dalyell that began 'As you are responsible for science, I wonder if I can ask … ?', to which a response had to be made; he also made use of Parliamentary Questions to press the government.[167]

External bodies

So far I have reviewed the work and roles of the bodies that made science policy within government – ministers and prime ministers, civil servants in Whitehall, chief scientific advisers, and the members and committees of Parliament. I have also noted the influence of industry through ACARD. Bodies external to Westminster were also important. These bodies included academies, campaigning groups and journalists. Many of these will be introduced in the appropriate place in the following chapters.

The Royal Society, the elite academy of science, worked to promote and protect the interests of UK science while also working, in parallel to the Foreign Office, to support the UK's international scientific standing. For much of the twentieth century, the presence of the Royal Society close to Parliament and Whitehall (before the Second World War in Burlington House and afterwards on Carlton House Terrace looking over the Mall), enabled both formal and informal exchanges of views (the latter sometimes at the Athenaeum, a private club with shared membership of senior civil servants and fellows of the Royal Society) with government. By the 1980s this cosy, gentlemanly relationship had become less straightforward.

In May 1986 the President of the Royal Society, George Porter, facilitated by Lord Rothschild, secured a meeting with Thatcher. Porter used the access to pitch two ideas. The first was that there should be a new 'National Science Advisory Council' of independent 'practising' scientists. This should be chaired, he thought, by the Prime Minister, and able to take a considered, long-term view, a 'proper perspective of the dependence of the nation's prosperity' on science and technology; it would thus provide 'a channel for scientists to make an input to Government policy-making at a high level'.[168] The second idea was higher salaries to ward off the 'brain drain' of talented scientists to the United States. Both ideas, but especially the first, were given short shrift. Significantly, one reason given by Fairclough, then Thatcher's GCSA, was that Porter was unaware of the new mechanisms of science decision-making that had been put in place (see Chapter 3).[169] The Royal Society's knowledge of science policy-making was not necessarily intimate nor complete.

However, the episode also reveals other aspects about science policy-making. First, Porter had been moved to write to Thatcher partly because of the establishment of the campaigning group, Save British Science, launched, as we saw in Chapter 1, in January 1986 to protest against cuts in university science; it had received the support of 100 Royal Society Fellows.[170] Porter had hoped that a National Science Advisory Council 'would improve morale generally in the scientific community, and take off some of the pressure generated by such movements as "Save British Science"'.[171] The 'real concern lying behind Sir George's proposal is his feeling that the presentation of the Government's science policy is not as good as it should be', one Thatcher's advisers noted, adding that a 'great deal of good work is going on, but it is difficult to publicise it, and the "Save British Science" lobby is accordingly able to capture support'.[172] This occasion was one of the very few moments that the arguments of Save British Science were raised – and even then only obliquely and mediated by the Royal Society's own interests, at the heart of government. Yet Save British Science, from the point of view of an external witness to the politics of 1980s science, was highly vocal in the public sphere. It is surprising how little it was heard from within the decision-making centre of the state. Second, Fairclough, in conceding that there was a presentational problem, and that 'we need to quieten this lobby if we are to be able to get a sensible discussion' on science issues, did note that one 'obvious vehicle for presenting the Government's science policy' would be a prime ministerial speech. This thought, underlined by Thatcher, may have been one origin of her famous 1988 speech on science, environment and curiosity-driven research.

The Royal Society did shape debate on specific issues, such as acid rain (see Chapter 7). There were plenty of other organisations, large and small, that sought to influence UK science policy. The impact of campaigning groups, such as the UK branches of Greenpeace and Friends of the Earth, will be shown in the chapters on nuclear and environmental issues. Individuals, too, could occasionally pitch science policy ideas that reached Number 10, such as the case of David Horrobin, who argued that 'any reading of scientific history indicates that leading experts can almost never be trusted to make decisions about what basic science is worth supporting' and wanted to ditch research councils for prizes. His idea reached the Prime Minister after the Vice Chancellor of the High Court happened to be having dinner with Thatcher at Lincoln's Inn.[173]

Yet in general, gaining the ear of the Prime Minister was very hard; those around her acted as gatekeepers, who might only open the gate when outside views aligned with hers or theirs. In Chapter 6 I will show that Charles Powell was the gatekeeper for missile defence research. George Guise, in the Number 10 Policy Unit, was also a gatekeeper. It was Guise, for example, who shot down Horrobin's plan, but also chose to forward her the science funding statistics of Terence Kealey because they were seen as undermining the pleas for more government funding made by Save British Science.[174] And it was Guise who, as I will show, outmanoeuvred ACARD, ABRC and the GCSA to reshape science policy in the late 1980s.

Finally, contemporary reportage and commentary on science policy was the product of highly able academic science policy units, at the universities of Sussex and Manchester in particular, and especially of a growing number of skilled science journalists. *Nature* and the *Times Higher Education Supplement* carried news and editorials on the state of British science, the BBC's *Horizon* television programme made occasional but important interventions and *New Scientist* was in a golden era of reporting on the intersections of politics and science. Their output was sometimes more than the first draft of history[175] – on several crucial issues we will see central government absorbing and responding to media commentary on British science and science policy.

Notes

1. PREM 19/241. Hunt to Thatcher, 16 May 1979.
2. PREM 19/241. Thatcher, handwritten note on Hunt to Thatcher, 16 May 1979.
3. John Campbell, *Margaret Thatcher. Volume 1: the Grocer's Daughter*. London: Jonathan Cape, 2000, pp.35–6. Penny Junor, *Margaret Thatcher: Wife, Mother, Politician*. New York: Henry Holt, 1983, p.12.
4. Russell Lewis, *Margaret Thatcher: a Personal and Political Biography*. London: Routledge & Kegan Paul, 1975, p.12.
5. One example is Thatcher's acceptance of the invitation to launch the Women Into Science and Engineering (WISE) travelling exhibition bus in 1984. Robin Nicholson suggested it was a good way of following up initiatives on science education and her Lancaster House Seminar on Science and Technology (see Chapter 3), and at the launch she said that 'the future of this country depends on ... [making] full use of the human talent at our disposal; and among other things this means persuading more girls to take up careers in science and engineering'. The chair of WISE, Baroness Platt of the Equal Opportunities Commission, considered it 'important' that 'young women ... should see the Conservative Party taking an intelligent interest' and Thatcher's involvement would show them 'that the Conservative Party is on their side. They are the women of the future'. PREM 19/1934. Nicholson to Thatcher, 10 January 1984. Memorandum on words at launch of WISE bus, July 1984. Platt to Thatcher, 30 May 1984.
6. Georgina Ferry, *Dorothy Hodgkin: a Life*. London: Granta Books, 1998, p.41.
7. Hugo Young, *One of Us: a Biography of Margaret Thatcher*. London: Macmillan, 1989, p.16.
8. PREM 19/3028. Handwritten comment by MS (Scholar), on Walker to Jenkin, 11 December 1980.
9. PREM 19/3028. Walker to Kilpatrick, 8 May 1982, notes evidence published in the *Lancet* of an association of the chemicals with soft-tissue sarcomas. For farmers' and other experts' knowledge of 2,4,5-T, see Alan Irwin, *Citizen Science: A Study of People, Expertise and Sustainable Development*. London: Routledge, 1995.
10. In the words of the report: 'The customer says what he wants; the contractors does it (if he can); and the customer pays'.
11. *The Organisation and Management of Government R&D* (Rothschild Report), published alongside Council for Scientific Policy (CSP), *The Future of the Research Council System* (Dainton), in

the Cabinet Office green paper *A Framework for Government Research and Development*, Cmnd. 4814. London: HMSO, 1971.

12. Neil Calver and Miles Parker, 'The logic of scientific unity? Medawar, the Royal Society and the Rothschild controversy 1971–72', *Notes and Records of the Royal Society* 70 (2016): 83–100. Jon Agar, 'Thatcher, scientist', *Notes and Records of the Royal Society* 65 (2011): 212–32. Miles Parker, 'The Rothschild report (1971) and the purpose of government-funded R&D – a personal account', *Palgrave Communications* 2 (2016). Online only, paper reference 16053 DOI: 10.1057/palcomms.2016.53. Peter Collins, *The Royal Society and the Promotion of Science since 1960*. Cambridge: Cambridge University Press, 2016, pp.113–16.
13. A partial exception is Richard Vinen, *Thatcher's Britain: the Politics and Social Upheaval of the Thatcher Era*. London: Simon and Schuster, 2009, p.5: 'It seems to me that events … probably did more to communicate Thatcherism than the speeches of Sir Keith Joseph'.
14. Agar, 'Thatcher, scientist', *op. cit.*, p.226.
15. Agar, 'Thatcher, scientist', *op. cit.*, p.226.
16. Stephen Davies, 'Rothschild reversed: Explaining the exceptionalism of biomedical research, 1971–1981', *British Journal for the History of Science*, published as a FirstView article online, 28 August 2018. DOI: 10.1017/S0007087418000523.
17. Young, *op. cit.*, p.16.
18. Interview with Sir Crispin Tickell by Malcolm McBain, 28 January 1999, p.20. British Diplomatic Oral History Programme (BDOHP). https://www.chu.cam.ac.uk/media/uploads/files/Tickell.pdf.
19. John Lloyd, 'The green light', *Sunday Times Magazine* (26 February 1989), pp.49–50, p.49. PREM 19/2657/1. Slocock to Bright, 31 October 1989.
20. PREM 19/1217. Chester to Thatcher, 5 June 1984. PREM 19/1471. Nicholson to Thatcher, 27 July 1984.
21. John Maddox, 'Mrs Thatcher asks for decisiveness', *Nature* 316 (18 July 1985): 178. Thatcher criticised the bureaucracy of research councils, and said that 'scientists must appreciate that money for research can come only by the creation of wealth, for which they have a responsibility'. She praised the University of Salford (under John Ashworth) and Cranfield Institute of Technology (under Henry Chilvers), both being run by her former advisers, for 'creating jobs', while labelling Manchester and Liverpool, in contrast, universities with which 'we have huge problems'.
22. Collins, *op. cit.*, p.292. Thatcher was a Statute 12 election, an honorary designation reserved mainly for the Royal Society's political friends. Serving politicians are no longer elected.
23. Thatcher MSS (Churchill Archive Centre): THCR 3/2/126 f3. Thatcher to Huxley, 31 October 1983. Thatcher MSS (Churchill Archive Centre): THCR 3/2/117 f121. Thatcher to Huxley, 12 May 1983, shows that there was an earlier delay because of the 1983 General Election.
24. Mary Warnock, *A Memoir: People and Places*. London: Duckworth, 2000, p.184. Thatcher's biographer Charles Moore called the episode the 'locus classicus for the conflict between Mrs Thatcher and the intellectuals'. Charles Moore, *Margaret Thatcher. The Authorized Biography. Volume Two: Everything She Wants*. London: Allen Lane, 2015, p.654.
25. PREM 19/585. 'Prime Minister's reception for inventors, Monday 26 January 1981. Notes on invitees', undated (1981).
26. For reporting of Calvert prize, see *Studies in Design, Education, Craft and Technology*. https://ojs.lboro.ac.uk/ojs/index.php/SDEC/article/viewFile/1014/982.
27. Also present were civil servants, including Ashworth, ministers (including Howe, Joseph, Brittan, Lawson and Macfarlane) and Members of Parliament.
28. PREM 19/585. 'Prime Minister's reception for inventors – 26 January 1981. Opening remarks', undated (1981).
29. PREM 19/585. 'Note of a discussion at the Prime Minister's Reception for Innovators and Financiers at Downing Street on Monday 26 January', 28 January 1981.
30. Thatcher's thoughts about inventors, and what she learned from the reception, formed the substance of her speech to the Parliamentary and Scientific Committee, 25 February 1981. Partial copy of text in ED 273/111. Full text at: http://www.margaretthatcher.org/document/104575.
31. PREM 19/2254. Thatcher, handwritten note on Howe to Thatcher, 31 December 1987. NLW to Thatcher, 22 April 1988.
32. PREM 19/3028. Thatcher, handwritten note on Alderton to Morris, 30 March 1990.

33. PREM 19/3028. Thatcher, handwritten note on Morris to Thatcher, 'Pesticide policy', 2 April 1990.
34. Of these officials, 18,246 were classed as part of the Scientific Civil Service in 1979. CAB 134/4415, 'Review of the Scientific Civil Service', STO(80)3, 23 April 1980. The total non-industrial civil service was 566,059 strong. In addition to the 18,246 were 566 statisticians and 259 psychologists.
35. *Review of the Framework for Government Research and Development (Cmnd. 5046)*, Cmnd. 7499. London: HMSO, March 1979.
36. See Chapter 3 for discussion at E(A) and the subsequent establishment of E(ST). STO was formerly STP and there was also later a narrower ST(CS), 'CS' standing for chief scientists. Membership lists of STO can be found in CAB 134/4654.
37. *First Joint Report by the Chairmen of the Advisory Council for Applied Research and Development (ACARD) and the Advisory Board for the Research Councils (ABRC)*, Cmnd. 8957. London: HMSO, July 1983, p.7.
38. Supply estimates 1979–80, quoted in Philip Gummett, *Scientists in Whitehall*. Manchester: Manchester University Press, 1980, pp.54–6.
39. CAB 134/4673. Ministry of Defence, 'The transfer of technology from defence R&D to the civil sector', September 1982, provides a good breakdown of what this research and development entailed.
40. Gummett, *op. cit.*, p.126 for 'largest'.
41. David Edgerton, *Warfare State: Britain, 1920–1970*. Cambridge: Cambridge University Press, p.322.
42. T 499/6. Barclay (Number 10) to Gieve (Treasury), 6 August 1984. See also Chapter 3.
43. Five Requirement Boards existed in 1982 (reduced from nine): Electronics and Avionics, Materials and Chemicals, Mechanical and Electrical Engineering, Metrology and Standards and Textiles and Other Manufactures. Department of Industry, *Research and Development. Report 1981–82*. London: Department of Industry, 1982.
44. Peter Hennessy, *Whitehall*. London: Macmillan, p.432.
45. T 499/6. Tebbit to Howe, 24 September 1984.
46. Gummett, *op. cit.*, p.56.
47. Gummett, *op. cit.*, p.28, on Zuckerman's use of the term in 1971. For a tendentious view, see: David Edgerton, 'The "Haldane Principle" and other invented traditions in science policy', *History and Policy*. http://www.historyandpolicy.org/papers/policy-paper-88.html.
48. Section 2(1) reads: 'The [Secretary of State] may, out of moneys provided by Parliament, pay to any of the Research Councils such sums in respect of the expenses of the Council as he may with the consent of the Treasury determine, and so far as relates to the use and expenditure of sums so paid the Council shall act in accordance with such directions as may from time to time be given to it by the Secretary of State'.
49. T 447/398. Turner to Brown, 'Science and government', 12 February 1982.
50. T 499/6. Andren to Gordon, 15 October 1984. In practice SRA covered 'quite a bit of pretty routine spending on North Sea drilling'. Haigh to Allen, 22 October 1984.
51. T 499/6. Whiting to Corlett, 31 October 1984.
52. T 447/398. Treasury reply to Select Committee on Science and Technology Sub-committee I – Science and Government, undated (late 1981).
53. Interview with Sir Crispin Tickell by Malcolm McBain, 28 January 1999, p.20. British Diplomatic Oral History Programme (BDOHP). https://www.chu.cam.ac.uk/media/uploads/files/Tickell.pdf.
54. Francis Tombs, chair of ACOST, criticised the lack of experience of departmental scientific advisers, see PREM 19/2836. Tombs to Thatcher, December 1989.
55. Susan Owens, *Knowledge, Policy, and Expertise: the UK Royal Commission on Environmental Pollution 1970–2011*. Oxford: Oxford University Press, 2015.
56. The ACARD terms of reference were: To advise Ministers and to publish reports as necessary on i) applied R&D in the United Kingdom and its deployment in both the public and private sector in accordance with national needs; ii) the articulation of this R&D with scientific research supported through the Department of Education and Science; iii) the future development and application of technology; iv) the role of the United Kingdom in international collaboration in the field of applied R&D. The terms of reference were revised in 1982.
57. PREM 19/241. Hunt to Thatcher, 16 May 1979.
58. PREM 19/241. Thatcher, handwritten note on Hunt to Thatcher, 10 August 1979.

59. PREM 19/241. Pliatzky to Thatcher, 23 October 1979.
60. PREM 19/241. Thatcher, handwritten note on Armstrong to Thatcher, 25 January 1980. Robert Armstrong explained that GEC through its technical director, Sir Robert Clayton.
61. CAB 134/4674. ACARD, 'List of members' addresses and telephone numbers', 19 May 1983.
62. PREM 19/585. ACARD, 'Exploiting invention', 16 December 1980. The recommendation was prefaced by the following comment: '"Tomorrow's World", "Horizon" and similar programmes are successful in stimulating interest in developments in technology but deal very little with the practical problems of turning these into profit'.
63. PREM 19/585. Pattison to Bell, 22 December 1980.
64. ACARD, *Technological Change: Threats and Opportunities for the United Kingdom*. London: HMSO, 1979.
65. PREM 19/1805. Cabinet Office brief for Thatcher's meeting with Sir Francis Tombs, 22 July 1986. The government's statement was made in *Government Response to the Report 'Science and Technology' by the House of Lords Select Committee on Science and Technology*, Cmnd. 8591. London: HMSO, 1983.
66. ACARD, *Technological Change: Threats and Opportunities for the United Kingdom*. London: HMSO, 1979.
67. PREM 19/1514. ACARD and ABRC, 'Improving research links between higher education and industry', June 1983. For the public debate that ensued, see: Editorial, 'Guidelines for helping industry', *Nature* 304 (14 July 1983): 101–2. 'In search of missing links ... ', *THES* (8 July 1983). 'Not just an academic question', *New Scientist* (7 July 1983). Peter Large, 'Thatcher "should spend £30m to bridge college-industry gap', *Guardian* (29 June 1983): 5. Editorial, 'Science based society', *The Times* (6 July 1983). 'British R&D policy: a commonsense approach', *The Economist* (9 July 1983).
68. PREM 19/1514. 'Executive summary' of ACARD, 'Exploitable areas of science', October 1985. The underlining is Thatcher's.
69. PREM 19/2252. Fairclough to Thatcher, 17 February 1987.
70. PREM 19/2477. Fairclough to Thatcher, 24 April 1987.
71. PREM 19/2479. Guise to Thatcher, 18 March 1988. Guise to Thatcher, 25 May 1988.
72. PREM 19/2478. Tombs to Thatcher, 30 September 1987, and subsequent draft replies. Tombs to Thatcher, 22 December 1987, shows that the lesson was learned.
73. PREM 19/2479 has the list of people attending.
74. PREM 19/3155 has the table plan showing where ACOST members and Thatcher sat for the meeting. Guise is present, sitting to one side.
75. PREM 18/3156. Turnbull to Fairclough, 25 July 1990.
76. 'Sir John Ashworth. Edited transcript of a conversation between John Ashworth and Peter Collins at the Royal Society on 22 April 2009', Royal Society Library.
77. Department for Education and Science, *The Science Budget: a Forward Look 1982*, October 1982. *First Joint Report by the Chairmen of the Advisory Council for Applied Research and Development (ACARD) and the Advisory Board for the Research Councils (ABRC)*, Cmnd. 8957. London: HMSO, July 1983, p.25.
78. ABRC, *A Study of Commissioned Research*, 'Mason report'. London: HMSO, 1983, p.8, p.10.
79. ABRC, *A Strategy for the Science Base*, Cm. 30. London: HMSO, November 1987.
80. PREM 19/3155. Gray to Crowne, 2 January 1990.
81. Solly Zuckerman, *From Apes to Warlords: the Autobiography (1904–1946) of Solly Zuckerman*. London: Hamilton, 1978.
82. Solly Zuckerman, *Monkeys, Men and Missiles: an Autobiography*, 1946–88. London: Collins, 1988, pp.189–92.
83. Roy Jenkins, speaking at Zuckerman's Memorial Service, quoted in: P. L. Krohn, 'Solly Zuckerman Baron Zuckerman, of Burnham Thorpe, O.M., K.C.B. 30 May 1904–1 April 1993', *Biographical Memoirs of Fellows of the Royal Society* 41 (1995) 41: 576–98, p.565.
84. Zuckerman, *op. cit*. (1988), p.194.
85. Zuckerman, *op. cit*. (1988), pp.211–21, p.215.
86. Zuckerman, *op. cit*. (1988), p.409.
87. John Sheail, 'Torrey Canyon: the political dimension', *Journal of Contemporary History* 42 (2007): 485–504. Zuckerman, *op. cit*. (1988), pp.406–9. In an echo of Zuckerman's wartime operational research days, the effectiveness of bombing the stricken oil tanker was a crucial calculation.
88. Susan Owens, *op. cit*. Zuckerman, *op. cit*. (1988), pp.413–15.

89. Zuckerman, *op. cit.* (1988), pp.418–422.
90. Attributes 1–7 are extracted from: Solly Zuckerman, 'Scientists, bureaucrats and ministers', speech to the Royal Institution, copy in CAB 134/4854, STO(CS)(84)33, 24 October 1984. Attributes 8–9 are extracted from: CAB 164/1642, Zuckerman to Robert Armstrong, 12 February 1981.
91. CAB 164/1642. Zuckerman to Robert Armstrong, 12 February 1981.
92. Completing a PhD at Birmingham University during wartime, Cottrell remained there for 19 years before taking up a post at Harwell, applying his metallurgical knowledge to understanding radiation damage of metals for the postwar atomic reactor projects. In 1958 he moved to lead and reinvigorate the Department of Metallurgy at the University of Cambridge, during which time he provided the theoretical underpinning of methods to assess the likelihood of failure in cracked engineering structures. In 1964 Zuckerman had asked Cottrell to become his Assistant Chief Scientific Adviser at the Ministry of Defence, working alongside William Cook. Soon after Zuckerman became GCSA in 1966, Cottrell, in late 1967, became deputy GCSA. R. E. Smallman and J. F. Knott, 'Sir Alan Cottrell FREng. 17 July 1919–15 February 2012', *Biographical Memoirs of Fellows of the Royal Society* 59 (2013): 93–124, 95.
93. Smallman and Knott, *op. cit.*, p.111.
94. PREM 16/903. Select Committee on Science and Technology, Science Sub-committee, 'Memorandum by the Lord Privy Seal', May 1976.
95. Jon Agar, '"Future forecast – changeable and probably getting worse": the UK government's early response to anthropogenic climate change', *Twentieth Century British History* 26 (2015): 602–28.
96. Smallman and Knott, *op. cit.*, p.112.
97. Smallman and Knott, *op. cit.*, p.111.
98. *Framework for Government Research and Development*. London: HMSO, Cmnd 5046, 1972 (White paper).
99. Smallman and Knott, *op. cit.*, p.111. Departments with chief scientists working in this mode in the mid-1970s were: Defence, Industry, Environment, Health, Agriculture, Energy and the Home Office. See: PREM 16/903. Select Committee on Science and Technology, Science Sub-committee, 'Memorandum by the Lord Privy Seal', May 1976.
100. For example, Cottrell, 'Research, risk and reward', the Sir Clavering Fison Lecture, on the 'anomaly'. 'If industrial research is good for industry why have we, who have spent so much on research, done so poorly industrially?', *Chemistry and Industry* (17 January 1981): 42–6.
101. CAB 178/15 for 'Security of nuclear materials, targets and facilities', 1974–1975. CAB 178/16–18 for radioactive wastes, 1974–1975. TBA CAB 178/21 for 'Uranium supplies for the nuclear power programme', 1968–1976.
102. CAB 164/1642. Zuckerman to Robert Armstrong, 12 February 1981.
103. PREM 16/903. Hunt to Wilson, 2 October 1975.
104. Handwritten note, initialled by HW signifying Wilson's approval, on PREM 16/903. Hunt to Wilson, 2 October 1975.
105. PREM 16/903. Dalyell to Wilson, 27 December 1975. Arthur Palmer (Select Committee on Science and Technology) to Wilson, January 1976.
106. Pearce Wright (Science Editor, *The Times*), 'Cabinet needs a chief scientific adviser', *The Times* (2 December 1975), p.3, reports Hodgkin's speech. Hodgkin invoked an imaginary rabies epidemic to show what would be missed without a GCSA. Rothschild responded, privately, by circulating a very odd satirical essay to Hunt and Wilson. Before diagnosing senior politicians of various imaginary diseases (Mr Michael Foot – 'glottal paralysis'), Rothschild makes one serious, cutting point: 'who says there is no Chief Scientific Adviser to the Government? The discerning reader will have noted … Solly Zuckerman's name … What is he doing to the Cabinet Office – and he is certainly there when he not elsewhere – if he is not giving the Government scientific advice?'. PREM 16/903. Lord Rothschild, 'TOP SECRET, BANNED AND REJECTED', undated (December 1975).
107. PREM 16/903. Hunt to Wilson, 19 January 1976. Wilson's response, handwritten on Hunt's note, is unexpectedly revealing about the Prime Minister's science policy: '1. I have no intention whatsoever of changing my mind about the appointment suggested. 2. I have some doubts about the science set up, which might be met by setting up a science min. [ministry]. (DES is useless here, but then the scientists are now – I started this in the 60s – where they belong in a scientist-consumer based relationship. 3. I agree the cosmetic proposals … '.

In other words Wilson considered, albeit briefly, a Ministry of Science in 1976, and claimed that the customer-contractor idea originated with him.

108. Biographical details from a press release issued because, in the words of the Cabinet Secretary, 'For news of the appointment to seep out gradually would not be fair to Professor Ashworth and could only be a source of capital to those who continue to press for the appointment of a full Chief Scientific Adviser'. PREM 16/903. Hunt to Stowe, 30 September 1976. See also: 'Sir John Ashworth. Edited transcript of a conversation between John Ashworth and Peter Collins at the Royal Society on 22 April 2009', Royal Society Library.
109. CAB 184/617. John Ashworth, 'Think tanks: a response to the need for administrative change', April 1982.
110. ACARD, *The Applications of Semiconductor Technology*. London: HMSO, 1978.
111. Ashworth was part of the team producing *Review of the Scientific Civil Service*, report of a working group of the Management Committee for the Science Group (CSD). London: HMSO, 1980.
112. Jon Agar and Brian Balmer, 'Defence research and genetic engineering: fears and dissociation in the 1970s', in Charlotte Sleigh and Don Leggett, eds, *Scientific Governance in Britain, 1914–1979*. Manchester, Manchester University Press, 2016, pp.122–43. For climate change, see Agar, *op cit*. (2015).
113. John Campbell, *Margaret Thatcher. Vol. 2: The Iron Lady*. London: Jonathan Cape, 2003, pp.642–3. Charles Moore, *Margaret Thatcher: the Authorized Biography. Vol. 1: Not for Turning*. London: Allen Lane, 2013, p.426.
114. PREM 19/1047. Armstrong to Thatcher, 9 April 1981.
115. Royal Society, 'The provision and coordination of scientific advice to government. Evidence submitted by the Council of the Royal Society to the House of Lords Select Committee on Science and Technology', March 1981.
116. PREM 19/1047. Armstrong to Thatcher, 9 April 1981. Thatcher, earlier, had spelled out why she was against a Ministry for Science: it would divorce 'those responsible for applied R&D from those concerned with formulating and implementing the policies to which their R&D related', and there would be an 'unwelcome division between responsibility for higher education and for the scientific community supported by the Research Councils'. PREM 191/1339. Thatcher to Lloyd, 20 August 1979.
117. Thatcher's handwritten note on PREM 19/1047. Principal Private Secretary (G. A. Whitmore) to Thatcher, 10 April 1981.
118. PREM 19/1047. Principal Private Secretary to Armstrong, 7 May 1981.
119. PREM 19/1047. Armstrong to Thatcher, 21 September 1981. For Nicholson and Biogen, see: 'Scientists in the market-place', *Financial Times* (19 November 1981).
120. 'Sir Robin Nicholson (FRS 1978). Edited transcript of a conversation between Robin Nicholson and Peter Collins at the Royal Society on 6 May and 18 May 2009', Royal Society Library.
121. Thatcher was also lukewarm about ACARD. Hunt thought that after 'a shaky start ACARD now seems to be doing a useful job', producing 'well received reports'. Thatcher in fact thought ACARD 'totally unnecessary', and the production of reports irrelevant if they had no 'practical effect'. PREM 19/241. Hunt, 'Science', 16 May 1979, with Thatcher's handwritten response at top, and Hunt, 'ACARD', 10 August 1979.
122. House of Lords Select Committee on Science and Technology, 'Science and government', 1981. PREM 19/1339. Armstrong to Thatcher, 8 January 1982. The Select Committee also called for a minister to be designated to speak for science and technology and the development of ACARD into a Council for Science and Technology among other requests. The STO (renamed STP committee) supported the strengthening of the role, recommending a 'Chief Scientist, Cabinet Office' who remained a member of the CPRS. PREM 19/1339. STO, 'Science and government', 27 April 1982.
123. PREM 19/1339. Draft command paper, 'Government response to the report by the Select Committee of the House of Lords on Science and Technology entitled "Science and government"', 27 April 1982.
124. The appointment is announced in PREM 19/1045, press notice, undated (June 1983). Thatcher's earlier rejection of the title is marked by the handwritten note 'NO' on PREM 19/1339. Armstrong to Thatcher, 21 May 1982. A cartoon, with Thatcher 'inventing' Nicholson in a lab setting accompanied Malcolm Rutherford, 'Who is Robin Nicholson?', *Financial Times* (5 March 1982).

125. 'Sir Robin Nicholson (FRS 1978). Edited transcript of a conversation between Robin Nicholson and Peter Collins at the Royal Society on 6 May and 18 May 2009', Royal Society Library.
126. PREM 19/1339. Armstrong to Thatcher, 7 September 1983. This response was described as 'largely the work of Robin Nicholson', and was agreed with by Thatcher.
127. T 471/45. Nicholson to Joseph, 17 December 1982.
128. PREM 19/1339. Nicholson to Barclay, 4 September 1984.
129. CAB 184/679. Nicholson to Thatcher, 6 July 1983.
130. PREM 19/1369. Turnbull, 'The science budget', 13 January 1984.
131. PREM 19/1369. Nicholson to Thatcher, 6 March 1984.
132. PREM 19/1740. Walker to Thatcher, 2 March 1984.
133. PREM 19/1740. Turnbull to Thatcher, 5 March 1984.
134. PREM 19/1740. Department of Energy, 'Review of the UK Atomic Energy Authority. Report by a Departmental Working Group', August 1984.
135. PREM 19/1740. Nicholson to Thatcher, 12 October 1984. By this point, Nicholson is signing himself off as 'Chief Scientific Adviser'.
136. PREM 19/1740. Owen to Thatcher, 12 October 1984.
137. PREM 19/1740. Walker to Thatcher, 1 May 1985.
138. PREM 19/1740. Turnbull to Dart, 27 May 1985.
139. PREM 19/1740. Nicholson to Turnbull, 31 May 1985.
140. PREM 19/1740. Thatcher to Nicholson, 17 June 1985.
141. Pearce Wright, 'Sir John Fairclough', *Guardian*, 11 June 2003: 23.
142. Martin Campbell-Kelly, 'Sir Maurice Vincent Wilkes. 26 June 1913–29 November 2010', *Biographical Memoirs of Fellows of the Royal Society* 60 (2014): 433–54, p.443.
143. PREM 19/2835. Fairclough, Speech to RAWB, 20 February 1989.
144. David Edgerton and Kirsty Hughes, 'The poverty of science: a critical analysis of scientific and industrial policy under Mrs. Thatcher', *Public Administration* 67 (1989): 419–33. They cite: John Fairclough, 'Setting priorities for science and technology', *Midland Bank Review* (Winter 1987), pp.18–23.
145. Rebecca Boden, Deborah Cox and Maria Nedeva, 'The appliance of science? New public management and strategic change', *Technology Analysis and Strategic Management* 18 (2006): 125–41, p.130.
146. PREM 19/3659. Willetts, 'The role of the Prime Minister's Policy Unit', draft from 1987, subsequently published as the RIPA Haldane prize essay in *Public Administration* 65 (1987): 443–54. I quote from the version that was shown to Thatcher.
147. Guise's own account can be found in a slim memoir *Inside the Tank*. Epsom: Bretwalda Press, 2015, and the very similar 'Margaret Thatcher's influence on British science', *Notes and Records of the Royal Society* 68 (2014): 301–9.
148. 'The conventional view is that the Civil Service gives "objective" advice to Ministers and that the Policy Unit because of its close personal loyalty to the Prime Minister cannot be objective. Historians and philosophers of science have invested enormous energy in investigating notions of objectivity in their own disciplines but this has not been fully reflected in the Whitehall formulations of the role of a civil servant. Whlist not advocating any jejeune relativism, it is clear that one can't just treat objectivity as unproblematic. First, there is no such thing as policy advice resting solely on objective facts … '. Willetts, *op. cit.*, p.15. I quote this Willetts passage at length partly because it is a step in an elegant argument that sought to retain objectivity for the Unit's ideological crusade, but also because it was perhaps Thatcher's closest encounter with the history and philosophy of science.
149. PREM 19/2252. Guise to Thatcher, 10 March 1987.
150. Boden, Cox and Nedeva, *op. cit.*
151. PREM 19/2252. Guise to Thatcher, 10 March 1987.
152. PREM 19/3157. Pearce Wright, 'Sir John Fairclough's contribution', undated (September 1990): 23.
153. David Fishlock, 'Top man takes "green" view', *Financial Times* (25 September 1990).
154. PREM 19/3155. Thatcher, handwritten comments on Gray to Thatcher, 22 December 1989.
155. PREM 19/3156. Turnbull, 'Note for the record. Meeting between the Prime Minister and Professor Stewart', 28 September 1990. Specifically: '(i) ACOST had a tendency to spawn new sub-committees … The Prime Minister hoped Professor Stewart would not only prevent any further proliferation but also prune back the existing apparatus. (ii) Although ACOST's terms of reference asked it to advise on priorities, it sometimes took the easy way out of recom-

mending higher spending rather than confronting the issue. (iii) The present arrangements for scrutinising applications for research funding were very bureaucratic ... (iv) The tendency for big scientific projects to absorb large sums of money needed to be watched carefully. (v) Although the balance of evidence pointed towards global warming, there were still many uncertainties and gaps in the scientific evidence ... '.

156. 'Sir William Stewart (FRS 1977; Chief Scientific Adviser, 1990–5). Edited transcript of a discussion between Bill Stewart and Peter Collins at the Health Protection Agency London office on 8 January 2009', Royal Society Library.
157. PREM 19/3156. Stewart to Major, 19 December 1990, a note the Prime Minister described as 'very helpful. I should meet him before too long'.
158. Conservative Party, *The Conservative Manifesto 1979*, 1979. Labour Party, *The Labour Way is the Better Way*, 1979. Liberal Party, *The Real Fight is for Britain*, 1979.
159. Conservative Party, *The Conservative Manifesto 1983*, 1983. Labour Party, *The New Hope for Britain*, 1983. Liberal–SDP Alliance, *Working Together for Britain*, 1983.
160. Conservative Party, *The Next Moves Forward*, 1987. Labour Party, *Britain Will Win with Labour*, 1987. Liberal–SDP Alliance, *Britain United: the Time Has Come*, 1987. The Liberal-SDP Alliance also promised to support the British Technology Group, set up a new Department of Education, Training and Science, and reverse cuts.
161. Hansard, 'Research and development. ACARD/ABRC Report', 10 February 1984.
162. Hansard, 'Research and development. ACARD/ABRC Report', 10 February 1984, p.1393.
163. PREM 19/1514. Armstrong to Thatcher, 11 November 1983. *Government Response to the Report 'Science and Technology' by the House of Lords Select Committee on Science and Technology*, Cmnd. 8591. London: HMSO, 1983. *First Joint Report by the Chairmen of the Advisory Council for Applied Research and Development (ACARD) and the Advisory Board for the Research Councils (ABRC)*, Cmnd. 8957. London: HMSO, July 1983.
164. CAB 134/4654. Nicholson and Fraser, response to House of Lords report, 19 March 1982.
165. PREM 19/1932. Armstrong to Thatcher, 29 July 1986.
166. Tam Dalyell, *A Science Policy for Britain*. London: Longman, 1983.
167. MAF 453/6. Dalyell to Thatcher, 9 March 1981, on privatisation of bacterial collections, was one such example. PREM 19/1339. Parliamentary Question, 26 October 1981, raised the general question in the Commons.
168. PREM 19/1933. Addison to Smith, 30 May 1986. Porter to Thatcher, 1 July 1986.
169. PREM 19/1933. Fairclough to Addison, 17 July 1986. The President of the Royal Society was subsequently invited onto ACOST.
170. Collins, *op cit.*, p.119.
171. PREM 19/1933. Addison to Smith, 30 May 1986.
172. PREM 19/1933. Addison to Thatcher, 4 July 1986.
173. PREM 19/2835. Browne-Wilkinson to Turnbull, 21 April 1989.
174. PREM 19/2835. Guise to Thatcher, 7 May 1989. Kealey has claimed to have had an audience with Thatcher on science policy in 1985. https://www.telegraph.co.uk/news/politics/thatcher-conference-liberty/10909494/Margaret-Thatcher-was-wrong-about-one-thing-science-doesnt-need-Nobel-prizes-to-thrive.html.
175. Martin Ince, *The Politics of British Science*. Brighton: Wheatsheaf Books, 1986. The bulk of the book was written by Jon Turney before he moved from the *Times Higher Education Supplement* to the Advisory Board of the Research Councils, where more discretion was needed.

3
The central debates on science and innovation

In this chapter I will show that there was a reversal in science policy and industrial strategy in 1987. The result was deep cuts in government funding of 'near-market' research and, in what I demonstrate was a flip side of the same coin, celebration of non-market-oriented 'curiosity-driven' research. The change came after years of growing tensions around government schemes to support research-intensive industry, cuts in public expenditure and their effects (especially on academic science), the perception of a lack of entrepreneurial spirit, the preponderance of defence research and onerous subscriptions to European collaborative programmes, specifically CERN. It is a long chapter, partly because all of these issues were intimately intertwined but need to be unpicked in detail. While the change in science policy was noted by commentators then and since, its causes and protagonists have been misunderstood – precisely because opposing policy aspirations have been conflated. Remarkably, Thatcher's change in policy was largely the result of influence of a single political adviser and made in the teeth of opposition from her science advisers.

Industrial strategy

Thatcher's government inherited numerous schemes that supported, directly and indirectly, what came to be called 'near-market research'; further proposals were made and adopted in the early 1980s.[1] Furthermore, in other promising areas, such as biotechnology, discussed below, Thatcher's government continued, and even expanded, 'pump priming' – in other words, the provision of considerable public funds as part of a strategy to create conditions for new industry to grow. However, one sector where Thatcher's approach to industrial policy was tested early was microelectronics. Under the previous Labour administration a policy of selective support had been adopted. The world microelectronics market was expected to be £10 billion per annum by 1985 and to be a substantial manufacturing sector in its own right; it would be pervasively applied. Yet, as a Department of Industry analysis, supported by ACARD, showed, only 5 per cent of British firms were active in microelectronics applications (notably Plessey, Ferranti and GEC), while 50 per cent declared that they were 'not sufficiently aware' even to assess opportunities and threats.[2] In response, in 1978 Labour had announced a Microprocessor Applications Project (costing £55m over three years), which aimed to encourage firms to apply microelectronics, and a Microelectronics Industry Support Programme (£70m over five years), which aimed to encourage the development and manufacture of microelectronic devices in the UK. Furthermore, the National Enterprise Board had funded Inmos, a company formed in 1978 by the British ex-Elliott Automation computer scientist Iann Barron with two American semiconductor experts, to the tune of £25m and rising.

Keith Joseph, then Secretary of State for Industry, wrote to Thatcher a month after she took office. 'Micro-electronics is of crucial importance to our future industrial and economic performance,' he argued, adding that in its way 'it is likely to be of the same sort of importance as was the steam engine with the difference that (a) it will be even more pervasive and (b) we are not in the forefront of the development'.[3] While he was 'in principle strongly opposed to support of this kind', Joseph thought the continued injection of considerable public funds was 'justified'. Thatcher's response was hostile. She demanded evidence that British industry would not embrace the technology as rapidly as competitors ('!!! Who said?'), and objected strongly to Joseph's proposal that the schemes 'be allowed to continue on broadly its present lines' ('No'). The 'whole area of policy' was to be discussed.[4]

Two responses are interesting, partly for what they reveal about paths not taken. First, on the back of consultancy produced by the Stanford Research Institute for the Department of Industry, civil servants wondered aloud whether a Californian model was worth pursuing:

> why not … convert [MAP] project support into a special fund for stimulating start ups and small firm growth in micro applications – the UK version of Californian venture capital. It is no use saying this is for the City or someone else in the financial world. This is not happening now and time is of the essence.[5]

The suggestion was not pursued. Second, when Kenneth Berrill, the economist and head of the CPRS, offered his advice to Thatcher, he introduced her to a new term: 'information technology':

> The government needs to discuss microelectronics in a wider context. The potential market for microelectronic equipment is very large and growing rapidly, but the interesting point is the breakdown of this market. No less than 60 per cent of sales are expected to be in 'information goods' ... It is not for nothing that the French talk of Information Technology rather than microprocessors. Approaching advanced electronics in terms of information technology creates a new perspective and the MAP/MISP/Inmos group of policies appears as only one part of the required response. If we have bottlenecks on the widespread use of the new 'information technology', our success in the microelectronics field is bound to be severely limited.[6]

It was necessary, said Berrill, to consider under this 'broader "information technology"' approach such things as the 'crucial importance of having an adequate communications network inside the United Kingdom (and hence the vital role of the Post Office)', the securing of access to frequencies and satellites, the 'many domestic and international issues in the fields of privacy, copyright, compatibility in data transmission', and the use of IT in the public sector.

On the specific subject of microelectronics, Thatcher's attitude hardened:

> She had very grave doubts about assisting the production of micro-electronics devices, and in particular the INMOS project. But she also questioned the support for applications. Her own view was that British industry was very ready to apply this technology, and that finance was not a constraint; where industry was not applying it, it was because of trade union opposition.[7]

Her doubts were shared by her chief scientific adviser.[8] INMOS continued amid arguments over the location of any manufacturing facility (Cardiff, Bristol or a venue in Scotland) and an opportunistic proposal from Lord Weinstock that it might be absorbed as a 'cottage industry' in the 'GEC Empire'.[9] It produced the innovative but loss-making 'transputer' chip.

However, 'information technology' became a major frame and object of government promotion in the 1980s. 1982 was named 'Information

Technology Year', and programme of events under the IT82 banner was marked at its end by a speech at the Barbican Centre by Thatcher, who noted that the proportion of the population that had heard of 'information technology' had increased over the year from 2 in 10 to 6 in 10.[10] In the same year the home computer market boomed, with consumers buying products from not only US but also UK manufacturers, not least Sinclair's ZX81 and, under the BBC's Computer Literacy Project, Acorn's BBC Micro. While the broadcaster's educational project had been conceived independently of the government's plans, it nevertheless, as Tom Lean notes, 'fitted well'.[11] Microcomputers entered government offices as well as schools with 600 microcomputers being delivered across Whitehall.[12]

During IT82, and in response to the Japanese announcement of its Fifth Generation computer initiative, the government also commissioned and received a report from a group, mainly industrialists, chaired by the BT board member John Alvey. In the summer of 1982 the Alvey Committee recommended massively supporting IT research in universities in collaboration with industry (which would provide some matching funds) – £350 million would be spent by 1987.[13] This was a clear industrial strategy, informed by experts: committees (SERC and ABRC had identified information technology as critical, another expert group, Alvey had set out a detailed plan), supported by joined-up government actions (one government department, Keith Joseph's Department of Education and Science, created new posts and a supply of trained talent, while another, the Department of Industry, worked with its clients to exploit them: 'Our hope and expectation', the minister for the first had written to the minister of the second, 'is, naturally, that our push will be matched by your pull').[14] The GCSA, Robin Nicholson, supported the plan (so long as the new posts were not tenured, which was also Thatcher's concern).[15] In January 1983 the IT Advisory Panel, concerned that the government had not announced its response to Alvey, urged the Prime Minister that the 'national interest urgently needs the formulation, publication and vigorous implementation' of a 'national strategy for Information Technology', if the UK was to keep up with the Japanese and the French.[16]

Consultation on the Alvey proposals continued over the next few months. Lord Weinstock was concerned that his mighty company, GEC, would not benefit as much as smaller competitors. Nicholson used this intervention to frame the Alvey programme to Thatcher in an acceptable way: GEC and large companies might lose out, but that would mean small entrepreneurial companies and innovative academic units would gain.[17]

Yet big industry pressure allowed multinational companies to partake in the Alvey programme, which was finally agreed by the Prime Minister and her ministers in March and April 1983. Brian Oakley, secretary of SERC, was appointed the director (interestingly the future Chief Scientific Adviser John Fairclough was one of four other candidates considered). Thatcher was not enthusiastic, but permitted this sizeable government injection of cash into industrially relevant research and development to go ahead. Her main complaint was that the staffing of the directorate was too large (a 'bureaucrat's paradise').[18] Even the free-market *Economist* welcomed the Alvey idea, while noting the 'scepticism' of 'that former research scientist, Mrs Thatcher', in an editorial titled 'Government can help':

> The sort of government money to avoid is the kind that tries to pick winners, or, as happened with the Inmos microchip company, tries to catch up on the world with one great subsidised bound. Government helps best in technology by getting a starter or two in time for the race.[19]

So in the early 1980s an industrial strategy of government funded near-market industrial research directed at areas selected by experts was acceptable to Thatcher and her administration.

During IT82 Kenneth Baker, as Minister of Information Technology, framed what was happening in terms of the 'knowledge industry' and the 'information revolution'.[20] Work would become cleaner, more flexible, less subject to unionisation. 'Flexi-working' hours would be shorter and interspersed with leisure time and continuing education. Politically, the Orwellian surveillance society could and would be avoided, argued Baker:

> Will the day come when the technology will not only allow the sending of messages along the cable to the home but also the spying on the recipient without his knowing it? ... Such a state of affairs however will only come about if people allow it to come about. It can't happen surreptitiously and it can and must be resisted. Data privacy legislation which we will be introducing as soon as possible is essential ... We should also engender a less deferential attitude to the state.

This in turn was linked, by Baker, to a necessary hostility to nationalisation and the welfare state:

> We should be jealous of the inquisitive nature of the State; we should resist the encroachments of the State into the economic and social domains that are best left to the individual; we should reject the all too popular and lazy reaction that the State should take upon itself an ever increasing responsibility for the welfare of the citizen. We should enhance the opportunities of private ownership for what the State owns it has to control.

In other words the information technology revolution, and a broad industrial strategy, could be hitched to the broader Thatcherite programme of privatisation and the freedom of the individual. Baker stated that he did not 'share the Orwellian nightmare because the microchip revolution through its pervasiveness and its cheapness can increase the area of human awareness and choice'. Books could be burnt or banned, but it was 'rather more difficult to cut off the wave bands'. Stopping the 'free flow of information' would be 'ruinous', if possible at all. In conclusion, so long as the political system could be robust enough not to succumb to 'push-button politics' or a 'continual series of referenda', Baker promised a rosy digital future:

> The Information Society will be better informed and also I suspect more relaxed, less formal, more mobile, less enamoured with structure, more skilled and less ridden with class and social difference and full of scope for more individuality.
>
> Yes, on balance, it will be a better place.

Commercial exploitation of academic science

A second area of science policy of considerable tension and attention during the 1980s, in which questions of industrial strategy and the role of the state were central, was the commercial exploitation of academic science and inventions. University scientists were allowed to charge for consultancy and to exploit their research for commercial gain (a right extended to local authority colleges and polytechnics in 1984,[21] this decision being taken after two years of discussion, in which the desire to 'encourage enterprise and economically useful enquiry' was balanced against the 'accountability safeguards').[22] However, individual entrepreneurial activity was affected by the presence of an alternative model for exploiting academic science. Under Attlee a collective instrument had been organised: the National Research Development

Corporation (NRDC), which had the right to first refusal to patent Research Council-funded inventions, and received further public funds to exploit them.[23] The NRDC's record was patchy, with failures (hovercraft, computers) and successes (pyrethrin), one problem being that the NRDC had little control over its commercial partner firms' choices: time and after time the firms took a narrow, understandably self-interested, short-term approach to development.[24]

In 1980 Margaret Thatcher visited Cambridge. She returned deeply angry. She had been told two stories that she read as catastrophic failures of commercial exploitation: an image intensifier developed at the Mullard radio astronomy observatory and monoclonal antibodies, discovered at the MRC's Laboratory of Molecular Biology. She asked her Chief Scientist, CPRS, John Ashworth for immediate advice. César Milstein and Georges Köhler had pioneered the monoclonal antibody technique in 1975 by fusing antibodies with a myeloma cancer cell to produce an immortal hybridoma that would continue to secrete antibodies when placed in a mouse – thereby turning the mouse into an immensely productive antibody factory. Monoclonal antibodies would subsequently become the workhorses of biomedical diagnostics and therapeutics.[25] The NRDC had been offered the chance to ensure the patenting of Milstein and Köhler's process, but had declined on the grounds that it was being published. However, a more assertive NRDC could have delayed such publication while patenting took place. Milstein later said that he did not regret the NRDC's decision, since, without a patent, 'it allowed him greater freedom to publish and share his results, and to get on with his research'.[26] Nevertheless, in 1979 *Nature* had reported the issue, asking pointedly whether 'Britain [had] lost large potential royalties through a failure to recognise the commercial potential of antibodies'.[27]

'First there is no doubt', wrote Ashworth in response to Thatcher's request, 'that either Cesar Milstein, the MRC or the NRDC (or some combination thereof) failed lamentably when they omitted to file a patent.'[28] But in terms of lessons to be learned, Ashworth's suggestions were mainly targeted at NRDC. As also shown by its attitude to biotechnology, the 'fundamental problems' were, first, that NRDC had 'monopoly rights over the results of the Research Councils' and, second, that by restricting its role to that of 'honest broker' it had failed to provide 'a 'technology transfer' service.[29] NRDC may make a profit, but that was 'a consequence of their cautious and risk averse policies'. Ashworth suggested the establishment in the UK of 'an entrepreneurial company along the lines of those established in Europe and the United States … which make a commercial business out of technology transfer'; it would have a 'privileged relationship with the MRC', which would 'de facto break the NRDC's monopoly'. This would be a twist on Celltech, the government-backed biotech firm already announced. With respect to individual research workers, Ashworth praised the Wolfson Industrial Units – a 'private initiative' that not only benefited universities financially, but was also interesting for the effect on other university researchers.

> The financial benefit of such units to the Universities can be considerable (I know that the units at the University of Southampton earned an income of over £1 million last year, for example) but even more impressive has been the effect on the attitudes of the staff of the University of seeing some of their colleagues engage in this kind of activity – and earn significant consultancy fees in consequence.

Such incentives of 'greater financial rewards to academic entrepreneurs' were good, wrote Ashworth, but there also needed to be a stick, perhaps 'discrimination against those who do not become entrepreneurial'. Perhaps if such sticks and carrots were in place then the radio astronomers, thinking about the other story told to Thatcher, 'would have gone off to a garage somewhere and set up their own little firm designed to sell image intensifiers'. 'The question to ask,' summarised Ashworth, 'is why the environment in Cambridge, Mass., encourages such behaviour and that in Cambridge UK inhibits it?'

ACARD, the voice of business in UK science policy, had urged similar measures 'to make it easier to found new businesses in the UK'.[30] In December 1980, in a report commissioned on Thatcher's request based on her hearing the two Cambridge stories, ACARD reiterated the point: 'the creation of an environment that favours entrepreneurial activity requires, we think, more radical changes, particularly in the attitude to business found in parts of United Kingdom society (notably higher education) and in the worth of intellectual property compared with physical property'.[31] ACARD offered criticism (more muted than Ashworth's) of the NRDC: it should continue, with government loans where necessary, but in return its monopoly right should be removed.[32]

Another immediate response was the organisation of a reception for inventors and innovators, held in January 1981. Thatcher addressed a gathering of small-scale garage inventors and owners of small tech businesses. A few had been assisted by the NRDC, but most had not. The message was that this latter class of innovator had her government's support.

The NRDC was merged with the National Enterprise Board in 1981 to form the British Technology Group (BTG). Initially BTG continued to have the monopoly of first refusal on patent rights arising from publicly funded research; it took under its wing, for example, the intellectual property on magnetic resonance imaging (MRI). But Thatcher remained unhappy. In Spring 1983 she mooted the abolition of BTG, while Patrick Jenkin, her Secretary of State for Industry pushed back, arguing that 'a body is still needed in the public sector to deal with the difficult and staff intensive task of identifying ideas, occasionally providing pre-development finance and of patenting and licensing those which have the promise of commercial success'.[33] The Conservative Party manifesto for the June 1983 election contained a commitment to 'accelerate the transfer of technology from the University laboratory to the market place'. Returned to power with a resounding victory at the polls, BTG reform was soon on the agenda. The removal of BTG's monopoly rights were part of a wider vision in which private capital would directly respond to new inventive ideas from public sector, allowing the BTG to focus either on technology transfer (not large-scale investment) or wither away as the market took hold – or at least act more commercially as it was forced to compete unprotected by monopoly rights. The status quo was not an option. The Chief Scientific Adviser Robin Nicholson told his Prime Minister that

> In the past BTG has shown all the sloth and leaden-footedness characteristic of a state-owned monopoly. To convert it into an enterprising and dynamic organisation is a formidable, but not impossible, task. But BTG will need to be freed from the previous restrictions in order to exploit fully its skills and expertise. ...
>
> BTG's eventual reward for success must be privatisation rather than elimination ...[34]

In Autumn 1983 ministers 'announced with a great flourish' that BTG's first refusal of patents would be removed, while at the same time Thatcher 'laid ... emphasis on the need for research activities to produce a commercial benefit to the UK economy'; these actions, worried the Treasury, created 'expectations of a brave new world where the results of research will show up bigger and faster than before'.[35]

The wrangle over NRDC/BTG prompted by Thatcher's response to the non-patenting of monoclonal antibodies took place in the context of the changing expectations of commercialisation of the life sciences associated with genetic engineering. The ways that the patenting of recombinant DNA techniques sparked a 'social transformation', in which molecular biologists 'formerly cloistered in academe, developed close ties with private industry as equity owners, corporate executives, and consultants', are familiar to historians of science.[36] The entrepreneurial culture had developed very rapidly – indeed the outrage felt in 1980 over Milstein's decision to not seek a patent for monoclonal antibodies in 1975, when non-patenting was unexceptional and unremarkable, is an indicator of the rapidity of this change. British biotechnology trailed the American lead, in terms of research, investment and industrial development.[37] In both countries a regulatory system was put in place in the late 1970s that permitted research to be conducted at different levels of biosecurity containment.[38] In April 1980 ACARD, the Advisory Board of the Research Councils (ABRC) and the Royal Society published a joint report (the 'Spinks report'), to which the government responded, in which many measures to support a biotechnology industry were accepted.[39] These included an expectation that private companies, large and small, would take the lead, while acknowledging that, given the long-lead time for development, some pump-priming and structural support was nevertheless to be provided by the government. Thus the overall aim might be:

> The Government's economic policies are designed to create a climate in which industry can take long-term risks with confidence. In this kind of environment companies are prepared to accept the high cost of investment in research and development and in the introduction of new technology and are able to generate the necessary funds to make this investment possible. The development of biotechnology will provide the private sector with new scope to exercise its enterprise and initiative. There will be opportunities for small and new companies as well as large and well-established ones.[40]

But in the interim, the NEB and NRDC would continue to invest (£1.5m up to 1980, with a further £1m under consideration, despite Thatcher's distaste for the two bodies).[41] By 1982 the level of BTG support was £13m, with the funds supporting (alongside private investment) 'nearly forty biotechnology initiatives, mostly small and university based'.[42] Of these initiatives Celltech was the largest enterprise (employing 100), and then Speywood (employing 40 scientists). These initiatives attracted public attention while the activities of large international pharmaceutical companies' activities (which included their own R&D teams as well as contracts in universities) were less visible.

The fraught issue of the commercial exploitation of academic science created some remarkable contradictions under Thatcher. We can see this clearly: first, in the paradoxical notion that free, individual academic recipients of public research council funds might be *directed* to be more entrepreneurial, and second, in Thatcher's insistence that arrangements such as the NRDC/BTG infringed individual researchers' rights.

Nicholson ordered research into whether a minister had 'powers to direct the nature and scope' of the Research Councils' research programmes. This constitutional investigation concluded that this breaching of the so-called Haldane rule was acceptable.[43] One Treasury official wondered if such 'stick [should be] applied to them because they had not done as much as they reasonably should of their own free will, to ensure that the economy in general and the taxpayer in particular can see some practical benefits for the funds he provides'.[44]

By 1984, after extensive discussions, Keith Joseph was ready with new proposals on the exploitation of Research Council funded inventions. Interestingly (given that he had the reputation as an ideologue), he tried to pull back from full liberalisation:

> I wondered if we should just have a free for all, allowing every researcher to do as he or she saw fit. Somewhat reluctantly, I think not – at least not yet. Public money is involved and there are statutory and other legal requirements to accommodate. Many researchers do not have the skills to pursue exploitation themselves; the incentive to develop such skills needs time and encouragement to grow.[45]

Likewise, Joseph said, he could not 'tell [universities] how to run their internal affairs'. Nicholson, this time, backed him up:

> Like the Secretary of State, I had originally been in favour of new arrangements in which authority and accountability were more fully devolved to the individual researcher. But, again like him, I have become convinced that such an enormous step from the present protective bureaucracy would not be wise and might, indeed, jeopardise the whole process of liberalisation of exploitation of research through the occurrence of a few 'scandals'.[46]

Thatcher was not pleased with this back-sliding. 'Why?' she scribbled next to Nicholson's comment, adding:

> No – I see no reason why an individual researcher should be denied the right to develop his own research in this country if he wishes. We can meet the public funds point by demanding a royalty.[47]

Joseph came back with the details of arrangements to succeed the ending of the NRDC/BTG monopoly in March 1985.[48] This draft reflected Thatcher's 'wish that [Joseph] should go further towards devolving rights in research to the individual researcher'.[49] Thatcher was now 'delighted'.[50]

In October 1987 the BTG chair Colin Barker proposed, and ministers approved, privatisation by means of a management-backed buy-out. Quite unexpectedly, in July 1988 Coopers & Lybrand, invited by the government to conduct a feasibility report, recommended that BTG stay in the public sector. The BTG management promptly commissioned their own report from Lazard Brothers, which, less surprisingly, came to an opposite view. The DTI brought forward the formal proposal for privatisation in December 1988, and it was eventually privatised under the Major government in 1992. Nicholson's prediction to Thatcher that privatisation, for BTG, would be the 'eventual reward for success' came belatedly to pass.

Cuts in the science base

The Thatcher administration's ambition to cut public spending notoriously impacted on British science, but its causes and consequences have been misunderstood. Rather than a simple story of cuts provoking outrage and response (in the organised form of Save British Science, established in late 1985), this crucial episode of British science and science policy was formed from several intersecting issues – notably the proper balance of support for defence and civil science, the relationships between science and innovation and of universities to industry, the question of British membership of European scientific organisations and the forms of scientific advice, as well as broad aims of reducing public expenditure.

R&D was an early target of Derek Rayner's 'crusade against waste and efficiency', as Peter Hennessey describes the quick but deep reviews of civil service work conducted by the Marks & Spencer executive much admired by Thatcher.[51] This programme of scrutiny was outlined to Cabinet in 1980, trialled on statistical services and then turned on R&D in government laboratories, quickly followed by many others (130 by 1982).[52] Thatcher approved the review in January 1981, writing that she was 'glad we are to embark on this scrutiny', emphasising that investigators should pay special attention to 'return on investment … especially important in def[ence] case'.[53] After ministers put in special pleas for certain favoured laboratories to be avoided, Rayner chose his case studies and his team went to work. Covering two-fifths of all government R&D

staff, Rayner's team identified savings of 1,518 posts and £14 million a year, as well as overprovision of services, 'waste of land and buildings', 'unrealistic charges', 'lack of cost-awareness' and 'too much bureaucracy'.[54] 'Once again we have seen unnecessary bureaucracy and costs which call for a determined effort to demonstrate that economy and efficiency matter,' Rayner told Thatcher. Thatcher responded that she was 'appalled that after all our efforts such gross inefficiency exists'. As one minister put it, Rayner had 'found widespread opportunities to reduce costs in the support "tail" without damaging the research "teeth", and that there is scope to continue with excellent research at less cost'.[55] Yet while the Rayner review shows that government science was not being treated differently from other parts of the civil service, the wider aim of reducing public funding certainly did raise concerns about blunted research capability.

The Rayner cuts affected science spent by government departments in support of government work. A much more extensive and notorious set of cuts concerned public funds that were spent outside of government, especially in universities. At an official level, the impact of cuts in public spending on research and development was watched. The newly constituted Official Committee on Science and Technology, a monster of 28 representatives, chaired by Cabinet Secretary Robert Armstrong, noted in its first meeting that the 'aggregate effect ... on British science' of all the 'decisions taken by spending Ministers in the light of the reductions in planned public expenditure' were 'likely to be substantial', and inter-departmental study was required in order to inform ministers.[56] A 'quick, broad study', led by the Department of Education and Science, with contributions from the CPRS, was initiated, although not until May 1980.[57]

In February 1981 Thatcher could state in a speech to the Parliamentary and Scientific Committee that 'we have kept up spending on the Science Vote, which supports research through the Research Councils and the universities'.[58] Under the 'dual-support' system the Department of Education and Science channelled money through the research councils, which supported research projects, institutes and laboratories, and postgraduate training, as well as international subscriptions such as CERN. Under the other half of the duality the University Grants Committee funded the universities, in particular staff salaries and university facilities.

The major cuts (primarily in the UGC block grants) were brought in by the March 1981 budget.[59] The line of the Royal Society was to: 'accept the fact of the cuts, at least in public; to press for them to be implemented selectively so as to protect the best research; and to advocate maintenance of the dual-support system'.[60] Informally, the Society would put pressure on the Chief Scientific Adviser.[61] Nevertheless, by 1982, severe problems due to cuts were evident. As the White Paper on Expenditure, covering 1981–2 to 1983–4, explained, 'the Government wishes to give protection to the support of basic science by ring-fencing the real value of the five Research Councils'.[62] But the university funding, through the UGC, was to be cut substantially: 'savaged' was the description in *The Times*.[63] Research funded by the Leverhulme Trust found that spending cuts were severely hitting British universities, and that, along with other pressures (for example, more time absorbed in preparing grant proposals, the research councils becoming 'increasingly dirigiste'), was causing plummeting morale.[64]

In October 1982 the Advisory Board of the Research Councils, led by the vice-chancellor of Bristol University, the nuclear physicist Alec Merrison, called for an increase in research funding, targeted at biotechnology, remote sensing, information technology, marine resources and neuroscience.[65] (He was still smarting from being told by Thatcher to earmark a modest increase in funds, following the Falklands War, for the British Antarctic Survey at the expense of what he considered 'other and better science', discussed in Chapter 7.[66]) In response, Keith Joseph stated that money should be shuffled away from the Social Science Research Council and asked for a further report on setting priorities.[67] A joint ACARD/ABRC report agreed that research spending should be more selective.[68] Meanwhile, the *Times Higher Education Supplement* began to speak of a 'crisis of science'.[69]

On 28 February 1983 the BBC science series *Horizon* broadcast an episode titled 'British science – on the wrong track?'. Introduced by Gavin Scott, the programme asked why British science's successes (illustrated by footage of Aaron Klug and John Vane's recent Nobel awards) did not translate into commercial successes; the monoclonal antibodies story was rehashed, with both the cuts in universities and the preponderance of military research criticised.[70] Thatcher saw the programme and was 'much disturbed' (monoclonal antibodies hit a particular raw nerve); she believed that it had presented 'a biased and one-sided picture, and she wished the record to be put straight'.[71] Her dissatisfaction had 'intensified' when she learnt that the BBC had interviewed her Chief Scientific Adviser, Robin Nicholson, for three-quarters of an hour, but had not used the material. Thatcher instructed Bernard Ingham to 'let the BBC know of her reaction to the programme'. She also immediately decided to chair a public seminar and reception on science, technology and industry at Lancaster House, at which she would give a speech.

Ferdinand Mount, Head of the Number 10 Policy Unit and one of Thatcher's inner circle of advisers, warned that there were dangers in such a move – it 'must not be allowed to degenerate into an anti-Government rally' and 'both your colleagues and the participants are likely to try to use the seminar as a forum for extracting commitments for more Government money. However, Thatcher insisted, declaring 'I am not thinking of a mass meeting, only about 150–200 scientists'.[72] It was to be a direct encounter. Bumped from May because of the ultimately triumphant general election, the public seminar took place on 12 September 1983. With BBC and ITN cameras present, Thatcher's speeches topped and tailed the event, with talks by ministers and industrialists (including Clive Sinclair and Lord Weinstock) in between. She spoke of high levels of spending of science and the way that fundamental science led to applications, promised to protect intellectual property and remove monopolies (recalling BTG), recalled the 'long and brilliant record' of Britain's science and engineering, listing the names of Newton, Faraday, Darwin and Fleming, Stephenson, Brunel, Royce and Barnes Wallis, and urged her present audience of fundamental researchers to be 'alert to its possible applications'. She recollected the application of science in her early career, but also how this process had speeded up:

> Ours is not only an age of discovery. It is an age of application – devastating in its swiftness; enthralling in its surprises; remorseless in its competitiveness.[73]

Her final words of the day were:

> I stress the point of the seminar is positive from the beginning to enable us all to do two things, to create new business and industry and to expand existing business industry. And as Professor Kingman [chair of SERC] said, always, because many of us are scientists, to reach out to the unknown, to try and unlock the secrets of nature which we have not yet solved, and to try always to meet the challenge of our times, which is the creation of new wealth and new business.[74]

Thatcher's science seminar, which in turn had, remarkably, been prompted by her furious reaction to a television programme, re-energised science policy discussions within government. Robin Nicholson penned a wide-ranging response, summarising key points and necessary actions. One point was that the display of attention had been vital: there was 'unanimous agreement from both individuals and the media that the most important fact was that the seminar happened and that the Prime Minister and her senior colleagues, and senior people in industry, finance and academia demonstrated their interest in the subject'.[75] For 'morale', the message of success must be maintained. A second point, among many anxieties expressed, however, was that with regards to the otherwise 'strong UK science base', which was 'seen as an essential UK asset by industry', there was 'evidence that the strength is slipping'. These were bold words from science adviser to prime minister. Nicholson's proposal was: 'identify constraints on the science base and restore to health by better allocation of public funds and more use of private sector funds'.

Mount was correct in his prediction that colleagues would use the seminar to ask for more funds. The Secretary of State for Education, Keith Joseph, immediately wrote to Thatcher on the subject of 'maintaining the strength of the science base'.[76] (Officials in the Department of Education and Science were already concerned about the 'plight of research in UK basic science and the risk of having to withdraw from a major undertaking if the Government did not provide some real growth of funding, certain and sustained, over several years'.)[77] Informed by conversations with David Phillips, chair of the ABRC, Joseph argued that even after 'some economies and consequent redeployment of resources within the science budget' (which would be made after investigation by Sir Ronald Mason), there was 'need for some more money … to maintain the existing range of research' and, furthermore, 'modest extra resources to enable research into whole new areas of science recently opened up'.[78] One possibility, suggested Joseph, was that extra money should be transferred from the defence research and development budget.[79] In the meantime David Phillips commissioned research in to the 'decline of basic science' from the Royal Society.[80] The pressure to continue to cap the Science Vote came from the Number 10 Policy Unit[81] and the Treasury, which had threatened extreme measures – such as a Star Chamber – to reduce R&D expenditure.[82] As a basis for argument, the government began annual reviews of R&D, starting in 1983.[83] 'The 1984 Review will be an essential database for a general critique of Government spending on R&D,' the Cabinet Secretary had told Thatcher, adding 'the urgent need for which the Chief Secretary, Treasury spoke about at your meeting with the Secretary of State for Education and Science on 19 October [1983]'.[84]

Unresolved tensions over defence R&D and CERN

Between late 1983 and early 1985 the ministerial tussle over science policy focused on two components of the science budget: defence R&D and the UK subscription to CERN, around which there was a wider debate about choosing priorities, all under the heading of 'maintaining the science base'. Let us examine each focus in turn. The UK spent about £1,900m on defence R&D, of which the bulk (83 per cent) was classed as development. Of the £330m on research, £141m was spent in industry and only £9m in universities.[85] Michael Heseltine, Minister for Defence, immediately responded to Joseph's suggestion of redeploying defence funds. He was 'very happy to look at the scope for increasing the proportion … placed in the Universities', but any straight 'transfer from defence R&D to "pay for" increases in the science budget would in practice represent a cut in the defence budget and would have to be justified'.[86] Robin Nicholson, on the other hand, agreed with Joseph, and, indeed told Thatcher that the risk to UK science was very high:

> the excellence of our science base is starting to be eroded through insufficient funding … It is tempting, of course, to postpone the increase in the Science Vote until the year when savings can accrue. I believe that this would be a disastrous course of action. It will be much harder (and more expensive) to restore the quality of our science once it has started to erode rapidly; we must act quickly now to maintain its quality.[87]

Nicholson did think there was scope for 'offsetting savings', which he expected to find in the atomic and defence sectors ('it seems unlikely that it will be possible to find arguments to sustain the privileged position of the Ministry of Defence and the United Kingdom Atomic Energy Authority'). Ferdinand Mount agreed: 'Our whole economy is distorted by the present preponderance of research on defence'.[88] The Prime Minister and key ministers met on 19 October 1983. On the broad point of maintaining (or even increasing) the science base, Robin Nicholson recalled:

> the Prime Minister who had, I felt, decided against the bid before entering the room, on the grounds of the current state of the critical discussions on public expenditure, said: 'I believe in science and technology but they cannot be set up on a pedestal with a private pipette to the Treasury (her phrase). … '[89]

On the narrow point of defence research, it was agreed that universities should be able to compete with the private sector for work, 'on the understanding that they would be contractors'. On the general point, however, Joseph's plea for extra funds was rejected:

> In discussion it was argued that the Science Budget … was already very large; that if difficult choices on priorities had to be made, this was equally true in other areas both in the public and private sectors; that the UK had for years financed fundamental research generously but with poor results. The priority now was to boost the commercial exploitation and application of technology. …
>
> Summing up, the Prime Minister said that the case for expanding the Science Budget had not been made out. It should be possible to absorb the proposed additions within the existing Budget. The priority was to achieve commercial exploitation of technology rather than expand fundamental research.[90]

Joseph and Heseltine rejoined battle over scraps of defence R&D funds in the following year, when the ABRC was concluding its annual review of government R&D.[91] Joseph had persuaded Thatcher that a transfer of funds might be worth considering, on the new grounds that universities were more likely to offer a range of routes to further application and commercialisation. A figure of £20m was proposed.[92] Heseltine now rejected the idea: defence science was best done by defence scientists who knew the Services' needs, taking money out would damage long-term intramural research at defence research establishments and removing the work to universities would distance it even further from applications.[93] Joseph, unsurprisingly, was in favour of an addition to the Science Vote (albeit, he noted, 'relatively small sums') to support certain 'strategic research' disciplines (therefore not closely tied contracts) that together would lead to a 'reinforcement of the science base that can be expected to benefit defence, and other industry more widely, in ways that cannot be foreseen at the outset'.[94] He listed an indicative set of 'disciplines' (specialties, really). The final choices would be made after 'close collaboration between MOD and the Councils),' noted Joseph. Such an approach, he added, 'would accord well with the plans I am working with the UGC, the ABRC and the Research Councils, to bring about greater selectivity in the UGC funding of research'.[95] Henry Chilver reported his ACARD group's conclusion that the UK's comparatively high defence R&D spend carried perhaps unjustifiable opportunity costs.[96] The Chief Scientific Adviser also weighed in on the side of Joseph and Chilver, but again set the issue in a wider context of perilous trends in science policy. On current plans MoD would increase its spending on R&D by 1 per cent, while at the same time 'expenditure on advancement of science will have declined by 3% in real terms ... These trends are in the wrong direction'.[97] Nicholson continued:

> I support ACARD's view that there is a high opportunity cost associated with pre-empting an ever-increasing fraction of the nation's R&D resources in defence technology. The ratio of £1 spent on R&D for every £3 spent on purchase of equipment is absurdly high. I have the impression that the MoD is feeding a leviathan with an insatiable appetite for R&D resources ... This trend must stop eventually and I think there is a case for examining the consequences of a reduction of the MoD R&D spend to roughly half its present value over a period of 5 years ... and a switch of the R&D resources thereby released to areas with a greater influence on the economic health of the country.

In February 1983 Lord Whitelaw had drawn to the attention of Cabinet colleagues 'widespread concern on the health of basic and strategic research' following a debate in the House of Lords.[98] Thatcher had asked how such problems might be solved 'without spending more money'. This in turn had prompted the search for solutions from 'improved efficiency and selectivity in basic research'. Now, however, Nicholson urged something more radical: 'the answer lies in re-allocating funds from other parts of Government's R&D spend such as defence'. In general, he said, 'we are over-committed in R&D for defence, agriculture and nuclear energy and under-committed in basic research and in strategic research for areas such as the environment and manufacturing industry'. The government should transfer (not simply cut, but also not provide an overall increase, as per Thatcher's instruction) funds, otherwise:

> There is real damage being done to our University and Research Council research. Although it is true that the past excellence of this research seems to have had little influence on the economic performance of the country, one does not solve that problem by reducing the excellence of basic research. At a time when the Government's policies have led to encouraging progress in the application of our scientific and technological skills to producing marketable goods and services, it would indeed be ironic if the same Government was to damage irreparably the very source of those skills and so inhibit the development of a strong science- and technology-based industry.[99]

On the specific question of what to do next about transferring £20m in defence R&D funds, Thatcher followed Nicholson's advice.[100] Nicholson rejected Heseltine's view that there was 'minimal overlap' between MoD- and DES-funded work, while transferring funds would inject 'vigour and scientific competition', not least in the defence laboratories which he viewed as lethargic.[101] Accepting these arguments, Thatcher asked Nicholson 'with the aid of the Chief Scientific Adviser, MOD and the Chairman of the ABRC' to 'examine the proposal more closely' with a view to 'clarifying the way in which a transfer of responsibility for research could be carried out'.[102]

Between September 1984 and February 1985, the issue was further explored. The Ministry of Defence, stating that the cut would remove basic research from the defence laboratories, pushed back – commissioning, for example, a report on the Royal Signals Research Establishment's impact on the economy.[103] The scrutiny did reveal a patchy record of mid-1980s defence-civil research connections in the UK.[104] Nevertheless, the end result was that there was no transfer of £20m of funding; instead measures to improve links between MoD and the wider scientific community, such as collaborative research grants and joint MoD and research council activities, were proposed.[105] Nicholson regarded the response as inadequate.[106]

Yet the issue of the damaging externalities of the UK's commitment to defence research did not go away. For example, Thatcher was advised in February 1986:

> UK public expenditure on R & D as a proportion of GDP is similar to that of other major industrial countries ... but it is strongly skewed towards defence. Defence R&D in 1985–86 (at £2,300 million) accounts for 52.7 per cent of total Government R & D expenditure. The defence industrial complex appears to be largely insulated from the ordinary pressures of the market economy, and pre-empts scarce scientific resources, especially in electronics and information technology, to the detriment of the rest of the economy.[107]

An ad hoc Cabinet committee of ministers (MISC 119), dominated by Heseltine, proposed to resolve this by setting up a new ministerial committee structure to review priorities across all government science, supported by a substantial 'R & D Evaluation Unit' in the DTI, another of Heseltine's ideas. But Brian Griffiths of the Number 10 Policy Unit attached this plan 'to second-guess the private sector' as 'Heathite Corporatism'.[108] 'A more useful approach would be to distinguish between applied research done by the private sector and pure basic research funded publicly,' argued Griffiths. He went on to make four recommendations to the Prime Minister:

1. Defence R&D should be cut.
2. Tax incentives should be used to encourage innovation and commercial risk-taking in privately funded R&D.
3. Accordingly, public funds for R&D should be directed away from applied commercial research and towards basic research and initial support for diffusing information about new technology.
4. For grant-aided university research, a market-responsive system should be developed whereby the brightest talent is drawn to the most fertile areas, at the same time attracting private venture capital and industrial support.

Crucially Heseltine resigned in January 1986 over the Westland affair, and so the option of a 'Heathite' industrial strategy receded, while the Griffiths argument would be successfully taken up and pushed through by a new man in the Number 10 Policy Unit, as will be shown below. Somewhat ironically, a 'Science and Technology Assessment Office' under the Chief Scientific Adviser in the Cabinet Office was indeed set up (announced in July 1986), even though its immediate *raison d'être* disappeared.[109] For now it should be noted that the UK's over-commitment to defence research – as a percentage of GDP it was nearly half again more than the French and six times West Germany's – was a major tension in science policy under Thatcher.

The other topic of disagreement, the UK's involvement in CERN, was more controversial, but ultimately also of little apparent outcome. In October 1983, reviewing the implications of cutting defence R&D, direct civil R&D support (ie grants) or central facilities, John Kingman had considered the possibility of cutting the Spallation Neutron Source, just then being completed at the Rutherford Appleton Laboratory, Harwell.[110] He did not mention CERN. At the meeting of ministers in October 1983 Thatcher had (after declaring that science and technology must not be put on a pedestal with a 'private pipette to the Treasury'), taken a swipe at CERN:

> The Science Vote and the Research Councils have been protected for 10 years, but have done nothing to manage their cash limits. There has been no real shift towards <u>useful</u> science and money is still lavished

on grand but useless projects such as CERN. At the same time other nations have benefited from our science because our University scientists are too toffee-nosed to get involved in applications. We can no longer afford to do science for prestige, it must be science for economic benefit.[111]

Nicholson listed actions that DES could take, one of which was 'Make a serious study of withdrawal from CERN'. In January 1984 Keith Joseph raised informally with Thatcher the suggestion, now apparently originating in David Phillips and John Kingman (the chairs of ABRC and SERC, respectively), that the UK's £30m funding on high energy physics, primarily for CERN, 'could be spent more productively elsewhere within the science budget'.[112] Joseph claimed he was 'initially sceptical, suspecting that the Professors might have been putting forward the most controversial option for cuts', but had come round to the view that there should be a review, not least because 'there might be substantial support within the scientific community for this switch in emphasis'. The Prime Minister agreed: 'she felt that CERN, in common with many collaborative projects, was extravagant'. All this was only a year after one of CERN's greatest triumphs: the discovery of W and Z bosons as predicted by electroweak theory. The molecular biologist John Kendrew was assigned the task of the review.

CERN management were depressed by the news, as diplomats reported, fearing that it was a 'preliminary step to almost certain UK withdrawal', which in turn might encourage other countries to take similar action.[113] The view from Bonn was that withdrawal would be met with 'some dismay', as well as 'evidence of penny-pinching unenthusiastic attitude to collaboration within Europe'.[114] Since the UK had no indigenous facilities approaching CERN's power, Germany saw such an action as 'a confession by the UK that we rank ourselves with those smaller powers who can no longer afford to play a role in the significant science of the twentieth century'. Withdrawal, said diplomats, would also 'go down badly with France', not least with Mitterrand.[115] (Thatcher, writing on the telegram, was unimpressed: 'Of course [Mitterrand would complain] the Lep ring is largely on French soil'.) Withdrawal would mean unemployment for 330 British scientists and technicians, as well as loss of £5m in high technology contracts with British industry. Geoffrey Howe raised these European relations aspects directly with Joseph, who in turn copied Thatcher in.[116] At least some of the pressure to review, and perhaps end, the UK's financial contribution to CERN came from British non-high-energy-physics scientists. Reassurance was sought, and received, that any saving would be returned to the Science Vote.[117] At this point, Thatcher's advisers made comments. The Political Adviser Oliver Letwin, while stating that the Conservative administration 'should certainly contain – and, if possible, reduce – spending on science', agreed that the review should not be a vehicle for such an aim.[118] Likewise Thatcher's Chief Scientific Adviser, Nicholson, supported redeploying money 'towards priority growth areas in science', although he doubted the review would recommend UK withdrawal:

> Withdrawal from CERN must be contemplated as one option on completion of the study – it would be unreal to exclude it. Personally I doubt that it will come to that. More likely will be recommendations to improve the cost-effectiveness of CERN (you've seen the gold plating yourself) and, crucially, to slow down the pace and hence the rate of spend on this area of research. There is no reason why the tax-payers of Europe and the USA should have to fund a private race between two scientific cliques carried out at a pace determined largely by their own curiosity and arrogance.[119]

Kendrew's working party studied the issue between March 1984 and June 1985, and reported to the ABRC and SERC.[120] Kendrew recommended a 25 per cent reduction in expenditure on CERN (quite dramatically, but only after the completion of the LEP in 1989), and only UK withdrawal if such a reduction could not be negotiated.[121] By the international agreement governing CERN any reduction would have to be accepted by and applied equally across all member states. Prominent scientists attacked the Kendrew recommendations.[122] Nicholson complained to Thatcher that DES's summary of the report omitted 'the important conclusions that expenditure on particle physics is too high irrespective of the current financial problems of the Research Councils'.[123] After going through the accounts with CERN's director, Professor Herwig Schopper, Nicholson concluded that a 15 per cent cut was the maximum attainable.[124] Reaction from other countries was mixed, some signalling support for a review and others sceptical.[125] An international review working group was set up, chaired by the French physicist Anatole Abragam.[126] In November 1986 Thatcher's new Secretary of State for Education, Kenneth Baker, reported that the Abragam review was going slowly, which meant at least a year's delay; any UK withdrawal from CERN would be pushed back to 1988.[127] 'The time taken is so long that we can only think the delay is deliberate,' wrote Thatcher. 'It is grossly inefficient.'[128]

Meanwhile, in 1986–7 the UK was embroiled in lengthy diplomatic negotiations over the size of the first European framework for funding R&D. Thatcher was deeply sceptical and insisted on capping the budget at 4.2 billion ECU.[129] Other countries, as well as the European Commission, wanted a much larger budget. Thatcher's intransigence on the issue frustrated other ministers, who argued that the United Kingdom was a net *beneficiary* of such European research spending.[130] This European tussle, seen by Thatcher as a matter of financial discipline principle and by the other governments as a Eurosceptic 'imperilling [of] Europe's ability to match the growing high technology challenge', pushed the CERN question to the back burner.[131]

Here is where a shift in influence among the advisers was ultimately critical. Nicholson, who was often sceptical of CERN's benefits, moved on,[132] to be replaced by John Fairclough in June 1986. Furthermore, George Guise joined the Number 10 Policy Unit and was soon paying close attention to science and technology policy.[133] Guise had come from the business world, specifically Consolidated Gold Fields, where he had been an Executive Director since 1981.[134] The policy on CERN would ultimately be settled in the wake of a fundamental reframing of science and technology policy that occurred in 1987 as Guise's arguments prevailed over Fairclough's. I will now trace this reframing as all the frustrations and tensions of policy came to a head.

A third area of active policy-making that should be briefly mentioned before returning to the core argument concerned the provision of trained engineers for industry. Keith Joseph proposed a 'switch', spending £42 million over three years to produce 600 more graduates and 500 more postgraduates each year in engineering and technology, the 'skills of tomorrow'.[135] Peter Warry and Oliver Letwin in the Number 10 Policy Unit were sceptical since there was no increased market demand, as measured by university applications, the creation of new jobs or engineers' salaries.[136] Nicholson, however, supported Joseph, although he also argued that a condition must be that the private sector 'make a full contribution in cash and in kind'.[137] Consultation followed with, for example, Lord Weinstock of GEC writing to Thatcher about shortages not just in new graduates but also the need to fund re-training in a fast moving field.[138] The Engineering and Technology Programme was announced in March 1985. Thatcher hosted a meeting with industrialists on 21 May 1985, at which they were told their support 'in cash or kind' was expected. Summing up the meeting, Thatcher sounded underwhelmed by the industrialists' contributions:

industry sometimes needed to adopt a wider concept of self-interest. Unless private enterprise was willing to ensure enough was done to encourage engineering and technical education and training, Government would have to step in. But they could never do the job as well as industry itself. … The Government had already done a good deal, however, and nearly every school, including primary schools, now had a microcomputer … Of course, more needed to be done and that was where industry came in. The UK's record on the research side was a good one; it was the practical application of research which we needed to concentrate on now.[139]

Risk-averse industry?

Frustrations at the centre of government were now building up in three areas: a defence industrial sector that absorbed too much R&D resources and was unresponsive to the market, the lack of entrepreneurial spirit among researchers and the perceived lukewarm willingness to contribute by industry. The argument that British industry and commerce were risk averse when it came to exploiting R&D was supported by two more documents we know that Thatcher read. First was a blunt letter from Sir Henry Chilver, chair of ACARD, who, 'alarmed', wrote of

> UK companies … avoiding high risk investments such as developing new products from R&D programmes and introducing new manufacturing methods resulting from technological advances … despite the fact that innovative products and services are vital to ensure competitiveness in the longer term and adoption of best practice can give current products a competitive edge … Until we have correctly identified the inhibiting factors, and found solutions to them, Government must not assume that setting an economic climate which encourages companies to make a profit is sufficient to ensure national prosperity in the long term.[140]

The second document was written by Sir Douglas Hague, chair of the ESRC. Drawing on arguments from SPRU's Keith Pavitt, Hague argued that the expense of science demanded that it be tied to national objectives, particularly 'increasing the efficiency and competitiveness of the British economy', while the government must provide financial support because

> British firms appear to underinvest in basic research and training, because of relatively short time horizons, risk aversion, and (most important) the fact that the responses of competitors will make it impossible for them to obtain all the potential benefits of their investment.[141]

Thatcher underlined risk 'aversion' in both documents. Interestingly, Hague called for a clear, functional and locational separation of research and development: 'the rule should be that basic research be done in universities and Research Council laboratories, and development in firms, when this is concerned directly with the promotion of economic development'.

When in 1985 the DTI ministers Geoffrey Pattie and Norman Tebbit wanted to extend government support for industrial R&D, the proposals were attacked by the Number 10 Policy Unit: 'All discretionary financial support to industry is suspect. Little money tends to go to projects that would not otherwise have been undertaken, and many of the large projects are ill founded'.[142] Thatcher agreed: 'I am not enamoured with this idea of support for everything. There is a lot we should refrain from doing because the money is ill-spent'.[143] Meanwhile, Nicholson, ACARD and leading ministers (such as Geoffrey Howe and Keith Joseph) were all expressing alarm that UK R&D, compared to that of other countries, was being cut, generating 'a great threat to our future economic prosperity'.[144]

In early 1986 Nicholson began promoting the establishment of 'a scheme to promote better "pulling-through" of outstanding advances in our science and engineering research base to provide new products and services to be sold profitably by UK industry'.[145] The plan initially had Thatcher's support.[146] Fleshed out, it was a proposal, by the name of LINK, for all Departments with significant research programmes and all Research Councils to reallocate money to provide a pot (£210 million in early plans) that would be more than matched by contributions pledged by industry (£400 million). It would complement existing schemes in information technology and offshore energy technology, and in the first instance target advanced electronic materials and molecular electronics. Fairclough, the new Chief Scientific Adviser, supported the industrial strategy plan. He hoped to complement the programme with another seminar

> on Priorities for Science and Technology … which will be designed to get across to scientists, industrialists and the city the

> Government's commitment to research and development which will lead to greater wealth creation. The City in particular remains woefully short-sighted in its attitude towards investment in R&D.[147]

Guise, Fairclough and the 1987 reversal of UK science policy

George Guise, the new voice in science and technology advice, had a word in Thatcher's ear.[148] He supported the objectives but not the methods of LINK, and made two trenchant criticisms. First, the departments which paid for their own R&D, in particular defence, would pay it only lip service. 'Indeed,' he said:

> the lack of commercial spin-off from the enormous volumes of defence R&D which have already been spent is something of a national disgrace. In terms of the investments this country has made there should be a thriving computer industry, a silicon chip industry, extensive developments in solid-state physics such as lasers, and a strong radio industry. In fact, the radio industry has declined to nothing.[149]

Second, without 'effective people with strong commercial connections' governing LINK, its objectives would not be met. Fairclough, he advised Thatcher about her Chief Scientific Adviser, should 'be pressed' on this issue. Guise proposed Sir Alistair Frame (of Rio Tinto Zinc) as chair and his suggestion was successful.[150]

Fairclough himself, in February 1987, with eight months experience as Chief Scientific Adviser, now felt able to offer 'a considered judgement about the issues and opportunities we face in securing greater economic contribution from our research and development activities'. His aim was bold:

> We need a renaissance of our industrial prowess through the contribution from research and development to again become an effective competitor to Japan, Germany and the United States. I offer the thought that this goal should be championed by yourself and become central to your future policies as you have already championed the control of inflation. We need a culture change in industry, science and the Ministry of Defence ... Your personal identification with this goal will be critical to such a task.[151]

Specifically, he called for the role of ACARD to be 'extended to cover basic and strategic science in addition to its role in applied research and development'. It would advise and balance priorities. By examining the 'whole environment' it could help with problems such as underinvestment in research by industry, the choice of which areas to invest to make 'world class science' ('we cannot afford … to engage in world class science in every subject and every University') and encourage privatisation of defence research. Guise, whose advice on science and technology was beginning to receive closer attention from Thatcher than Fairclough's, supported the Chief Scientific Adviser's proposal to simplify the 'structural tangle' of R&D funding policy.[152] He observed that Fairclough 'is asking for an annual science plan, analogous to a nationalised industry's corporate plan, which would be settled annually and run for several years ahead … [the internal Cabinet committee] E(RD) and ACARD would become the overlords of Government R&D expenditure'.

Guise wanted to go further, however. In particular, going beyond Fairclough's call for privatisation of some defence research, 'there is a strong case for privatisation of the Government Research Laboratories which would take the contractor/customer principle to its logical conclusion'. (This is an important point. Some commentators have taken the Rothschild principle as merely an arrangement for relabelling the relationship between government department and its research. Here we see that the 'logical conclusion' of market language is private markets.) Guise, too, thought Fairclough's plan, if it was to succeed, needed Thatcher's personal strong backing.[153]

Thatcher backed Fairclough to the extent of asking for a more detailed proposal.[154] After consultation (with ministers, the Cabinet Secretary, Nicholson – now in the private sector at Pilkington – and Lord Dainton), Fairclough's proposals remained largely unchanged. However, he now had a name for the newly extended ACARD: the Advisory Council on Science and Technology (ACOST).[155] The proposals were made public in a response to a report of the House of Lords Select Committee on Science and Technology.[156] Again Guise praised Fairclough's plans as far as they went, but argued that they did not go far enough:

> There is much evidence that British R&D suffers from mismanagement … By far the greatest culprit is private industry which refuses to put risk capital into maintaining a modern technological base, *as long as the Government is prepared to do it for them*. …
>
> We cannot improve industrial R&D by central edicts about annual reports. What is needed is an efficient Government structure for allocating public funds which addresses the balance of responsibility between industry and Government. The first step

and one within the Government's grasp is a strong attack on the methods by which public research funds are allocated and subsequently managed …

Fairclough's paper does not address many of these issues. It concentrates on how Government can take a lead by getting its own house in order.[157]

At a meeting on 29 April 1987, with Fairclough, Brian Unwin, Robert Armstrong and Guise, Thatcher agreed to Fairclough's plan. It was a centralisation of the mechanisms for making science policy. A cabinet committee E(ST) – the Ministerial Steering Committee on Economic Strategy Sub-Committee on Science and Technology, a renamed, strengthened E(RD) chaired by Thatcher – would 'determine priorities in terms of functional spending'; this would extend even to the Research Councils, 'possibly to the extent of requiring expenditure of a particular sum in a particular area of research'.[158] E(ST) had its first meeting on 1 July 1987, soon after the general election. The new ACOST, which would retain Tombs as chair, would advise.[159]

Meanwhile, in July 1987 – at the invitation of E(RD) to consider the 'case for a stronger centralised management of the activities of the Research Councils'[160] and at a time when the Councils were regarded as being in 'serious difficulties'[161] – the ARC published *A Strategy for the Science Base*.[162] In public this science policy document was the one that was fiercely debated, especially its proposal to separate out a class of research-active universities that would receive most research funds.[163] While rejected, the subsequent compromise on university funding led to a beefed-up Research Selectivity Exercise (1989) as a mechanism for guiding the allocation of funds (a key moment in the trajectory that has led to the prominence of the cycle of Research Assessment Exercises and Research Excellence Frameworks in today's UK university system).[164] Nevertheless, despite this public acrimony, within the private discussions around Number 10 the ABRC's *Strategy for the Research Base* was also, ironically, the document that Thatcher's science policy would shift in opposition to and away from.

With the central machinery streamlined, George Guise could now pitch his big vision for science policy.[165] It would prove to be a deeply influential document. One small measure of this influence was the fact it started with an anecdote that Thatcher, having read it, would repeat in speeches and memoirs:

When Gladstone asked Faraday whether electricity might ever have a useful purpose, he replied 'Yes, Sir. One day you will tax it'. By contrast, Rutherford claimed in the 1930s that 'anyone who expects a source of power from the transformation of atoms is talking moonshine'. The prescience of Faraday is rare. Most people working in fundamental science, as well as those who fund them, have no idea what economic benefit the work will bring.[166]

At the heart of Guise's pitch was an attack on what he saw as a widespread and conventional understanding of the role of government (and especially public money) in supporting science:

> There exists in Government and industry a general mood which stresses the economic rather than the intellectual value of science and hence that resources should be shifted from pure science to technology. This attitude underlies the recent ABRC proposals and is endemic at the DTI, who continually confuse value for money with return on capital. It is actually a form of national short-termism and reeks of state intervention in industry. This philosophy misunderstands both the contribution of science to economic progress and the proper role of the public sector in stimulating it.

Guise then listed six examples that showed, he said, that the 'greatest economic benefits have always resulted from advances in fundamental knowledge rather than the search for answers to specific applied problems'. Specifically: transistors were not 'discovered by the entertainments industry seeking new ways of marketing pop groups', but by 'people working on wave mechanics and solid state physics'; computer logic circuits were not built by accountants; nuclear energy was not discovered by oil companies; the induction coils of in motor cars came from Faraday, not the transport industry; while electronics and electro-magnetic waves were the result of the work of Thompson [*sic*], Lorenz, Maxwell and Hertz, not the manufacturers of 'televisions and cellular telephones'.

Guise claimed that 'each' of these examples served to describe 'a basic scientific discovery whose application has proved, in the narrowest of economic terms, hugely profitable'.[167] Scientific inquiry 'without economic direction, over the past centuries has formed the bedrock of a modern economy'. Yet the consensus now was that 'public sector funds should be directed towards specifically applicable research projects'. (As well the DTI and the ABRC, Guise aimed his ire at the head of the ESRC, Douglas Hague, as an example of the class of economists who held this view. The Alvey and LINK programmes discussed above also exemplified it.) Guise wanted 'precisely the reverse of the policy'.[168] Alvey had been the flagship component of the early 1980s industrial strategy. Now, Guise said, any further Alvey work, if at all, should merely coordinate and not spend public money when industry should be paying for research.[169]

In short: Guise persuaded Thatcher that her government should cut funding from near-market research, because only then would profitable companies step up and fund more of their own R&D.[170] As Guise pointed out, this was the science policy that would truly resonate with Thatcherite policy; it was the one 'consistent with our long term goal of minimising State intervention'. For industrial research the 'proper role for Government', according to Guise, was one restricted to 'co-ordination, information dissemination and the reduction of bureaucratic impediment'. 'Basic research', on the other hand, was 'essential to long term national prosperity and its funding is a primary function of Government'.[171] It was 'organically part of the national interest and … the route to success is to back individuals and teams' (and to do so in an unashamedly elitist way; to spread the jam widely was 'a diluted form of socialism') and not to set 'remote goals which pre-judge the outcome of work'. The extent to which Guise's framework became the strategy for UK science policy can be seen in the first instance in the fate of Fairclough's new arguments for technology policy. Fairclough, the Chief Scientific Adviser, wanted to argue that Guise's paper was

> only about science and the Government role there. The Government also have a crucial role in funding technology … if scientific discoveries are to be properly exploited.[172]

In particular, a 'discovery becomes economically significant only when the science of how and why it works is sufficiently understood for it to be embodied in the design of an artefact or a process'. In other words, science was still needed in the 'intervening process between discovery and exploitation … the science phase does not come to an end and the exploitation stage take over'. Fairclough cited the 'the new warm super conductors … materials [that] could be as important as the transistor to economic progress' as a case in point. He wanted government departments to have the strength and the funding to 'focus on priorities for technology and sponsor selected new emerging technologies'.[173]

Fairclough's proposal that the Government's 'role to support the development of technology' be recognised (and funded) was immediately criticised by Guise. 'Government should fund basic science, but very little technology except where it is the user,' he wrote, as point 1 of his summary of key points.[174] Such a separation was essential:

> Fairclough mentions a 'transitional' phase where the science is 'becoming' exploitable technology. Once the science is properly understood then the routes to technological development are clear and can be costed. If economic benefit may be derived from pursuing such development then private sector business should fund it.[175]

Again the warm superconductor was a case in point (and it is telling that advisers could write to their Prime Minister about 'zero point energy' and assume that they would be understood):

> The warm superconductor ... is a phenomenon which is not yet fully understood. The concept of zero point energy, which comes directly out of fundamental quantum theory, has been known for decades and accounts for the superconducting behaviour of metals near the absolute zero of temperature. By contrast, the statistical behaviour of electron groups several hundred degrees higher, at room temperature, is not understood in fundamental terms. Therefore, the behaviour of certain ceramics which appear to display superconductivity at higher temperatures needs further work to establish fundamental theory.

Rather than step in and fund technology, the government should restrict itself to funding basic research. Even where government had a customer role or an 'operating function' (such as 'the provision of a health service, a defence service or an adequate road network'), work should not be placed intramurally – that was 'the old centralised control philosophy and has led to much inefficiency'. Instead it 'should be put into the private sector as fast as possible'.[176]

With the science policy reversed, the question of subscription or withdrawal from CERN could be resolved. Guise received a leaked summary of the Abragam report, via his friend, the Oxford particle physicist Christopher Llewellyn Smith (an adviser to Abragam who would also become Director General of CERN in 1994). Guise agreed with and repeated the arguments of Llewellyn Smith (and indeed Abdus Salam, who had visited Guise and the Number 10 Policy Unit) that the UK should now remain in CERN: it had pressed for the Abragam review which had delivered the assessment that CERN's administration and funding should be overhauled, and that remaining at the forefront of particle physics was a historic British necessity.[177] Fairclough also offered support for CERN, conditional on the Abragam proposals being followed through.[178]

Edgerton and Hughes argued that ultimately Thatcherite science policy was about the extension of control.[179] We can see elements of this, but ultimately it mistakes the part for the whole. There was a move to extend control, that is what Fairclough's reform of the machinery was about, extending ACARD's work into basic science (making ACOST), setting up an expanded, more powerful, internal cabinet committee E(RD) with the Minister in the chair. But this was merely a step to an end. The end was indeed ideological – the minimisation of State intervention – and this was expressed by the out-manoeuvring of Fairclough, once his work was done, and the implementation of the Guise reforms. The point is that research at CERN could be firmly classified as 'fundamental sc.[ientific] endeavour', as Thatcher wrote, and within the proper role of government to fund.[180] When the director general of CERN presented the arguments to Thatcher, she, now, would write back in relatively supportive terms.[181]

Thus when, in December 1987, the government heard of a threat by 'some, perhaps all, of the independent members' of the ABRC that they would resign if CERN was not funded, the result was not alarm since the issue was in fact more or less already resolved,[182] although the threat remained.[183]

What is becoming clear from my historical study of 1980s UK science policy is that there was a sharp shift in science policy – one that separated Thatcher's early and late years as Prime Minister. Early on, say between 1979 and 1987, there were increasing frustrations with the unresponsiveness of both civil and military science to markets, and rising anxieties among ministers about maintaining the state of the 'science base' as state funding was cut back. Then there was a crystallisation of policy: government funding for near-market research was abruptly curtailed (because private industry should step up), and, to balance this, the science base, especially 'curiosity-driven research' was heralded.[184]

The details of this history are convoluted, but the proximate steps towards the ascendance of 'curiosity-driven research' in UK science policy were as follows. In the early 1980s the common division of science into kinds or types had been threefold. As her Chief Scientific Adviser Robin Nicholson had briefed Thatcher in 1984:

> Basic research is that undertaken primarily to acquire new knowledge, without any particular application in view. Strategic research covers the area where basic concepts are established, but where it is not yet possible to identify specific products or processes. Applied research is directed towards a specific practical aim, such as the development of new products or processes.[185]

Curiosity in this first phase of Thatcher's administration was barely mentioned. When it was, indeed, the reference was as likely to be derogatory as otherwise – such as when, quoted above (p.84), Nicholson had argued that, in the context of arguments over CERN, 'the tax-payers of Europe and the USA should have to fund a private race between two scientific cliques carried out at a pace determined largely by their own *curiosity* and arrogance'.[186]

In December 1987 the eminent Cambridge molecular biologist Max Perutz laid into the ABRC report *A Strategy for the Science Base* in an article for *New Scientist* magazine titled 'How to stifle innovation'.[187] The attack received a warm and immediate reception from the advisers closest to Thatcher, notably George Guise, because it squared with the new science policy of curtailing near-market research. (It was nevertheless criticised, paragraph by paragraph, by others less close.[188]) Thatcher herself read the Perutz article, as we can tell by the blue ink.[189] It might have particularly provoked her with its mention of monoclonal antibodies – the exemplary case for her of British science's failure to make profits. Thatcher, again, underlined these words in blue. Perutz attacked mission-oriented science. He gave a list of great innovations, stating that they 'all arose from basic, curiosity-motivated research'.

Perutz's arguments in the *New Scientist* had been prefigured almost word-for-word in a September 1987 letter he had written to the leading figure of Save British Science, Denis Noble.[190] It has the same list (and more) of innovations, and the same claim that they 'all arose from basic, curiosity-motivated research'. Noble's Save British Science, as the advertisement that I quote at the beginning of this book shows, channelled anger among academic scientists that 'basic' science was in a crisis of underfunding. Save British Science's narrow focus on protecting basic science blinded it to the (probably unanticipated) rhetorical support its arguments might lend to those seeking the dismantling of support for near-market research. Fascinatingly, Perutz went on to hold up his own Laboratory of Molecular Biology, under his direction (from 1962 to 1979) as an exemplar:

> My laboratory is often held up as a centre of excellence, but this is not because I ever 'managed' it. I tried to attract talented people by giving them independence, listening to them and taking an interest in their work, helping them get what they needed for it and making sure they got the credit afterwards. … Had I tried to direct peoples' work, the mediocrities would have stayed and the talented ones would have left. The laboratory was never 'mission-oriented'.

> The brilliance of British science is one of the country's greatest cultural achievements, if not the greatest, but it is a fragile flower as I know from Austria, my country of birth. Once destroyed by bad politics it cannot be restored.

One of the extraordinary discoveries to have come out of the LMB under Perutz was Milstein's monoclonal antibodies. The story Thatcher was told about monoclonal antibodies at Cambridge, the one that she repeated so often to illustrate a failure in academic science–industry relations, was here presented as a parable of the freedom of the individual – which must, ironically, include the freedom to choose not to patent – which in turn led to the unfortunate decision on near-market research.

The rejection of government funded near-market research was made public in January 1988, when it informed the white paper *DTI – the Department for Enterprise*.[191] It was implemented on a department-by-department basis, for example they were known as the 'Barnes cuts' in government agricultural research after the person in charge.[192] They amounted to a removal of 30 per cent of funding by government departments for science.[193] A remarkable feature of this shift in science policy was that it was carried out in opposition to the direction of advice coming from the two, heavyweight advisory bodies, the ABRC and ACOST. When the ACOST (the successor, recall, to ACARD) had written its first communique to the Prime Minister, it seriously mistook its function. 'ACOST's starting point,' wrote the chair Francis Tombs, 'is that the overall total of UK expenditure on civil R&D must be raised over the next 5 years' – not least because 'other countries are well advanced in their plans for increased, targeted government expenditure and incentives to industry to capture growing world markets for goods and services based on advanced technologies'.[194] This, wrote a Number 10 staff member to Thatcher, was a '<u>most</u> unhelpful letter'. In no uncertain terms Tombs was told that his and ACOST's role was not to advise on overall funding levels, but on priorities. A chastened Tombs came back with a report on ACOST's work on priorities.[195] Likewise the ABRC had produced its report *A Strategy for the Science Base* in May 1987. From the outside this might have been taken as a statement of the direction of UK science policy (as the *Financial Times* and Edgerton and Hughes did).[196] From within Number 10, however, it was precisely the opposite. We have already seen Guise's complaint about the 'attitude [that] underlies the recent ABRC proposals' above, and how it galvanised the shift in science policy. This is one demonstration of how access to previously private primary sources can change what we thought we knew.

A second feature of the shift in science policy was that it was scaffolded by anecdotal history of science. It was George Guise who urged Thatcher that the end to government-funding of near-market research was the right approach to science policy. In March 1988 he was complaining to Thatcher:

> Despite the intellectual turnaround of the past few years, despite the public rejection of government-funded near market research in the recent DTI White Paper, and despite all the talk of ceasing to try to pick winners, there is still much muddled thinking both in the ACOST letter from Tombs … and in the ABRC strategy for the science base. If Ernest Rutherford couldn't anticipate nuclear fission as a practical form of power generation as late as the 1930s, how can ACOST or any other quango predict the economic benefit from basic science?[197]

He cited the JET fusion programme and the fast-breeder project at Dounreay as two examples where the 'real culprit was the intellectual arrogance of the whole centralising, long-range winner picking philosophy'.

The same features can be seen in the discussions over the proper role, location and funding of the new Interdisciplinary Research Centres (IRCs), which had been proposed by the ABRC's *A Strategy for the Science Base* and backed by ACOST and the Department of Education and Science. Thatcher disliked them in practice, considering them too bureaucratic; she asked why it 'should be necessary to set up a new mechanism to force scientific disciplines to work together' and 'why, if IRCSs were successful, they had not been introduced before?'.[198] Guise thought ACOST and DES displayed the same 'confused' and 'muddled' thinking. Despite the debate being 2,000 years old (Guise cited Plato's *Republic*),[199] he could see a clear resolution: government would fund curiosity-driven research (which could either be basic[200] or, as in the IRCs, 'strategic and exploitable'); technology would be funded by industry; industry would not shape (but could fund, in a hands-off manner) the former and government would keep out of the latter. IRCs should be centres of clusters, but industry should not seek to direct research, and certainly not 'seek specific answers for an immediate benefit':

> It is … good that industry be encouraged to support IRCs and it is in industry's own general interest to do so. A strong national science base, well managed as opposed to economically directed, automatically contributes to high industrial productivity and national return on capital. The proximity of the top research establishments

> in California to Silicon Valley and in Massachusetts to Route 126 [*sic*] are not coincidental. Much academic research is funded by industry and many of the industrial leaders have their educational roots in proximate academies. This partnership did not, however, flourish through industry attempting to pre-ordain the outcome of the research and pushing funds into what some committee foresaw to be commercially exploitable! It was based on identifying and supporting high quality team leadership, setting a budget, and then leaving well alone – the Perutz approach rather than that of David Phillips and Francis Tombs.[201]

Even Silicon Valley, Guise wrote, implausibly, was the result of curiosity-driven research.[202] The critical point was that Guise and Thatcher regarded State intervention as deeply undesirable, and this included public funding for near-market research. The ideological desire to remove the State's role from funding much applied research was the obverse of the new enthusiasm for 'curiosity-driven research'. They were two sides of the same science policy coin. 'Curiosity', especially since the late 1980s, in contrast to how it is often perceived – a neutral, child-like motivation common to scientists – had become a term wielded for political purpose.

Thatcher's new policy was fully expressed in her famous Royal Society speech of 27 September 1988. In the crucial months leading up to the speech, it is clear that the Chief Scientific Adviser, John Fairclough, had far less access to and influence on the Prime Minister on science policy matters than had Guise; he was 'excluded, apart from written comments' for a crucial six months.[203] Thatcher's speech, which took place in the Fishmongers' Hall in the City of London rather than at the Society's headquarters, is remembered today primarily for her call to arms on anthropogenic climate change. (That, discussed in Chapter 7, by the way, was another abrupt turn for Thatcher; there is documentary evidence to suggest she was a leading sceptic in 1979).[204] But the other important announcement was on curiosity:

> Of course, the nation as a whole must support the discovery of basic scientific knowledge through Government finance. But there are difficult choices and I should like to make just three points.
>
> First, although basic science can have colossal economic rewards, they are totally unpredictable. And therefore the rewards cannot be judged by immediate results. Nevertheless the value of Faraday's work today must be higher than the capitalisation of all the shares on the Stock Exchange!

> Indeed it is astonishing how quickly the benefits of curiosity driven research sometimes appear. …
>
> Second, no nation has unlimited funds, and it will have even less if it wastes them. …
>
> So what projects to support? Politicians can't decide and heaven knows it is difficult enough for our own Advisory Body of Scientists to say yea or nay to the many applications. I have always had a great deal of sympathy for Max Perutz's view that we should be ready to support those teams, however small, which can demonstrate the intellectual flair and leadership which is driven by intense curiosity and dedication.[205]

She concluded:

> Mr. President, this country will be judged by its contribution to knowledge and its capacity to turn that knowledge to advantage. It is only when industry and academia recognise and mobilise each other's strengths that the full intellectual energy of Britain will be released.

It is this speech that gives us the modern prominence of curiosity-driven research, a survey of a large corpus of literature shows a sharp inflexion in frequency of use on the late 1980s and a subsequent tenfold rise.[206]

After the reversal

In late 1988, science policy discussions in Number 10 continued to be a pattern of Tombs (as chair of ACOST) and Fairclough (as GCSA) seeking influence while Guise acted as a critical gatekeeper to Thatcher. Topics under debate included the Isis neutron facility at the Rutherford Appleton Laboratory, defence research and priorities in the science base. On Isis, Fairclough gave an upbeat report on the Isis facility (previously Spallation Neutron Source), which had been opened by Thatcher in 1985. It was, he wrote, 'a world class instrument … acknowledged as the best pulsed neutron (and muon) source in the world … supporting basic research in the core sciences. It is backing good people and ideas, with no pre-set selection by subjects … [while] the bulk of the work is in those strategic areas of chemistry, physics and materials science which in the long term could be of interest to industry'.[207] Users were scientists from the UK (nearly 60 per cent), but also from other countries, which contributed financially. He added that demand for beamtime outstripped supply. But to Guise this was evidence that 'foreign users do <u>not</u> pay their own fair share':

> If ISIS were a facility in the private sector with such an enormous backlog of unsatisfied demand, there would probably be a clamour for Government to regulate it in order to prevent monopoly profits! The back of commensurate foreign revenues at ISIS is typical of the reluctance of scientists to think about financial efficiency. ... The whole history of ISIS shows how naïve Britain can be when turning its leadership to financial advantage.[208]

On defence R&D (last an issue troubling Number 10 in the proposal for a modest reallocation from defence to civil budgets and in the SDI proposals), Tombs reported ACOST's latest analysis.[209] The concern again was how to obtain greater benefits for the civil economy. But now the argument focused on the proportion of 'dual-use' R&D, the meaning of which here was 'enabling' research that could equally well find civil as defence applications. ACOST had found that only one-fifth of UK defence R&D was dual-use in this sense. More would benefit the civil economy (through spin-offs) but also the military (who would benefit from competition from more providers). There was also a problem in the 'absence of an organisation to provide and be accountable for the technological oversight of the overall process of translating Services' needs into operational systems'.

On the science base, Fairclough sought a meeting with Thatcher to discuss various next steps (rearranging and even unifying research councils, encouraging the civil research establishments to be more competitive and flexible, allowing ACOST to 'develop a more responsive role to Government in order to balance its other more wide ranging and long-term work').[210] Guise was critical and, given that Thatcher had 'made the Government's position crystal clear', doubted even that a meeting between Prime Minister and her Chief Scientific Adviser was a 'necessity'.[211] He declared that ACOST 'continues to live in a world favouring centralised action by UK Ltd to ensure that the country is competitive in its innovation policy'. It hankered after government support for technology. And it barely commented on science. 'Perhaps it should be renamed ACOT', he joked. 'ACOST,' he went on, 'ought to be thinking of questions like:

> (i) In which sciences in Britain ahead of the world?
> (ii) How is such a ranking measured
>
> - by usefulness?
> - by Nobel Prizes?

- by number of papers published?
- by numbers of university departments?

(iii) How do we reconcile the Perutz principle of supporting individuals and small teams who are making breakthroughs with the principle of concentrating funding in areas where Britain is already ahead?

(iv) How can the value for money in basic science be measured if commercial exploitation is not a key parameter? (Has ACOST ever thought about how many telescopes British astronomy should support? The astronomers will want one for every hill in Hawaii while the non-astronomers will regard all of them as a colossal waste of money!)

(v) How do we ensure that organisations like the Rutherford Appleton laboratory or CERN or the Laboratory of Molecular Biology at Cambridge which produce excellent science are run efficiently?[212]

On the civil research establishments, and Fairclough's call for supporting more managerial flexibility, Guise responded that it was 'well intentioned but naïve':

> It is fine to set up certain specific activities as prototype businesses, and research establishments are ideal candidates, provided that <u>the ultimate goal is privatisation</u>. However they should become agencies only for a clear defined bridging period and not as a perpetual limbo where their costs will go out of control. The litmus test of whether an organisation is truly subject to the raw, real forces of competition is whether it can go bankrupt. All else is mimicry of the true market world of the private sector.[213]

All these points were made in the run up to E(ST) Cabinet committee meetings, the central forum for decision-making in UK science policy in this period. Guise worked hard, even when Thatcher wobbled. In January 1989 he wrote

> you expressed your concern about Government policy on science … You were also worried that by championing basic science and totally renouncing near market research, Government may further harm the nation's competitiveness. My belief is that we have already done harm to British innovation by removing much of the R&D burden from industry since the war. That which is not paid for,

> or earned, is rarely valued and the poor status of engineers today is in part because industry has not had to pay for and nurture them in order to stay ahead in the market place.
>
> Meanwhile, the engineers and would-be technical innovators have formed cabals whose prime purpose is to get money out of Government for technical research. …
>
> Unless the managements of our businesses face the raw, real forces of competition, and accept that they cannot win without developing and paying for technical research themselves, Britain will continue to underperform.
>
> It is a vicious circle. The more Government featherbeds technical research, the less top management will value it and the lower our innovative standing will be. Those in ACOST and ABRC and the endless quangos who lobby Government for technical support will continue to present this as evidence that more money is needed. The cure has been exacerbating the illness![214]

Guise urged Thatcher to be resolved. 'Among the basic science community your standing is extremely high,' he said, adding that this was because 'You are regarded as an ally against the forces of bureaucracy and centralisation'. Against ACOST and ABRC was, suggested Guise (again going back to Perutz), the individual:

> We have all fallen into a logical type confusion about 'picking winners'. What we should foreswear is the picking of winners by ranking the potential economic benefit of basic work … We shall always have to pick individuals and decide how their efforts can best be financed. It is this kind of individual winner that you need to see …

The Cabinet Secretary attempted to heal the rift between the sides of Thatcher (and Guise) on the one hand and ACOST, ABRC and Fairclough on the other. 'The Government – you, John Fairclough and ACOST – are struggling to make a big change in the Government's R and D priorities; and by past standards big changes are being made,' he noted, inclusively; ACOST were

> not, in my view, 'picking winners' in the sense of super-imposing their view on those of the scientists and enforcing their view by the allocation of funds: they are recommending priorities and encouraging industry and research to come closer together with a more deliberate aim.[215]

'Fairclough and ACOST want very much to help,' he emphasised, urging Thatcher to meet with her Chief Scientific Adviser. ('He is the only Chief Scientist we have for the time being and if he is to be effective he needs to know, from you, how you think.' chipped in Thatcher's Principal Private Secretary.)[216]

Thatcher did begin to meet again with her GCSA, Fairclough and her chair of ACOST, Sir Francis Tombs. She was briefed by Fairclough about ACOST and met Tombs, prior to Thatcher herself chairing a meeting of ACOST, held on 1 February 1989. She raised with Tombs a number of concerns: she felt money was going to the 'big battalions' rather than individuals ('the people most likely to make scientific breakthroughs'; the IRCs were 'too bureaucratic'; 'industry was still not undertaking enough near-market and product research' (despite the 'salutary effect that could often follow when Government-funding was reduced'); Britain was falling behind on Nobel Prizes; and whether enough research was being undertaken on the environment.[217] Tombs replied that 'ACOST would welcome suggestions from the Prime Minister on issues and areas of work that she would like to be pursued'. Thatcher said she would give it further thought, but did make one suggestion (ironically one that would be the centre of a major controversy for her successor): 'we seemed to be entering a period of much greater difficulty over bacterial disease in food production and preparation'.

ACOST offered its first set of advice on national priorities in 1989, although it took the government until the following year to respond. The topics addressed energy R&D, increasing the level of civil R&D, increasing the number of science and mathematics teachers, deepened analysis of industrial R&D and manpower, a more active role for the DTI, continued support for LINK and EUREKA and more support for global environmental research. Many of the suggestions were incremental, and even then some of these were declined in the government's response.[218] The response was very much in line with the government's core science policy strategies of reining in government funding of near-market research, discouraging government departments to take interventionist leads into matters appropriate for private industry and reducing bureaucracy. For example, ACOST's request to deepen surveys of industry's expenditure on 'technologies and on qualified scientific and engineering manpower' – information necessary to a more interventionist industrial strategy – was rebuffed with the explanation that the government did not want to increase the 'form-filling burden on firms, particularly small businesses'.

Thatcher also complained of bureaucracy and waste in another project. In March 1989 Kenneth Baker wrote to her expressing his support for a major new capital project from the Medical Research Council. This project was to move the Clinical Research Centre to the Royal Postgraduate Medical School, creating a new National Centre for Clinical Research and Training, built in Hammersmith, London, at an estimated cost of £48.5 million. The MRC considered this weaving together of basic science, clinical research and teaching essential to reverse a declining trend. 'The amalgamation would produce more than the sum of its parts,' argued Baker, because it would 'bring together on one site in multidisciplinary groups and in a clinical research environment the top scientists, clinicians and teachers', and thereby 'enhance the cross-fertilisation of basic research and work with patients and speed up the clinical application of scientific advances'.[219] The influence would be long term. It would 'mould our future leaders in academic clinical medicine who will in their turn train our next generation of doctors and researchers, who will then go on to disseminate the new approach and establish centres of excellence throughout the UK'. The ABRC also backed the plan.

Thatcher hated it. 'A total and deliberate waste of money designed to pre-empt a decision which must not be taken', she wrote of Baker's allocation of seed money, adding:

> No – this will be £2m wasted. The excellence of research does not depend on the extravagance of the building. You will be stopping an overwhelming amount of research by spending this amount of money on bricks and mortar. What a waste.[220]

The proposal was subsequently watered down, although the Prime Minister remained hostile.[221]

Distaste for bureaucracy also probably contributed to Thatcher's approval of a rearrangement of the research council system in the late 1980s. In 1988 the ABRC had begun a review to examine overlapping responsibilities, initially restricted to the biological sciences, which were seen as of growing importance. However the outcome, the 'Morris' report of April 1989 (named after its chair, the head of the engineering company Brown and Root), went further. It called for a single National Research Council, with six only 'semi-autonomous' divisions, 'overseen and co-ordinated by a holding Board and Director-General'.[222] In parallel, the suggestion that the AFRC and NERC be merged was made by the House of Lords Science and Technology Select Committee. Since the single National Research Council would require substantial legislation, while parliamentary time was scarce, the Morris proposal was not followed (if it had been, then something very similar to today's UKRI would have been established). Instead an alternative plan was offered: slim down the ABRC (from 26 to 14 members), but strengthen the secretariat and give the whole a 'more explicit remit to improve coordination and joint working' among the research councils, and approve the merger between AFRC and NERC.[223] Subject to her stipulation that NERC's distinctive polar and climate model work was protected, Thatcher approved the reconstitution of the ABRC and, perhaps significantly, gave the reduction in membership an emphatic blue tick.[224] The merger of AFRC and NERC however immediately 'produced much steam from Departments', and Guise advised that Morris's call for a single National Research Council had 'connotations of centralisation and enhanced bureaucratic control' – even (and here he quoted Max Perutz) 'Kremlinisation'.[225] Nevertheless it represented a 'healthy piece of fresh thinking', for shaking up science's bureaucracy. Guise also suggested that Sir David Phillips, as the continuing head of the reconstituted ABRC, should be 'given a private but clear remit to develop within two years practical proposals for pushing down the allocation of research funds nearer the workface'.

Thatcher chaired another meeting of ACOST in March 1990, and in preparation carefully read and annotated reports on 'adaptive biology' (how evolution of organisms might be affected by climate change), for a meeting that also discussed the science base and advanced manufacturing technology. While relationships between Prime Minister and advisers settled down in 1990, it was still the case that the Number 10 Policy Unit and ACOST were, in a sense, competitors in shaping science and industrial policy. A case in point was with advanced manufacturing. ACOST pointed to the low increases in productivity in the UK since the 1960s and called for modest initiatives to enhance industry awareness of advanced manufacturing technologies, but also advocated for the establishment of a national centre.[226] Guise thought these were 'the same old thinly disguised arguments for near market research funding by Government'. He backed this up by circulating to Thatcher the views of the professor of manufacturing systems at the University of Warwick, S. K. Bhattacharyya – closely aligned to those of Guise.[227] Thatcher visited Bhattacharyya and Guise continued to hold him up as an exemplary figure, one who 'doesn't want or use any Government funding'.[228] (For Guise on climate change, see Chapter 7.)

In July 1990 Thatcher met Lord Tombs, the chair of ACOST again. Fairclough and Guise were both present. Tombs pushed for an extra £100 million for basic science, half to renew increasingly 'badly run down' equipment, and half to be used by heads of department to support young scientists.[229] Fairclough described the case for the extra cash as 'compelling', but also noted that ACOST 'did not face up to the question of where the money should be found for this top priority if the public expenditure situation did not permit additional funding'.[230] He thought it should be transferred from defence R&D and from the Department of Energy's nuclear research budget. Guise supported such a transfer, because it represented a 're-allocation to basic science from applied and defence R&D'. But he was much more concerned about continued foot-dragging on the broader issue of principle. 'There remain backwoodsmen in the DTI,' he informed Thatcher, 'who never really accepted the policy of Government being the principle supporter of basic science with industry funding near market research.'[231] Later in the year Guise could be heard complaining of ACOST's continued backing of government funding for advanced manufacturing research ('the whole things read like Wilsonian plans from the sixties'), while warning that an initiative from Peter Lilley to assist innovation in small and medium-sized businesses 'smells like near-market research'.[232] Perhaps, wondered, Guise, ACOST should abolish itself?

By September 1990 Fairclough had departed. He was replaced as GCSA by Professor William Stewart. Tombs also went, and in returned Robin Nicholson, the ex-GCSA now chair of ACOST. On 28 November 1990 Thatcher, too, had gone. Guise followed her soon after.

Conclusion

The major finding of this chapter is that there was a profound shift in science policy and industrial strategy in 1987, in which near government-funded near-market research was cut and curiosity-driven basic science championed. This was driven by the arguments of the Number 10 Policy Unit, specifically those of George Guise, in opposition to traditional, long-standing sources of advice. The contrast is stark. At the public seminar on science, technology and industry on 12 September 1983, Thatcher had proudly spoken in her opening speech of how 'we [the government] spend money to stimulate the development of science-based products, and to help bring them to the market-place. Spending on this has increased by nearly 20 percent in real terms over the past four years'.[233] Tension over science policy had built up through the 1980s and had played out in fierce internal arguments over issues such as defence research, cuts to the 'science base' and CERN. By 1984 there was a feeling of stalemate. The Chief Scientific Adviser considered the science base to be in danger of fast erosion, the debate over small sums of defence research for universities had become a distraction and there was a growing conviction, exemplified in Thatcher's mind by the case of monoclonal antibodies, that the root problem was a failure to translate basic research into commercial products. For example, at a meeting on the science budget on 3 May 1984 it was recorded that the 'Prime Minister said it was unsatisfactory that Britain had made such advances in basic science but had failed to develop profitable applications'.[234] The new research policy after 1987 resolved some of this tension, while being aligned to broader Thatcherite ideological commitments to privatisation and a reduced role of the state.

All this happened largely behind closed doors. The *public* debate centred on the decreasing morale of scientists, risks of a new brain drain and the threat to 'basic' research that came from the cuts in university funding. This public debate peaked between the rebellion of the scientists under the Save British Science banner in January 1986 and opposition to the ABRC's *A Strategy for the Science Base* in 1987. Narrowly focused on protecting basic science, and with a unsophisticated model of innovation, this movement provided unwitting cover for the reversal in science policy – and occasionally, as in the case of Perutz, lent it valuable ammunition. The institutions that sought entry to express the scientists' voice at the heart of government – whether elite and traditional, as in the Royal Society, or grass roots, such as Save British Science – struggled, or in the case of the latter were rebuffed.[235] Likewise the public debate around the reorientation of university funding – the University Grants Committee was replaced in 1988 by a University Funding Council, 'a body numerically dominated by people from commerce and industry, with only a sprinkling of academics'[236] – focused on a perception of centralisation and direction, when in fact the private decisions show that possibility of industrial direction of basic science was being denied.

It is important to note that *alternative* models were articulated in competition right at the heart of government. One example was ACARD's plan, the product of several years' work, put forward in its 1985 report on 'exploitable areas of science'.[237] This plan called for the prioritised areas to be supported and efforts 'to guide a fairly high proportion of that part of the national scientific resource paid for by the taxpayer'. Technology was the 'bridge' between science and the production of goods; it needed to be studied in depth, supported and built well. Therefore ACARD called for a centre for comprehensive science and technology assessment to provide and share 'a broad view of the inter-relation of developments in scientific knowledge and market trends' to provide the knowledge for such wise steersmanship. Such centres, it was claimed, could be found in economic rivals Japan, France and the United States. Academics at SPRU produced working examples of what foresight would look like.[238] The then Chief Scientific Adviser, Robin Nicholson, had supported the plan to forecast and guide, and had tried to reassure Thatcher that it was '*not* about bureaucrats telling scientists what basic research to do', it was '*not* about "coordination" to remove decision-taking responsibilities from where they properly belong' and 'most emphatically, it is *not* about bureaucrats "picking winners"'.[239] He protested too much. It was precisely this comprehensive industrial strategy for exploiting science that, once Nicholson had gone, his successor John Fairclough sidelined and that, once Guise had Thatcher's ear, would be rejected.

Notes

1. For example, a proposal from 1982 in which the Department of Industry paid 50 per cent of a company's participation in joint university-industry research in optoelectronics, with the Science and Engineering Research Council paying the academic costs. T 494/149. Chapman to Chivers, 2 August 1982. It became the Joint Opto-electronics Research Scheme (JOERS). Also, Department of Industry, 'Support of collaborative research between industry and universities in specific technologies', 29 July 1982.
2. PREM 19/269. 'Microprocessor Applications Project (MAP)', 7 May 1979. The ACARD finding was mentioned in Biffen to Thatcher, 2 July 1979.
3. PREM 19/269. Joseph to Thatcher, June 1979.
4. PREM 19/269. Lankester to Duguid, 11 June 1979.
5. PREM 19/269. Solomon to Lankaster, 3 July 1979.The SRI report is Michael A. Placko, 'A status report on microprocessor technology and its implications for the United Kingdom. Interim draft summary', undated (1979).
6. PREM 19/269. Berrill to Thatcher, 12 July 1979.
7. PREM 19/269. Lankester to Duguid, 16 July 1979.
8. Ashworth recalled: 'There was an amazing row in government about an organisation called INMOS which was our pathetic attempt to create an Intel, a chip manufacturer. We thought we might be able to take on Intel or some such fantasy'. 'Sir John Ashworth. Edited transcript of a conversation between John Ashworth and Peter Collins at the Royal Society on 22 April 2009', Royal Society Library.
9. PREM 19/269. 'Note of a telephone conversation between the Secretary of State for Industry and Sir Arnold Weinstock, concerning INMOS', 21 March 1980.
10. Tom Lean, *Electronic Dreams: How Britain Learned to Love the Computer*. London: Bloomsbury, 2016. The programme for the end of IT82 conference is in T 471/45. Thatcher's Keynote Address was followed by the Vice President of the Commission of the European Communities speaking on 'The information era – made in Europe?'
11. Lean, *op. cit.*
12. Again mostly from UK manufacturers, although including Apple II and IIIs, despite abiding by GATT rules. T 471/45. Press release, Treasury, 'Microcomputer contracts for government departments', 25 January 1983.
13. Brian Oakley, 'Industry has lessons for academics', 2 July 1987.
14. PREM 19/2116. Joseph to Jenkin, 30 November 1982.
15. PREM 19/2116. Nicholson to Joseph, 17 December 1982. Nicholson wanted the 'new blood posts … to be used to start a new era for our universities', in other words 3–6 months' notice and no tenure.
16. PREM 19/2116. IT Advisory Panel to Thatcher, 13 January 1983
17. PREM 19/2116. Nicholson to Thatcher, 17 March 1983.
18. PREM 19/2116. Thatcher, handwritten comments on Jenkin to Thatcher, 22 April 1983. Also includes several particularly emphatic 'NO's.
19. Editorial, 'Government can help', *The Economist* (9 April 1983), p.16.

20. T 471/45. Kenneth Baker, 'Towards an information economy', speech to the British Association for the Advancement of Science, Liverpool, 7 September 1982. In his later speech at the end of IT82, Baker noted 'When I was appointed Minister 2 years ago and persuaded the PM to designate 1982 as IT Year, some people even suggested that I had invented IT. I didn't, but I am still a solitary figure as the only Minister of IT in the world'. Baker, Department of Industry press release of text of speech at IT82 Conference, 9 December 1982.
21. T 494/105. Press notice, Department of Education and Science, 'Plans to let further education sell its expertise', 27 March 1984.
22. T 494/105. Note by Allan on Sallow-Smith to Allan, 2 July 1984. Allan tells an anecdote of a further education lecturer who spent research grants and bank loans on a fully equipped pig farm (with students as farmworkers) as evidence of the need for accountability safeguards.
23. The monopoly rights did not cover UGC-funded research, although university-based inventors nevertheless offered potential patents to the NRDC.
24. Wilkie, *op. cit.*, pp.52–3. John Hendry, *Innovating for Failure: Government Policy and the Early British Computer Industry*. Cambridge, MA: MIT Press, 1990.
25. Lara V. Marks, *The Lock and Key of Medicine: Monoclonal Antibodies and the Transformation of Healthcare*. New Haven, CT: Yale University Press, 2015.
26. Marks, *op. cit.*, p.33.
27. David Dickson, 'California set to cash in on British discovery', *Nature* 279 (1979): 663–4. Marks, *op. cit.*, p.28. Soraya de Chadarevian, *Designs for Life: Molecular Biology after World War II*. Cambridge: Cambridge University Press, 2002, pp. 356–7. In fact the wide applicability, indeed usefulness, of Mabs was not at all apparent in 1975.
28. PREM 19/585. Ashworth to Sanders, 'Innovation in the UK', 2 September 1980. The underlining is Thatcher's.
29. Again underlining is Thatcher's. The examples Ashworth could give of such technology transfer services were Battelle and Stanford Research Institute, both in United States.
30. ACARD, 'Industrial innovation', 1980.
31. ACARD, 'Exploiting invention', 16 December 1980. This report was written by a working party of Alfred Spinks (ex-ICI, and lead author of the 'Spinks report', also under ACARD auspices, into biotechnology), and two others: G. Taylor (TDC Development Ltd) and J. C. Duckworth (IDJ Investment Services Ltd and managing director of NDRC, 1959–70), and assisted by Ashworth. The evidence for this report being directly seeded by Thatcher's Cambridge experience is in PREM 19/585. Pattison to Bell, 22 December 1980.
32. ACARD, 'Exploiting invention', *op. cit.*, p.6. ACARD (p.12) also thought that the BBC TV shows *Tomorrow's World* and *Horizon* could do more to publicise how 'developments in science and technology' might practically be turned to 'profit' – such advice 'could be given in an unobtrusive way (rather as farming advice used to be given in every episode of "The Archers")'.
33. CAB 184/679. Jenkin to Thatcher, 9 May 1983.
34. CAB 184/679. Nicholson to Thatcher, 6 July 1983.
35. T 494/142. Kelley to Nicholson, 28 February 1984.
36. Susan Wright, 'Recombinant DNA technology and its social transformation, 1972–1982', *Osiris*, 2nd Series, 2 (1986): 303–60.
37. Geoffrey Owen and Michael M. Hopkins, *Science, the State and the City: Britain's Struggle to Succeed in Biotechnology*. Oxford: Oxford University Press, 2016.
38. Susan Wright, *Molecular Politics: Developing American and British Regulatory Policy for Genetic Engineering, 1972–1982*. Chicago: University of Chicago Press, 1994.
39. ACARD, ABRC and Royal Society, *Biotechnology* (Spinks Report). London: HMSO, 1980. For government response, see PREM 19/3154. An official committee on biotechnology was launched following the Spinks Report. Thatcher was never enthusiastic about its existence, and only support from Robin Nicholson (in 1982) and Nicholson and Norman Tebbit (in 1984) preserved it from extinction.
40. CAB 134/4415. 'Draft white paper on biotechnology', 2 July 1980. White paper published as Cmnd 8177.
41. 'Why do we need a NRDC and a NEB, don't say anything that perpetuate their separate identities'. PREM 19/3154. Thatcher, handwritten comments on Morgan to Pattison, 30 December 1980.
42. CAB 134/4564. Official Committee on Biotechnology, 'Report on biotechnology', April 1982.

43. T 494/142. Kelley to Nicholson, 28 February 1984. See also PREM 19/1369. Nicholson to Turnbull, 21 February 1984. In the previous autumn, Nicholson had called for an investigation in 'to what extent Councils' charters really inhibit them from making usefulness as strong a criterion as scientific quality when funding research'. ED 273/111. Nicholson to Hancock, 24 October 1983. Thatcher publicly aired the question of more direction for research councils in an interview in *Nature* in 1985 in the context of the CERN debate discussed later in this chapter: 'Mrs Thatcher ... thinks it may have been a mistake to leave the judgement of what lines of research would be supported exclusively to the research councils ... now she wonders why the research councils are spending so much in this field [of high energy physics]'. John Maddox, 'Mrs Thatcher asks for decisiveness', *Nature* 316 (18 July 1985): 178.
44. T 494/142. Kelley to Nicholson, 28 February 1984.
45. PREM 19/1619. Joseph to Thatcher, 19 December 1984. 'Public money ... understanding' with Thatcher's blue underlining.
46. PREM 19/1619. Nicholson to Thatcher, 21 December 1984.
47. PREM 19/1619. Thatcher's handwritten comment on Nicholson to Thatcher, 21 December 1984.
48. PREM 19/1619. Joseph to Thatcher, 29 March 1985.
49. PREM 19/1931. Addison to Thatcher, 9 April 1985.
50. PREM 19/1931. Handwritten comment on Addison to Thatcher, 9 April 1985.
51. Hennessey, *Whitehall*, p.592.
52. It is nevertheless notable that the first two Rayner reviews concerned knowledge production.
53. PREM 19/772. Soames to Thatcher, 8 January 1981, and Thatcher's written notes on this document.
54. PREM 19/772. Rayner to Thatcher, 9 June 1982. Also Management and Personnel Office report, 'Review of support services in research and development and allied scientific establishments', June 1982. Rayner highlighted the 24 years' supply of embossing tape. According to Hennessey the most notorious finding was the £30 rat bred at Ministry of Agriculture's veterinary laboratory at Weybridge (compared to a commercially available £2 rat). Hennessey, *op. cit.*, p.601.
55. PREM 19/772. Baroness Young to Whitelaw, 29 June 1982.
56. CAB 134/4415. 'Cabinet. Official Committee on Science and Technology. The consequences for British science of the projected public expenditure reductions', by Department of Education and Science, 2 November 1979.
57. CAB 134/4415. Minutes, STO(80)1st meeting, 2 May 1980. The delay happened after some departments expressed 'reservations'.
58. Margaret Thatcher, Speech to Parliamentary and Scientific Committee, 25 February 1981. She ended this speech by quoting Richard Feynman at length: 'We are very lucky to live in an age in which we are still making discoveries. It is like the discovery of America – you only discover it once. The age in which we live is the age in which we are discovering the fundamental laws of nature, and that day will never come again. In this age people are experiencing a delight, the tremendous delight that you get when you guess how nature will work in a new situation never seen before. What is it about nature that lets this happen, that it is possible to guess from one part what the rest is going to do? That is an unscientific question: I do not know how to answer it, and therefore I am going to give an unscientific answer. I think it is because nature has a simplicity and therefore a great beauty.'
59. Specific examples of the difficulties in individual departments caused by cuts can be found in ED 273/111. Denning to Brown, 18 October 1983, and Kurzfield (SERC) to Tanner, 21 October 1983. Myelnikov is entirely right when he argues that earlier cuts, such as the 1981 massive reduction in funding for the Animal Breeding Research Organisation, stemmed from earlier and independent decisions. Dmitriy Myelnikov, 'Cuts and the cutting edge: British science funding and the making of animal biotechnology in 1980s Edinburgh', *British Journal for the History of Science* 50 (2017): 701–28.
60. Peter Collins, *The Royal Society and the Promotion of Science since 1960*. Cambridge: Cambridge University Press, 2016, p.67. The Royal Society President, Andrew Huxley, wrote to *The Times* (23 July 1983) outlining the Society's position.
61. Nicholson recalled: 'there was a continuous rumble between me and some senior Officers in the Royal Society, and I remember particularly with George Porter, about whether my role should be promoting science and expenditure on science at all costs within the Cabinet Office. And I said no, that is not my role. I said, the elected government of the day has decided, had been elected on the basis of a reduction in public expenditure that's applied across the board

and my role as scientific adviser is to limit any reductions which affect science and to promote the better use of that scientific spend'. 'Sir Robin Nicholson (FRS 1978). Edited transcript of a conversation between Robin Nicholson and Peter Collins at the Royal Society on 6 May and 18 May 2009', Royal Society Library.

62. Although Keith Joseph did cut the 1982–3 budget of the Social Science Research Council by £1.1m, against ABRC advice for level funding. 'Publish the advice, says ABRC', *Times Higher Education Supplement*, 17 September 1982.
63. 'Squeeze on science budget', *The Times*, 12 February 1982.
64. Geoffrey Oldham reported in Jon Turney, 'Science research "in danger"', *Times Higher Education Supplement*, 3 September 1982. Oldham, 'The future of university research in Britain', Papers in Science, Technology and Public Policy no.2, Imperial College/SPRU, 1982.
65. ABRC, *The Science Budget: a Forward Look 1982*, October 1982. 'Science chiefs want more investment in new projects', *Financial Times*, 28 October 1982.
66. PREM 19/1505. Merrison to Thatcher, 18 August 1982.
67. Joseph's reply, of 1 December 1982 was published as part of a DES press release on budget allocations, 1983-84. T 494/134. Phillips to Joseph, 29 April 1983, encloses ABRC's requested report on priorities, 'Scientific opportunities and the Public Expenditure Survey', April 1983.
68. *First Joint Report by the Chairmen of the Advisory Council for Applied Research and Develoment (ACARD) and the Advisory Board for the Research Councils (ABRC)*, Cmnd. 8957. London: HMSO, July 1983.
69. 'Crisis of science', *Times Higher Education Supplement*, 28 January 1983. This criticised the modest 'new blood' funding provided for fresh academic posts.
70. Programme date from http://genome.ch.bbc.co.uk/70c18733b7b34bda989ba147b5b16968. The programme was reviewed by Bernard Dixon in 'Big money and old wives scrutinised', *New Scientist*, 10 March 1983, who noted the 'vivid portrayal of the unprecedented butchery visited on academe', but bemoaned Scott's lack of historical perspective, or 'any visible grasp of the reports and councils, White Paper and Green Papers, which have reconnoitred this terrain before and illuminated the thorny issues at the heart of science policy'. A second *Horizon* programme on UK science policy was also brought to the Prime Minister's attention in November 1984. See: PREM 19/1619. Nicholson to Thatcher, 16 November 1984.
71. PREM 19/1370. Scholar to Nicholson, 9 March 1983.
72. PREM 19/1370. Mount to Thatcher, 11 March 1983, and handwritten comment by Thatcher.
73. PREM 19/1370. Transcript of opening speech, by Thatcher, 12 September 1983.
74. PREM 19/1370. Transcript of closing speech by Thatcher, 12 September 1983.
75. PREM 19/1370. Nicholson to Thatcher, 23 September 1983.
76. PREM 19/1369. Joseph to Thatcher, 15 September 1983.
77. ED 273/111. Tanner to Phillips, 21 June 1983. The document considers the tactics of writing a memorandum that would seek to 'interest the PM and enlist some sympathy and support, without her feeling cornered by the science lobby. The aim is some more money sometime'.
78. The figures requested in total for this extra money after redeployment were £35m in 1984–5, rising to £55m in 1985–6 and £75m in 1986–7. Phillips stated to Joseph: 'Quite simply, there is an urgent and overriding need for additional funds for science in the years immediately ahead. Coupled with some indication of longer term Government intentions'. ED 273/111. Phillips to Joseph, 15 September 1983. Phillips' argument to Joseph was to stress the 'prevailing sense of excitement' in science and to note the pressures on the science budget (the need to find redundancy money, rising costs of international subscriptions (but not mentioning CERN), the falling success rate of alpha-rated grant applications and defensible capital needs. The reasons to be excited by science are interesting: 'First, new insights in quite fundamental areas – the nature of matter and the forces that hold it together, the processes governing the formation of the universe, the molecular processes governing life, and ways of thinking scientifically about the obscure (but introspectively real) boundary between body and mind – are felt to be amazingly fertile in helping to answer long-standing questions concerning the nature of the universe, the causes of diseases such as cancer, the possibilities of creating artificial intelligence. The second such advances also look exceptionally fertile for new technology, as the examples indicate'.
79. This might not necessarily mean reductions in overall defence research. Joseph floated the suggestion that new defence research contracts should be placed with universities rather than at government research establishments. He had in mind, therefore, an expansion of military research on campus.

80. Collins, *op. cit.*, p.119. Peter Swinnerton-Dyer, chair of the UGC, also did an analysis, see: ED 273/111. Swinnerton-Dyer to Richard (unknown surname), enclosing 'The effect of the cuts on university research', 10 October 1983. Swinnerton-Dyer made use of various indicators, the data of which were collected at UGC, such as expenditure on consumables (ie research supplies), salaries and wages, and books and periodicals.
81. Permanent Secretary at DES, David Hancock, informed his minister of informal intelligence: 'The No. 10 Policy Unit have advised the Prime Minister that the extra bid for science is out of the question unless full offsetting savings can be found from the Defence Budget or elsewhere'. ED 273/111. Hancock to Joseph, 18 October 1983.
82. The Treasury, in its frustration over bilateral negotiations, came very close to implementing a general science policy (for the first time in UK history), if Robin Nicholson is to be believed (and there is little reason to doubt him): 'I gather from the Treasury (in strict confidence) that the Chief Secretary came close to rejecting all R&D in his bilaterals, bringing the bids together and setting up a Star Chamber to prioritise across Departments to the extent of funds available'. ED 273/111. Nicholson to Hancock, 24 October 1983. In the context of the quotation 'all R&D' would have included both defence and civil R&D. According to Peter Collins (personal communication), before 1965 (and before the implantation of the Trend report), many research bodies, including the Royal Society, had to negotiate directly with the Treasury for their budgets.
83. The reviews were prepared by the Cabinet Office's Science and Technology Secretariat, working under the guidance of the Sub-Committee of Chief Scientists, chaired by Nicholson.
84. PREM 19/1514. Armstrong to Thatcher, 11 November 1983.
85. Note the small amount of defence R&D placed in universities. In the view of John Kingman this was because of reluctance on both sides: 'Oddly there has been no real attempt to commission research in universities perhaps because the academic reluctance to engage in classified research clashes with [or rather complements] the chronic military tendency to over-classification'. ED 273/111. Kingman to Hancock, 13 October 1983.
86. PREM 19/1369. Heseltine to Thatcher, 4 October 1983.
87. PREM 19/1369. Nicholson to Thatcher, 14 October 1983. Nicholson evidence, drawn from his experience as a member of SERC and ABRC was threefold: the substantial number of 'alpha' proposals going unfunded by research councils, other countries moving faster in 'the exciting new areas of science: neuroscience, biomolecular electronics, artificial intelligence, remote sensing, robotics, novel materials, medical diagnostic techniques', and a 'renewed "brain drain" amongst young people'.
88. PREM 19/1369. Mount to Thatcher, 18 October 1983. However, Mount also disliked the precedent of shifting money from the MoD to DES budget outside of the established annual Public Expenditure Survey mechanism.
89. ED 273/111. Nicholson to Hancock, 24 October 1983. The continuation of Thatcher's comments (which mentioned CERN) are discussed below.
90. PREM 19/1369. 'Maintaining the strength of the science base', Turnbull to Hodgkinson, 19 October 1983. Copy with 'financed fundamental research generously, but with poor results' with red underlining and exclamation mark in ED 273/111.
91. Cabinet Office, *Annual Review of Government-funded Research and Development*, 1984. As with the 1983 review, a published version issued with classified information on future defence expenditure deleted. Full versions, with defence figures, can be found in PREM 19/1369. The defence figures are significant, since they forecast a rise, and a portion of this increase would provide the proposed £20m transfer. When categorised by 'primary purpose', the estimated changes in expenditure between 1983–4 and 1986–7, in real terms, were: advancement of science -3 per cent, support for policy-making –4 per cent, improvement of technology +9 per cent, support for procurement decisions +4 per cent. PREM 19/1369. 'Annex', July 1984. For Nicholson's views on the picture emerging from the 1983 and 1984 annual review, see: PREM 19/1619. Nicholson to Bailey, 15 August 1984. For Treasury views, see T 499/6. 'Annual review of R&D', undated (August/September 1984).
92. CAB 164/1790. Barclay to Mottram, 23 May 1984.
93. PREM 19/1369. Heseltine to Thatcher, 25 June 1984. Copy also in CAB 164/1790. Amusingly, he cited Rothschild's recommendation that government research laboratories continued to undertake 10 per cent of basic research, in support, noting that the defence research establishments were already only doing about 5 per cent.

94. PREM 19/1369. Joseph to Thatcher, 2 July 1984. Copy also in CAB 164/1790. 'low dimensional structures, including molecular electronics', bio-sensors, protein engineering, 'systems design and control, specifically the application of computers to mechanical engineering', lasers, information technology and a 'UK-led space science mission'. 'Strategic research' had been raised as a point of concern by Willie Whitelaw on 22 February 1984.
95. This might look like ministers seeking greater control over direction of research, but remember that in the context here the alternative is closely tied, contract research.
96. PREM 19/1369. Chilver to Armstrong, 11 July 1984.
97. PREM 19/1369. Nicholson to Thatcher, 27 July 1984.
98. CAB 184/4853. Minutes, ST(CS), 18 April 1984, records departmental chief scientists' discussion on the debate, which was prompted by the first joint report of ACARD and ABRC chairmen. 'Viscount Whitelaw had been impressed by the debate and identified 4 major points which needed attention' the first of which was 'effect on research of cuts in University funding'. In discussion it was noted that 'Many Committees attended by Chief Scientists had discussed the effects that reductions in Government funding for Universities ... were having on the UK science base. A consensus was emerging that the resulting problems could only be solved if more money was made available. It was difficult for members of the sub-committee to reconcile departmental policies to constrain expenditure with this apparent need for increased expenditure'. Thatcher was informed of the debate by William Waldegrave. Waldegrave to Thatcher, 22 February 1984. The Royal Society's survey of the health of basic science, commissioned by David Phillips, noted above, was also part of the context of these discussions.
99. Note that the main Edgerton finding of no correlation between R&D and national economic growth is unexceptionally taken as read here. It would not have been news to Nicholson.
100. More generally, Nicholson's view at this time was that 'Government-funded R&D is substantially out of balance: over-committed on defence, agriculture, and nuclear energy; and under-committed on basic and strategic research, especially in manufacturing'. PREM 19/1619. Barclay to Thatcher, 2 August 1984.
101. PREM 19/1369. Nicholson to Barclay, 25 July 1984.
102. PREM 19/1369. Barclay to Mottram, 30 July 1984. Copy also in CAB 164/1790.
103. CAB 164/1790. RSRE, 'The impact of RSRE on the UK economy. Draft paper to CSA', undated (November 1984). RSRE was formed by amalgamating the larger Royal Radar Establishment (RRE) at Malvern, with the smaller Services Electronic Research Laboratory (SERL), Baldock and the Signals Research and Development Establishment (SRDE), Christchurch.
104. The documents of this policy discussion are a valuable record of such connections. In 1984, the MoD spent £9m on 650 contracts for research in 72 universities and institutions (a small amount, compared, for example, to the £164m spent extra-murally in industry and research associations. CAB 164/1790. 'Record of a meeting to discuss Maintaining the Strength of the Science Base, 4 December 1984. There were significant differences between in the armed services. The Army Personnel Research Establishment, Farnborough, commissioned £850k at places such as Loughborough, Cranfield, Cambridge Consultants and CAP Scientific. RSRE funded £850k per year on 11 radar contracts at 11 universities, and a further £870k on materials. CAB 164/1790. 'Record of a meeting to discuss Maintaining the Strength of the Science Base, 28 September 1984. The report cited above has more detail on specific projects. Yet, compare with the view of the Head of Marine Geology and Geophysics, Institute of Oceanographic Sciences, T. J. G. Francis's view that 'no such relationship existed' (in sharp contrast to the state of affairs in the United States and France) in oceanography. Francis, 'A personal view', *Underwater Technology* 10(2) (1984): 1.
105. CAB 164/1790. Nicholson to Thatcher, 2 January 1985. Barclay to Hodgkinson, 8 January 1985. Barclay to Nicholson, 8 January 1985.
106. CAB 164/1790. Nicholson to Barclay, 22 February 1985. PREM 19/1932. Norman (CSA, MoD) and Phillips (chair, ABRC) to Thatcher, 21 November 1986 reported on progress.
107. PREM 19/1805, Unwin to Thatcher, 18 February 1986.
108. PREM 19/1805. Griffiths to Thatcher, 18 February 1986.
109. PREM 19/1805. Cabinet Office brief for Thatcher's meeting with Sir Francis Tombs, 22 July 1986.

110. In opening the SNS was renamed ISIS. ED 273/111. Kingman to Hancock, 13 October 1983. On defence R&D, Kingman wrote: 'If I were viewing the scene from the Kremlin, I would want to bounce the West into spending more on defence and less on civil research, on the grounds that the resulting increase in military strength would be outweighed by the long-term weakening of the economy. Conversely, the interests of national security would paradoxically be furthered by a modest shift from the Defence to the Science Vote'.
111. ED 273/111. Thatcher's words, speaking on 19 October 1983, and quoted by her Chief Scientific Adviser in Nicholson to Hancock, 24 October 1983.
112. PREM 19/1369. Turnbull to Hodgkinson, 13 January 1984. The 1983 Forward Look of the ABRC had raised the possibility of 'withdrawing completely from a major area of scientific activity' in the light of the grave problems faced by the Research Councils. In 1983 the two potential areas had been high-energy physics and satellite astronomy. Of the two, HEP had fewer immediate applications, 'although, remembering Rutherford's view of the relevance of his research, we would be wash to state that no application will be found in the future'. Extract, 1983 Forward Look of the ABRC, submitted to Secretary of State for Education and Science, 25 November 1983.
113. PREM 19/1369. Telegram, Geneva to FCO, 20 January 1984.
114. PREM 19/1369. Telegram, Bonn to FCO, 20 January 1984.
115. PREM 19/1369. Telegram, Paris to FCO, 20 January 1984.
116. PREM 19/1369. Howe to Joseph, 2 February 1984. Joseph to Thatcher, 5 March 1984.
117. Joseph: 'we owe this appraisal to people in other parts of UK science who believe they can give better value for money and are pressing for a frank discussion of issues and options'. These scientists are not named. Elsewhere on the document Thatcher agreed with the comment that without the 'assurance [that money saved would be spent on science], the scientists will not weigh alternatives frankly'. PREM 19/1369. Joseph to Thatcher, 5 March 1984.
118. PREM 19/1369. Letwin to Turnbull, 6 March 1984.
119. PREM 19/1369. Nicholson to Thatcher, 6 March 1984. Commenting on this advice, Thatcher repeated her interpretation of French interests: 'Of course France is <u>concerned</u> – the new LEP ring is largely on <u>her</u> territory and <u>we</u> get little from it'.
120. Membership: John Kendrew (chair), Douglas Hague (chair ESRC), Eric Ash (a late replacement for the busy Peter Hirsch, chair UKAEA), Jack Lewis (chemistry, Cambridge), John Mason (recently retired director-general of Met Office), Ken Pounds (space science, Leicester), Francis Tombs (various business and financial interests). Christopher Llewellyn-Smith (theoretical physics, Oxford) was added as an adviser when Howe raised concerns that none of the group were specialists in high-energy physics.
121. PREM 19/1931. Joseph to Thatcher, 18 June 1985.
122. For example Abdus Salam, the Pakistani physicist at Imperial College and Nobel Prize winner for his contributions to theoretical physics, circulated a paper 'Why the Kendrew recommendations must be scrapped', which began 'So the Kendrew Report has decided for British science to kill its own creation – the subject of Particle Physics, just when it has discovered its systematising principle', and repeated the argument in Salam, 'Particle physics: will Britain kill its own creation', *New Scientist* (3 January 1985).
123. PREM 19/1931. Nicholson to Thatcher, 18 June 1985.
124. PREM 19/1932. Nicholson to Thatcher, 14 November 1985.
125. PREM 19/1932. Unwin to Thatcher, 17 December 1985. The Netherlands and West Germany 'welcomed' the budget review, Belgium, Austria, Denmark and Italy 'favoured' it, while 'scepticism' was expressed by Switzerland, Norway and possibly Sweden; France was 'not averse'. Waldegrave negotiated the terms of reference.
126. Details of the membership can be found in PREM 19/1932. Walden to Secretary of State, 29 April 1986. The UK representative was the chemist, VC of Keele University and ex-Director of Institut Laue Langevin, Brian Fender.
127. PREM 19/2477. Baker to Thatcher, 25 November 1986. Supported by Fairclough to Thatcher, 1 December 1986.
128. Handwritten note by Thatcher on PREM 19/2477. Bearpark to Thatcher, 3 December 1986.
129. PREM 19/2252. Pattie to Howe, 4 December 1986. Pattie to Howe, 3 March 1987. Pattie to Howe, 25 March 1987.
130. PREM 19/2253. Howe to Tebbit, 30 April 1987. Interestingly, Tebbit tried to suggest that there was a party political angle, suggesting that opposing European science funding 'will be natural grist to the "alliance" mill'. PREM 19/2253. Tebbit to Thatcher, 29 April 1987.

131. Newspaper cutting, 'Britain faces EEC conflict on research plan veto', April 1987.
132. In addition to the advice noted above, see also Nicholson's sceptical comments about Merrison's case for UK participation in CERN. PREM 19/1369. Nicholson to Thatcher, 18 May 1984.
133. PREM 19/2046. Cole to Griffith, 18 July 1986, asks about Guise's access to documents before his PV (personal vetting) was complete, suggesting he had started work by then.
134. 'Bett to Leave BBC for British Telecom', *Financial Times* (17 March 1981), p.25.
135. The money was not new. £31 million was transferred from other departments (notably DTI and Department of Employment), and £12 million was 'UGC money redeployed from other purposes'. Joseph's decision followed industry representations through the Engineering Council and the IT Skills Shortages Committee. PREM 19/1931. Briefing on 21 May 1985 meeting.
136. PREM 19/1619. Warry and Letwin to Thatcher, 5 February 1985. 'Queues for arts places at university are stronger than those for science places', 'The bulk of new jobs are created in the service sector rather than electronics', 'The market does <u>not</u> appear to favour engineers. Their salary rates may be rising, but are still lower than accountants'. Warry later went into nuclear higher management (during privatisation) and became a professor at Warwick University.
137. PREM 19/1619. Nicholson to Thatcher, 5 February 1985.
138. PREM 19/1931. Weinstock to Thatcher, 19 April 1985.
139. PREM 19/1931. Addison, minutes, 'Meeting with industrialists on 21 May 1985', 21 May 1985.
140. PREM 19/1932. Chilver to Thatcher, 2 July 1985. Chilver said the problem was widespread, not confined to companies of particular sizes, sectors or locations. The issue in medium- and large-sized companies was likely 'attitudes in board rooms and in City institutions, and amongst City commentators', while for smaller companies it was still access to capital (despite the growth of venture capital, the Unlisted Securities Market and wider holdings of company equity. Thatcher has underlined 'aversion'.
141. PREM 19/1932. Douglas Hague, 'Notes on public support for science and technology', paper for House of Lords Select Committee on Science and Technology, for 19 March 1986. Hague took aim at the equipment-driven politics of Big Science (such as supercomputers and also the unhelpful growing sophistication of models, including the Treasury Model).
142. PREM 19/1804. Warry to Turnbull, 15 March 1985.
143. PREM 19/1804. Thatcher, handwritten comment on Warry to Turnbull, 15 March 1985.
144. PREM 19/1805. Howe to Thatcher, 8 July 1985: 'our national commitment to Research and Development shows a relative and significant decline'. Joseph to Thatcher, 4 July 1985: Councils must be supported 'if we are not to suppress – to our lasting damage – intellectual creativity or to risk driving our best scientists abroad'. Chilver to Thatcher, 26 June 1985: 'Expenditure on R & D by UK private industry appears to be declining ... We are concerned that current Government mechanisms make it difficult to manage and fund R & D programmes of national importance'. Nicholson to Thatcher, 26 July 1985: 'You may wish to conclude that R &D trends in the UK are alarmingly out of line with those of our industrial competitors and this carries a great threat to our future economic prosperity'.
145. PREM 19/2252. Cunningham to Bearpark, 14 November 1986. Nicholson used the cabinet committee system, forwarding a report from MISC119 to E(A) for discussion on 19 February 1986. PREM 19/1805. Unwin to Secretary of State for Trade and Industry, 15 April 1986, shows that one plan (resisted) was to raid the fast-breeder reactor programme for cash to support Nicholson's 'pull-through'.
146. At this E(A) meeting, 19 February 1986, chaired by Thatcher, she said that Nicholson's proposal should be given 'urgent consideration' by the new E(RD) committee, chaired by the Secretary of State for Trade and Industry (Paul Channon). Internal to government, E(RD) worked to 'reshape Departments' spending patterns on R&D so as to give greater emphasis to programmes with substantial industrial participation, which offer the best prospects for greater wealth creation'. PREM 19/2252. Channon to Thatcher, 24 November 1986.
147. PREM 19/2252. Fairclough to Thatcher, 14 November 1986. Fairclough was working on the seminar idea with Phillips and the new chair of ACARD, Sir Francis Tombs. Fairclough viewed the 'pull-through project as a test case of the new E(RD) machinery and particularly whether the Departments can agree a common goal and make a contribution towards a realisation of that goal'. PREM 19/1805. Fairclough to Addison, 1 August 1986.

148. For Guise's own brief and partial overviews of his work on science policy, see: *Inside the Tank*. Epsom: Bretwalda Press, 2015 and 'Margaret Thatcher's influence on British science', *Notes and Records of the Royal Society* 68 (2014): 301–9.
149. PREM 19/2252. Guise to Bearpark, 17 November 1986.
150. PREM 19/2252. Bearpark to Thatcher, 27 November 1986.
151. PREM 19/2252. Fairclough to Thatcher, 17 February 1987.
152. PREM 19/2252. Guise to Thatcher, 10 March 1987. Guise cited Fairclough's experience at IBM, which had an 'outstanding track record for effective R&D', where he 'supervised the work of 9,000 research staff, reporting to him through six layers of management'.
153. 'What is certain is that Fairclough has an impossible task in making any of this happen without your strong personal support for his plan. He will not be asking for this tomorrow but he will be expecting some view of whether you think this is where we should be going. I think it is'.
154. PREM 19/2252. Norgrove to Thatcher, 12 March 1987.
155. PREM 19/2477. Fairclough to Thatcher, 24 April 1987. E(RD) would be renamed E(ST), and Fairclough recommended Thatcher chair it, since having a 'Minister from one of the spending Departments in the chair will always be awkward'.
156. PREM 19/2253. Fairclough to Thatcher, 2 July 1987, contains the response to the House of Lords Select Committee on Science and Technology.
157. PREM 19/2477. Guise to Thatcher, 28 April 1987. My emphasis.
158. PREM 19/2477. Norgrove to Fairclough, 29 April 1987.
159. But advice to increase science spending was frowned upon. When ACOST requested an increase in civil R&D spending, it was told in no uncertain terms that it was 'most unhelpful' and told that it should never again start a letter to the Prime Minister with comments on overall funding. PREM 19/2478. Tombs to Thatcher, 30 September 1987, plus comments on letter and subsequent draft replies.
160. PREM 19/1805. Channon to Thatcher, 11 August 1986, contains the instructions from E(RD).
161. ED 273/89. Baker to Thatcher, 11 March 1987: 'All of the Councils are in serious difficulties ... SERC will probably have to cut new research grants from 900 to under 300, and reduce postgraduate studentships as well. MRC will reduce their research grants by at least 15% and their capital allocations to MRC establishments by 25%. The great majority of their establishments can expect not to receive any capital equipment allocation at all. The AFRC is already deep into redundancies.' This was at the moment when, as discussed in Chapter 7, Baker was committing to issue a direction to NERC, overriding its scientific council decision.
162. ABRC, *A Strategy for the Science Base*, Cm. 30. London: HMSO, November 1987.
163. Mary Fagan, 'ABRC's research plan meets united assault', *New Scientist* (5 November 1987), p.19, details the attacks (from BAAS, the CVCP, AUT, SBS) on the ABRC proposal for three tiers of universities, ordered by research activity: R (research), T (teaching) and X (mixed). Research funds would only go to R-class and some of the X-class institutions.
164. Ben Martin and Richard Whitley, 'The UK Research Assessment Exercise', in Richard Whitley, Jochen Gläser and Lars Engwall, eds, *Reconfiguring Knowledge Production*. Oxford: Oxford University Press, 2010, pp.51–80, p.56. The first Research Selectivity Exercise (1986) was largely driven by causes internal to science funding bodies rather than being driven by external demands for greater accountability and selectivity. See Norma Morris, 'Authority relations as condition for, and outcome of, shifts in governance: the limited impact of the UK Research Assessment Exercise on the biosciences', in Whitley, Glaser and Engwall, eds, *Reconfiguring Knowledge Production*. Oxford: Oxford University Press, 2010, pp.239–65, p.245.
165. Guise subsequently thought that E(ST) was designed from the start as the means of reallocating funds from applied and defence R&D to basic research. PREM 19/3156. Guise to Thatcher, 24 July 1990. While this may have been in his mind, the reorganisation of machinery that produced E(ST) immediately preceded the pitch for the new policy.
166. PREM 19/2477. George Guise, 'Public expenditure on science', 24 July 1987. Margaret Thatcher, *The Path to Power*. London: HarperCollins, 1995, p. 176.
167. 'The economic benefits [of electricity] would greatly exceed the cost of all fundamental scientific research undertaken in the UK since the time of Newton!'.
168. PREM 19/2477. George Guise, 'Public expenditure on science', 24 July 1987. Guise makes the same point later: 'This philosophy [of industry expecting government to fund industrial research] led to the Alvey programmes, to Inmos, and to the billions of pounds that have gone on industrial support over the past decades'. PREM 19/2835. Guise to Thatcher, 12 January 1989.

169. PREM 19/2116. Guise to Thatcher, 1 May 1987.
170. The exceptions would be where the state is the customer.
171. One might ask why fund basic research at all? Why not rely on other countries? Guise dismissed this suggestion, since without well-funded basic research in the UK 'there would be nobody here who understood what was being discovered or [would be] invited to participate in the idea flow … The argument that we can leave everything to the United States is as fallacious on research as it is on defence'.
172. PREM 19/2478. 'Priorities for science and technology', drafted as Fairclough to Thatcher, 16 September 1987. Some (possibly Thatcher) has written 'why?' against 'funding'. Faiclough also calls for several other policy changes: the DTI should have a proper Chief Scientific Adviser (rather than a mere Chief Scientist and Engineer who was essentially a line manager), the ABRC move towards multi-disciplinary research centres at selected universities should be supported, funding for large-scale science should be separately budgeted, CERN's administration should be renegotiated on efficiency lines and the financing of energy R&D needed reconsideration in the light of privatisation.
173. The same commentator as in the above note has written 'How will they judge?' against this point about departments choosing priorities.
174. PREM 19/2478. Guise to Thatcher, 17 September 1987. The key points were: '1. Government should fund basic science, but very little technology except where it is the user. 2. We should be quite generous in defining the point at which scientific activity becomes technology. It should go to the point where the basic science is sufficiently well understood to make some machine based upon it. Beyond that point we are into the realm of commercial development. 3. Government should support the development of enabling technology, but principally by co-ordination rather than direct investment. 4. International collaboration is necessary, particularly in big science, but proper management methods must be imposed. CERN is an excellent case as Fairclough recommends. 5. Government funding for basic science should be strongly supported with the money concentrated intellectually and geographically on specific centres of excellence. The thin spreading philosophy must be abandoned because the funds are spent more on generating acrimony than science or technology!'.
175. PREM 19/2478. Guise to Thatcher, 17 September 1987.
176. PREM 19/2478. Guise to Thatcher, 30 September 1987.
177. PREM 19/2478. Llewellyn Smith to Guise, 24 November 1987. Guise to Thatcher, 25 November 1987. Also Guise to Thatcher, 5 November 1987. Llewellyn Smith: 'the tradition in particle physics goes back farther and deeper in the UK than in any other country in Europe'. Guise, following Salam: 'Since Newton, British scientists have been in the vanguard [of particle physics]. Atomic theory was given its first rigorous presentation by Dalton. J. J. Thompson discovered the electron. Rutherford discovered the atomic nucleus. Indeed it is no great exaggeration to describe particle physics as a predominantly British invention!'. Llewellyn Smith: 'Having set up the Abragam Enquiry, it would seem – and would actually be – incredible if the UK were to give notice … '. Guise: 'Having set up Abragam, largely upon UK initiative, it would be perverse to make an irrevocable departure decision … '.
178. PREM 19/2478. Fairclough to Thatcher, 18 November 1987.
179. David Edgerton and Kirsty Hughes, 'The poverty of science: a critical analysis of scientific and industrial policy under Mrs Thatcher', *Public Administration* 67 (1989): 419–33.
180. PREM 19/2478. Wilson to Thatcher, 25 November 1987.
181. PREM 19/2478. Schopper to Thatcher, 17 November 1987. Thatcher to Schopper, 31 December 1987.
182. PREM 19/2478. Baker to Major, 10 December 1987. The UK responded to the Abragam by issuing a statement to the CERN Council, saying that the UK could remain a member of CERN so long as major efficiency savings (in line with the Abragam recommendations) were followed. Baker to Howe, 16 December 1987, containing the message to CERN Council. This message was reiterated in the letter from the Prime Minister to the DG of CERN. Thatcher to Schopper, 31 December 1987. Baker reported positively on progress in PREM 19/2479. Baker to Thatcher, 24 November 1988. Fairclough and Guise supported this view, although Thatcher remained 'unhappy', but decided to 'reluctantly agree to continued UK membership'. PREM 19/2479. Gray to Thatcher, 28 November 1988.
183. PREM 19/2479. 'Note for the record. Meeting with ACOST', 30 March 1988, ends with a brief discussion of CERN, in which 'the Prime Minister noted that if rapid corrective action was not

taken it might be appropriate for the UK to give notice that its subscription would be stopped after December'.

184. Parts of the following pages were previously published in Jon Agar, 'The curious history of curiosity-driven research', *Notes and Records of the Royal Society*, published online 20 September 2017. DOI: 10.1098/rsnr.2017.0034.
185. PREM 19/1369. Nicholson to Thatcher, 2 May 1984. This categorisation of science was used by bodies such as the ABRC.
186. PREM 19/1369. Nicholson to Thatcher, 6 March 1984. My emphasis.
187. Max Perutz, 'How to stifle research', *New Scientist* (10 December 1987), p. 57. Tam Dalyell asked the government for a response via a question in Parliament in January 1988. ED 34/311. Tam Dalyell question to Secretary of State for Education and Science, 19 January 1988. Perutz's anger was directed at the Advisory Board for the Research Councils (ABRC) report, *A Strategy for the Science Base* (May 1987).
188. PREM 19/2479. Jeffrey to Gray, 26 January 1988. Phillips to Fairclough, 21 January 1988. Fairclough to Gray, 27 January 1988. David Phillips wrote a rebuttal on each paragraph. Fairclough was also critical. Both made the argument that Perutz, by insisting that science's creativity came from individuals, had failed to grasp the role of collaboration, networks of local support and resources and the role of management (even for accounting for the success of his own institution, the Laboratory of Molecular Biology). Phillips deploys plenty of history of science in his rebuttal, both recent (the development of the LMB) and less so (quoting Rutherford on 'There is physics and stamp collecting', for example). Phillips also attacks Perutz's interpretation of the monoclonal antibodies story (again underlined by Thatcher, and again reminding us that the importance of this case): 'Monoclonal antibodies were invented by two people working full time at the MRC Laboratory of Molecular Biology, among the strongest concentration of talent in the world and with access to resources that are unparalleled in the UK. To represent this as characteristic of the creativity of individual university research workers is ridiculous'. Note the individual vs collective at play here.
189. PREM 19/2478. Annotated copy of Max Perutz, 'How to stifle research', *New Scientist* (10 December 1987), p.57.
190. Perutz to Noble, 11 September 1987. Campaign for Science and Engineering (CaSE) archives.
191. Department of Trade and Industry, *DTI – the Department for Enterprise*, Cm. 278. London: HMSO, 1988.
192. Miles Parker, 'The Rothschild report (1971) and the purpose of government-funded R&D—a personal account', *Palgrave Communications* 2 (2016). Online only, paper reference 16053 DOI: 10.1057/palcomms.2016.53. Nicholas Read, 'The "near market" concept applied to UK agricultural research', *Science and Public Policy* 16(4) (1989): 233–8.
193. Parker, *op. cit.*, citing House of Lords, *Select Committee on Science and Technology Session 1988–89 3rd Report* (HL Paper 24). HMSO: London, 1989.
194. PREM 19/278. Tombs to Thatcher, 30 September 1987. Tombs was chair of Rolls-Royce plc. Francis Tombs, *Power Politics: Political Encounters in Industry and Engineering*. London: I. B. Tauris, 2010.
195. PREM 19/2478. Tombs to Thatcher, 22 December 1987. The talk now was of 'redirecting' funds rather than increasing the total civil R&D spend. In April 1988 Thatcher chaired a meeting of ACOST dedicated to the discussion of priorities. The priorities fell in four areas: advanced materials, biotechnology and life science, control and communications, and design and advanced manufacturing. Present at this ACOST meeting were: Tombs, Andrew Bain (chief economic adviser, Midland Bank), Sir John Collyear (chair, Technology Requirements Board), Professor Gordon Edge (Chief Executive, Scientific Generics Ltd), Antony Gill (chair, Lucas Industries), Terry Harrison (chair, Northern Engineering Industries), Dr Graham Hills (Principal and VC, Strathclyde University), Professor Leonard Maunder (Newcastle), David McMurtry (chair, Renishaw), Professor Stan Metcalfe (Manchester), Dr Dennis Oliver (ex Pilkington), Professor Sir David Phillips (chair, ABRC), Professor Keith Peters (Cambridge), Sir George Porter (President, Royal Society), Dr Charles Reece (ICI), Dr Alan Rudge (BT), David Smith (ex Esso Chemicals UK), Professor Desmond Smith (Heriot Watt) and Professor Sir Peter Swinnerton-Dyer.
196. Michael Prowse, 'Why bigger many not be better', *Financial Times* (21 December 1988). Prowse gathered quotations from David Phillips, Sam Edwards, Ben Martin (SPRU), Denis Noble (Save British Science), Edgerton and Donald Braben (head of venture research at BP).

Prowse states that Phillips (ABRC), with Fairclough (CSA) 'was the driving force behind the new science policy'. Guise's central influence was hidden and missed.

197. PREM 19/2479. Guise to Thatcher, 18 March 1988.
198. PREM 19/2835. Minutes, ACOST, 1 February 1989.
199. PREM 19/2479. Guise to Thatcher, 25 May 1988. 'In Plato's Republic, Socrates asks Glaucon if money should be spent on astronomy and Glaucon replies that it ought to be because of its military and navigational value. Socrates replies that he is amused to see such fear of recommending useless studies. The debate is therefore over 2000 years old and still unresolved in the [ACOST and DES] papers we have today'.
200. For Guise the characteristic of basic science (and soon to be labelled 'curiosity-driven science') was that it was the ultimate source of wealth but for which 'economic benefit analysis is a waste of time because the greatest advances have always come from unexpected directions. The economic consequences of basic science is in principle unplannable'. Likewise: 'What appears in the short term to be useless is frequently that which has the greatest longest long term value. Electricity and atomic energy are obvious examples and there are many others. If the electronics industry was known as the quantum physics industry in the same way as the chemical industry is so-called, the point would be obvious to many who do not have the scientific education to perceive the direct, albeit long term linkage'. And finally: 'Development should always be drawn by the possibility of market demand whereas research must continuously extend the boundaries of the possible with little pre-judgement about usefulness and profitability'. All these points were prepared by Guise to guide Thatcher at the crucial E(ST) Cabinet committee.
201. PREM 19/2479. Guise to Thatcher, 25 May 1988.
202. PREM 19/2479. Guise to Thatcher, 25 May 1988. 'It is a great fallacy that research can be corralled into specific areas by economic forces alone. Unless the intellectual leadership is there, and intellectual leadership is always driven by curiosity, economic pressures will simply lead to second-rate work' (Thatcher's underlining). For the Cold War context for Silicon Valley, see: Stuart W. Leslie, *The Cold War and American Science: the Military-Industrial-Academic Complex at MIT and Stanford*, New York: Columbia University Press, 1994; Paul N. Edwards, *The Closed World: Computers and the Politics of Discourse in Cold War America*, Cambridge, MA: MIT Press, 1996. For a moderated view, see Christophe Lécuyer, *Making Silicon Valley: Innovation and the Growth of High Tech, 1930–1970*, Cambridge, MA: MIT Press, 2007.
203. PREM 19/2479. Guise to Thatcher, 15 November 1988. 'Fairclough is seeking to use the opportunity of your meeting with ACOST in mid-December to have a pow wow on science and technology policy in general. This is understandable in that he was excluded, apart from written comments, from the putting together of the Royal Society speech. Indeed he not had a formal discussion with you since the join meeting with Francis Tombs in mid-March'.
204. Jon Agar, '"Future forecast – changeable and probably getting worse": the UK government's early response to anthropogenic climate change', *Twentieth Century British History* 26 (2015): 602–28.
205. Margaret Thatcher, 'Speech to the Royal Society', 27 September 1988. Copy of the text (as prepared rather than as spoken) can be found in the online archives of the Margaret Thatcher Foundation. http://www.margaretthatcher.org/document/107346. Accessed 3 April 2017. Thatcher's third point is very pertinent to current Brexit times: 'My third point is that, despite an increase in the basic science budget of 15 per cent in real terms since 1979, the United Kingdom is only able to carry out a small proportion of the world's fundamental research and that of course is true of most countries. It is therefore very important to encourage our own people to be aware of the work that is going on overseas and to come back here with their broadened outlook and new knowledge. It is also healthy to have overseas people working here'.
206. Google Ngram conducted for the term 'curiosity-driven research', 2016, included in Agar, 'Curious history', *op. cit.*. For another view, see: Richard Phillips, 'The impact agenda and geographies of curiosity', *Transactions of the Institute of British Geographers* 35 (2010): 447–52.
207. PREM 19/2479. Fairclough to Thatcher, 25 August 1988.
208. PREM 19/2479. Guise to Thatcher, 12 October 1988. The underlining is Thatcher's. The second underlined passage also garnered a big blue tick.
209. PREM 19/2479. Tombs to Thatcher, October 1988.
210. PREM 19/2479. Fairclough to Thatcher, 11 November 1988.
211. PREM 19/2479. Guise to Thatcher, 15 November 1988. 'Our present policy is built on the belief that it is Government's job to finance science and industry's to finance technology, unless Government needs that technology for its own purposes such as defence'.

212. PREM 19.2479. Guise to Thatcher, 15 November 1988. 'It is alarming that high ranking impartial bodies like ACOST and the Royal Society are not even thinking about these issues'. He also complained that Fairclough barely mentioned the environment: 'We are told that closer collaboration on the matter is needed and that a further study will advise on a coherent UK approach. This does not sound like the determined start which will give Britain intellectual leadership on the big, long-term ozone and greenhouse problems'.
213. Guise's emphasis. Thatcher underlined 'privatisation' and many other parts of the document. Note that other officials were alarmed by Guise's interpretation. One checked with the PM: 'My understanding of Next Steps is that it is principally a way of managing better what for the foreseeable future will remain in Government. George's approach would stop the movement to agencies in its tracks.' PREM 19/2835. Note by AT (probably Andrew Turnbull) on Gray to Thatcher, 6 January 1989.
214. PREM 19/2835. Guise to Thatcher, 12 January 1989.
215. PREM 19/2835. Butler to Thatcher, 13 January 1989.
216. PREM 19/2835. Andrew Turnbull, handwritten note on Butler to Thatcher, 13 January 1989.
217. PREM 19/2835. 'PM's discussion with Sir Francis Tombs', 27 January 1989. On the Nobel question, in discussion 'it was suggested that former UK Nobel Prize-winners were perhaps not as industrious as they might be in putting forward nominations for new UK candidates'.
218. PREM 19/3155. Thatcher to Tombs, 9 January 1990, including 'Government response to ACOST's advice on national priorities for science and technology: 1989'.
219. PREM 19/2835. Baker to Thatcher, 20 March 1989.
220. PREM 19/2835. Handwritten comment by Thatcher, on Gray to Thatcher, 21 March 1989.
221. PREM 19/2836. Gray to Thatcher, 2 October 1989.
222. PREM 19/2836. DES, 'The future structure of the research council system', December 1989. See also: Andy Coghlan, 'Research councils face abolition threat', *New Scientist* (15 July 1989), p.24. The Morris working group consisted of Morris, three academics and the research director of Amersham plc. PREM 19/3155. Guise to Thatcher, 23 January 1990. The Morris report also called for the management of science to be professionalised through UK business school qualifications. Guise objected: 'Engineering "research" can probably be managed in this sense. Basic science certainly <u>cannot</u>. Who could have "managed" the work of Newton, Einstein or Francis Crick?'.
223. PREM 19/2836. McGregor to Thatcher, 21 December 1989. Among the detailed reasons for the merger given by DES was one point made by Sir David Smith, VC of Edinburgh: 'the forces of NERC and AFRC cannot now mount a national expertise on the greenhouse effect. Like soil science, neither subject used to be fashionable, revitalisation needs a unified approach'. McGregor to Thatcher, 21 December 1989.
224. Handwritten: 'NERC scientists would be worried about Antarctic/Arctic/Ocean Circl'n/Satellite Observation/Computer Climate Models. I don't think we could have a merger <u>unless</u> it works in detail and as agreed'. Fairclough shared Thatcher's concerns on this point. PREM 19/3155. Fairclough to Caines, 9 January 1990.
225. PREM 19/3155. Guise to Thatcher, 23 January 1990.
226. PREM 19/3155. 'Synopsis of AST(90)8. Advanced manufacturing technology', undated (March 1990).
227. PREM 19/3155. Bhattacharyya to Guise, 8 March 1990, including 'Centre for advanced manufacturing technology', undated (March 1990). Bhattacharyya: 'Where there has been a big input from public funds to equip universities and other organisations, they have never weaned themselves off continued public support ... Where companies themselves have taken the lead, for example at Warwick, the story is different ... I suggest that it is not ACOST's role to advocate propping up unsuccessful companies. I am personally surprised that ACOST has even tackled such a topic, which is properly the business of industry and not of Government. In short, the Working Party recommendation smacks of an academic-biased plea that "the Government should do something" and of DTI officials yearning for the good old days of intervention'.
228. PREM 19/3156. Guise to Thatcher, 24 July 1990.
229. PREM 19/3156. Turnbull to Fairclough, 25 July 1990.
230. PREM 19/3156. Fairclough to Thatcher, 24 July 1990.
231. PREM 19/3156. Guise to Thatcher, 24 July 1990.
232. PREM 19/3156. Guise to Turnbull, 30 October 1990.
233. PREM 19/1370. Transcript of opening speech, by Thatcher, 12 September 1983.

234. PREM 19/1369. Turnbull to Hodgkinson, 4 May 1984. This meeting was between Thatcher, Joseph, Peter Brooke (PUS, DES), David Hancock (permanent secretary, DES), Nicholson, Phillips (chair, ABRC) and Kingman (chair, SERC). Thatcher contrasted the UK situation to that of Japan. The meeting received some press coverage, for example: Paul Flather, 'Scientists called to Downing St', *The Times* (4 May 1984). This in turn brought in outside comment, for example Alex Merrison (President of the Council of CERN), 'CERN and the UK', 14 May 1985, which argued that just as the Rutherford and the nuclear atom eventually had far-reaching applications, so would the research at CERN.
235. Save British Science, with the momentum of its debut, won an audience with Keith Joseph in February 1986, and found the minister in an uncompromising mood. His successor, Kenneth Baker, declined invitations to meet. Mulvey, 'Report of the interim meeting on the future of SBS', 18 April 1986. CaSE archives.
236. Warnock, *op. cit.*, p.185. Universities were certainly, following the Jarratt Report (1985) and Kenneth Baker's Education Reform Act (1988), expected to be more managerial, cost-effective and accountable, but this modelling after business management is quite distinct from strategic industrial direction of university science.
237. ACARD, *Exploitable Areas of Science*. London: HMSO, 1985. Other alternative plans could be found in discussions in the Treasury, see T 199/6. Deloittes brought in to help in 1982 by the Treasury suggested science parks, specialist 'intermediaries' (in other words what would not be called support for 'translation'), the promotion of spin-off and diagnosed a fundamental dearth of entrepreneurs with imagination. See T 494/134.
238. PREM 19/1514. 'ACARD report – exploitable areas of science. A synopsis of the approach by reference to communications', 1985, including the 'capsular' diagram.
239. PREM 19/1514. Nicholson to Thatcher, 3 October 1985.

Coda
AIDS and bioethics

In her first term as Prime Minister, Thatcher was confronted first by AIDS, a new, devastating and rapidly spreading disease for which there were few, if any, precedents and second by a crisis point in how society should regulate biomedical research, especially the use of, and research on, human embryos. In both the AIDS and bioethics cases Thatcher intervened only occasionally, and her responses were often moralistic in tone: a survey should be stopped because it offended and intruded into people's private lives, specific types of embryo research should be prohibited because she found them repulsive in character. In neither case did she invoke her reserved right to answer questions on 'science matters' – nor did her central advisers or external parties approach her, as they sometimes did, as I have shown, 'as a scientist'.

Commentators have suggested a connection between policies pursued and a broader Thatcherite ideology in both instances: in the case of AIDS in terms of traditional or even 'Victorian' family values, and in the case of bioethics in a rejection of self-regulating professional expertise. There is something in both of these positions, as I explore, but there is also a sense in which neither of these science-related issues was at all typical of how science policy was developed under Thatcher. While to the public AIDS and embryo research were perhaps the two most high-profile science-related issues of the 1980s, to the government, in the ways they were deliberated and handled, they were exceptional and atypical.

AIDS

The realisation that young, urban, homosexual men in the United States, presenting symptoms including various infections, loss of weight, fever, enlargement of lymph glands and rare cancers, were cases of a new disease happened in 1981. First labelled gay-related immune deficiency (GRID), the condition was given the name of Acquired Immune Deficiency Syndrome (AIDS) in August 1982. Once recognised, earlier cases began to be identified and, over time, a pattern of transmission from Africa was traced. Pictures of the spread of the disease are therefore time dependent. In 1985 it was thought that in 1979 there were 11 known cases, 10 in the United States and one in the United Kingdom. In Britain the Communicable Disease Surveillance Centre, a body with its roots in the late nineteenth century[1] and part of the Public Health Laboratory Service, began national surveillance of AIDS in 1982, part of an international network. In 1983 French and American scientists independently isolated a virus as the cause. In early UK official documents the virus was first referred to as HTLVIII (the name used by one of the American teams, under Robert Gallo). For clarity, however, I will use the name HIV (Human Immunodeficiency Virus) that became the international taxonomic standard in 1986.

In the mid-1980s it was thought that the first reported case of AIDS in the UK was in 1979, and that by July 1985 this had risen to 196, of whom 110 had died. The Chief Medical Officer, Donald Acheson, estimated that there were 10,000 people infected with HIV, while up to 2,000 cases of the fully developed disease were expected within three years.[2] 'The majority of these persons are in London and the number is increasing perhaps at the rate of 50–100 per week,' reported Acheson. 'People infected with [HIV] ... are usually free of symptoms for many months or years, are unaware of their infections, but are nevertheless infectious ... [An] exponential increase in the number of infected persons can be expected.'[3] London hospitals became the main centres of treatment, notably St Mary's (Paddington), St Thomas' (Lambeth), Middlesex (Fitzrovia) and St Stephen's (Chelsea). While homosexual men were the main group, other 'at risk groups' were identified as cases became known: haemophiliacs (also almost exclusively male) who were treated with repeated injections of the Factor VIII clotting agent, intravenous drug users, recipients of blood transfusions, female partners of homosexual and bisexual men, children of infected mothers and health care workers. There was no cure or vaccine. 'Plans should be based on the assumption that no means will be available to prevent the disease by immunisation in the next five years,' warned Acheson.

Measures instigated by the government in response to AIDS included formulating and distributing centralised expert advice, general public education and specific information campaigns, targeted biomedical research and a reorganisation of blood donation and transfusion services. The government also considered screening visitors from Africa and the United States, starting with students coming to the country on British Council schemes.

An Expert Advisory Group on AIDS was established in February 1985, chaired by Donald Acheson, to watch and comment on all aspects of the disease, while an interdepartmental group of senior civil servants was later established to advise a Steering Group of ministers on the wider implications ('for employers and employees, life insurance, education, certain occupational groups, and so on').[4] Examples of the issuing of advice from the centre included such acts as the Chief Medical Officer sending a letter to all doctors in England on matters of 'information on groups at risk, clinical presentation and diagnosis and measures to prevent the spread of the infection', or the Chief Nursing Officer contacting professional nursing organisations, and then on to members, about community care of AIDS patients.[5] Government funding went into general information campaigns, for example *Some Facts about AIDS* from the Health Education Council – a body that was independent of government, but had members that were appointed by the Secretary of State for Health and a budget that was largely supported by public money.[6] There were also pamphlets on specific aspects, such as being a haemophiliac or a blood donor.[7] Counselling, such as that developed and provided through the Terrence Higgins Trust, also received government funds, for example £10,000 in 1985 and £25,000 already given.

The biomedical research effort was coordinated and funded by the Medical Research Council, starting in mid-1983, with contributions from the Department of Health and Social Security. About £430,000 was spent by 1985. The MRC requested and received an extra £1 million to spend over 1987/88. Most AIDS research was funded and carried out by private pharmaceutical companies. However, it was also the case that research funding for AIDS was judged against other claims. In particular, when a small tranche of extra research funding (£15 million) was made available to British science in 1985, the ABRC advised

> That it would not be appropriate for any of the new money … to be diverted to the epidemiological research programme [on AIDS]. We understand that the additional £15m was secured to meet the twin objectives of sustaining strategic research of industrial relevance and helping to halt the brain drain of talented British scientists. The Board could not advise that the AIDS

epidemiological research would have a high priority against these criteria.[8]

While the point was partly tactical, to pressure DHSS to provide more of its own funds, it also reveals the middling priority given to AIDS research, compared to the issues of industrial strategy considered in the Chapter 3, even as the full-scale AIDS crisis was emerging.

Blood supply was systematically reorganised. Groups at risk were discouraged from donating blood. Intensive effort was made to find tests for HIV in order to screen donated blood passing through the National Blood Transfusion Service. These tests were expected to be in operation by October 1985, although the question of whether the results of tests should be made known to donors was yet to be resolved. Haemophiliacs, by their nature likely to be frequent recipients not only of donated blood, but also of extracted blood products such as the blood-clotting protein Factor VIII, were a particularly problematic group. In 1985 the plan was to heat treat Factor VIII for haemophiliacs, but also take steps to ensure that the UK was self-sufficient in blood products, necessitating a major £38 million expansion of the Blood Products Centre at Elstree. In 1987, against the wishes of John Major, then at the Treasury, the government offered compensation to haemophiliacs infected with HIV by the Factor VIII injections.[9]

Much of this government activity was led by the Minister of State for Health, Barney Hayhoe. Thatcher was kept informed, by reading documents and a briefing from Acheson, and she occasionally asked to see scientific papers on AIDS.[10] She also did not think that the administrative machinery needed to be brought into the Cabinet committee network, as Robert Armstrong had suggested, so that the Prime Minister might 'keep on eye on what is going on', but could be serviced at a distance from the centre by the Department of Health and Social Security.[11] Hayhoe's public stance was largely to be reassuring:

> The Government fully understands the public concern about AIDS. We are tackling the disease on a broad front and, with the continuing co-operation of those in the at-risk groups. I am hopeful that we will be able to control the spread of the disease and reduce the appalling suffering which accompanies it.[12]

'We have to walk a difficult tightrope between being accused of bureaucratic inertia and being so active as to whip up public hysteria,' David Willetts of the Number 10 Policy Unit advised Thatcher. 'Barney Hayhoe's announcement gets it about right.'[13] But Willetts also wondered whether Thatcher might want to open the new Elstree centre. 'It combines attractive themes,' he said, explaining that these included 'high quality British science, action to protect innocent [*sic*] victims of AIDS and spending on health infrastructure'. Mark Addison, Thatcher's private secretary, also equated haemophilia with innocence and purity (and therefore, implicitly, homosexuality with guilt and impurity):

> The lab will ensure that haemophiliacs can be supplied from our own pure sources with special blood plasma, to protect them from becoming innocent victims of AIDS. ... My own feeling is that the Prime Minister should stay clear of AIDS (!), even when it is a question of opening laboratories to help innocent victims. If she is going to do a medical visit, I should prefer to suggest opening a hospital, or a home for children with incurable diseases, etc.[14]

Thatcher's direct interventions into AIDS policy were not about the science, high quality of otherwise. Instead, they became more motivated, it seems, by moral revulsions, public perceptions of the role of government and her understanding of concerns about family and privacy. I will give two examples.

The first related to advertisements that Norman Fowler, a minister at DHSS, intended placing in Sunday newspapers in February/March 1986. They were 'explicit and distasteful', thought Willetts, but the AIDS 'problem [was] ... now so serious that we must do as he proposes'.[15] 'Do we have to do the section on risky sex?' queried Thatcher. 'I should have thought it could do immense harm if young teenagers were to read it.'[16] On this occasion Acheson, Chief Medical Officer, refused to back down. He argued that the 'passages ... contained the essence of the message that he needed to get across; and that in his professional judgement their inclusion in the publicity was vital'.[17] Thatcher then wondered if it might breach the Obscene Publications Act. Yet a conviction under this Act required that a publication had to tend to 'deprave and corrupt' and the Act also contained an exemption on grounds of 'public good'. The Home Office did not think the publication of the advert would be obscene. Thatcher responded again:

> I remain against certain parts of this advertisement. I think the anxiety on the part of parents – and many teenagers who would never be in danger from Aids – exceeds the good it may do. It would be better in my view to follow the 'VD' precedent of putting

> notices in public lavatories etc. But adverts where every young person could read and learn of practices they never knew about will do harm.[18]

Thatcher finally relented when the words 'and should be avoided' were added after the line 'Rectal sex involves the highest risk'. The adverts appeared in all national Sunday and daily newspapers on 16 and 17 March and 6 and 7 April 1986. Not a single public complaint was lodged. The 'Don't aid AIDS' adverts were followed by a leaflet, *AIDS: Don't Die of Ignorance*, which was sent to every household in the UK, at a cost of £2 million. Thatcher initially opposed this mail drop, which was delayed until January 1987.[19] When it eventually happened it was accompanied by poster and television campaigns, but Thatcher ruled out use of a Category 1 Ministerial broadcast – a rare public emergency measure, last used under Callaghan.

If some in government thought the government's own public education campaigns were distasteful, then this was revulsion was doubled when some of the more independent AIDS advice was reviewed. The completely explicit *National AIDS Manual*, produced by the National AIDS Trust (funded in part by the Home Office) and written by Peter Scott, was described by Professor Brian Griffiths of the Number 10 Policy Unit as 'pornographic … this material legitimises all kinds of deviant behaviour'.[20] Thatcher agreed that the 'borderline between the permissible and the pornographic would seem to have been crossed' and felt 'strongly that a publication of this sort should not be financed from Government funding'.[21]

The second intervention concerned a survey. In February 1989 the H(A) Cabinet committee had agreed that more information was needed about the population's sexual attitudes and behaviour to inform further AIDS campaigns. The Economic and Social Research Council (ESRC) and the Health Education Authority (formerly Council) explored what an arms-length survey might consist of: both bodies expected the results to make an important contribution. In July 1989 David Mellor, successor but one to Barney Hayhoe as Minister of State, Health, therefore proposed a large-scale survey, saying that government 'need[ed] information for forecasts of the likely future spread of the HIV epidemic'.[22] The expert committees, as well as Acheson, the Chief Medical Officer, agreed that there was a 'scientific case for a survey on the scale proposed … anything smaller would not include sufficient numbers from relevant sub-groups in the population to provide the information for forecasts which is needed'. The survey, judged Acheson, would significantly narrow statistical predictions of the number of AIDS cases, allow mathematical models to look further ahead, enable forecasts of HIV positives to be made ('none currently exist') and forecasts of heterosexual spread, again 'where none exist at present'.[23]

The government, Mellor proposed, because of the interest of the departments of Health, should directly provide £200,000 out of a total cost of £810,000. The rest would be research council funding and therefore the ESRC's responsibility. Mellor, recognising that there might be an adverse public reaction to the government asking 20,000 people about their sexual practices, suggested finessing the presentation, a sort of discretion by omission:

> Health Departments' interest in the survey would not be printed on the questionnaire, nor would people be told about it in the interview, unless they asked. If they did ask then they would be told that Government were providing some financial support for the survey and the reasons for this would be given. Respondents would also be assured that their individual anonymity would be guaranteed.[24]

Thatcher, however, was against the survey in any form. 'I doubt the need for this survey,' she wrote. 'I should have thought that there is so much information available now from the US that we could use that. Also – we have been severely criticised for some of the things we have done in this programme [to curb AIDS]'.[25] She also thought respondents' answers would be unreliable. When it was suggested that US data could not be read across to describe the UK's sexual behaviour, Thatcher gave her third reason to rule out the survey:

> I think people rightly would be deeply offended by questions of this kind and I do NOT think we are entitled to intrude into their privacy. Neither Government, nor Government money should be involved in any way – if this survey goes ahead.[26]

'The decision not to put any Government money into the survey will be very controversial,' noted a DHSS civil servant, 'and the announcement will require very careful planning and handling.'[27] Yet when it was announced, the Wellcome Trust stepped in to cover the missing funds. The Wellcome funding story was scooped by the *Independent* before the Trust's courtesy heads up had reached the Prime Minister, to the charitable foundation's embarrassment.[28]

Overall then the AIDS issue was one in which the Prime Minister did not invoke her right to respond on science-related matters. In the case of the public education campaigns she regarded the explicit detail with distaste, citing the innocence of teenagers. In the case of the survey, she intervened to prevent direct government money going towards the gathering of factual data that her expert advisers regarded as very important, although she made no attempt to interfere with the choices of the research council involved (it was therefore not a breach of the so-called Haldane principle, as has been implied).[29] Rights to privacy and concern for adverse public reaction (which would have been led by criticisms in the right-wing press) were the two reasons she gave. Despite the immensity and suddenness of the AIDS crisis it was not, for Thatcher, one of major science issues of her term.

Bioethics

Historian Duncan Wilson locates the origin of the term 'bioethics' – in the sense we use it today, as the 'ethical scrutiny of specific problems raised by medicine and the biological sciences' – to the work of the Dutch obstetrician André Hellegers and the political activist Sargent Shriver in the very early 1970s.[30] The persuasive view was that bioethics could not be left to medical practitioners or biomedical scientists alone, but rather that policy should be formed by a wider range of people. While it flourished in the United States and other countries, Wilson notes that bioethics, 'did not gain currency in Britain until the 1980s, when increasing numbers of philosophers, lawyers and theologians became actively involved in the public discussion of medicine and biology, the teaching of professional ethics and the development of regulatory guidelines'.[31]

The period of Thatcher's prime ministership is therefore the context for the entrenchment of bioethics in Britain. Indeed Wilson goes further and links the ideas and influence of two key bioethicists to Conservative values and projects. First, the academic lawyer Ian Kennedy, 'who was the most high-profile advocate of the approach he explicitly termed "bioethics" … was influential because [his endorsement of bioethics] dovetailed with the Conservative government's neoliberal belief that professions should be exposed to outside scrutiny to make them publicly accountable'.[32] Kennedy called for 'outside scrutiny' of medicine by a supervisory 'board of committee' in his BBC Reith lectures in 1980, and followed it up with the observation that 'only someone who is free from any claims which medical professional loyalty may make on his objectivity … can successfully examine the institution of medicine'.[33] Second, the philosopher Mary Warnock, who would lead the most important political and ethical intervention into regulating biomedicine, specifically embryo research and IVF, in Britain in the 1980s, also chimed with 'how the Conservative government prioritised "non-expert" involvement in public inquiries into science and medicine during the 1980s'.[34] Specifically Warnock, Wilson reveals, rejected the view that authority trumped arguments from moral preferences and sentiment, an argument that there 'cannot be moral experts' leading to a position that was aligned to 'the neo-liberal emphasis on individual autonomy and echoed Margaret Thatcher's belief that '"choice is the essence of ethics"'.[35] When writing his book, essential to understanding bioethics in Britain, in the early 2010s Wilson was able to access departmental and research council files – DHSS and MRC – but not the central government, notably the Number 10, files. So, with these files now available, we can explore how his argument fares in the light of new evidence.

When *in vitro* fertilisation (IVF) was first used successfully in humans with the birth of the 'test tube baby' Louise Brown in July 1978, the initial public response was very positive. However, by the early 1980s Conservative politicians and the right-wing popular press increasingly regarded the technique as intensely problematic, a threat to the 'traditional' nuclear family and wasteful of the life of embryos. But 'most controversial', judges Wilson, was research on embryos *in vitro*. IVF generated more embryos than were used in implanting, and experimentation on this material was of great interest and use to scientists. (The list of techniques being called into question compiled by the government overlaps, but was not identical with those highlighted by the press.)[36] Wilson regards the controversy over IVF as being not only a matter of the resurgence of 'Victorian values' under Thatcher, but also because 'criticism of these practices … reflected and bolstered growing calls for external involvement with scientific and medical ethics'.[37] This pressure for external scrutiny, a product of academic debate and Conservative scepticism of autonomous professional experts, was the reason why the appointment of the philosopher Mary Warnock to lead a wide-ranging inquiry into IVF was particularly significant.

Thatcher's 'initial view' was that 'some form of independent inquiry into these ethical issues' [of 'in vitro pregnancies'] was 'necessary, in view of the growing public concern'.[38] Shirley Williams, the ex-Labour minister and co-founder of the Social Democratic Party in 1981, had written to Thatcher suggesting that the form of such an inquiry should be a Royal Commission. Williams had, in the 1970s, once agreed that the medical profession should regulate itself, but had begun to move towards the principle of wider viewpoints shaping regulation. Writing to Thatcher, Williams argued that a Royal Commission 'would enable those with varying knowledge and experience to contribute to a significant assessment of the issues and to make recommendations for the future – and its members should be drawn not only from scientists and the medical profession, but also those with understanding of the law, theology and education'.[39] Leo Abse, the Welsh Labour MP and no friend of Thatcher's (he later wrote an unflattering psychoanalytical biography of her), made a similar suggestion, which he pursued through Parliamentary mechanisms.[40] Norman Fowler, minister at the DHSS, supported the idea of a wider inquiry, if not a Royal Commission:

> The Department's position until now has been that it was best to have the views of the medical bodies concerned – the General Medical Council, the British Medical Association, the Medical Research Council, and the Royal College of Obstetricians and Gynaecologists – before reaching conclusions on the nature of any wider enquiry ... But the fact is that the issues go way beyond purely medical questions, and involve much wider considerations, as well as very specific and detailed legal problems. It may well be, therefore, that further action should not wait on these bodies: there are indications that the profession themselves [*sic*] share this view.[41]

Fowler therefore instructed his officials to prepare advice on the form that a wider inquiry might take. This proposal was run past Thatcher, who, although she underlined sections, offered no specific comment. By April Fowler was convinced that an 'official' 'Committee with an outside Chairman and members' should be established, with the following terms of reference.

> To consider recent and potential developments in medicine and science related to human fertilisation and embryology; to consider what policies and safeguards should be applied, including consideration of relevant legal matters, and to make recommendations.[42]

Fowler thought the membership should consist of 'doctors, scientists, lawyers, persons with a background in marriage counselling and in theology as well as four or five non-experts', while as 'Chairman' a judge would fit the bill. At this point Thatcher did intervene: the inquiry did not call for a judge.[43] Baroness Young, a leading Conservative figure in the House of Lords, suggested amending Fowler's terms of reference from 'relevant legal matters' to 'relevant legal and ethical matters'; this change, she said, 'would put beyond doubt what we intend'.[44]

Perhaps this insistence on the ethical was influential, since by May 1982 the DHSS had decided that an ethics philosopher, Mary Warnock, was their favoured choice as chair, although they carefully ran the names of Warnock and a short list of another three candidates past Robert Armstrong in the Cabinet Office.[45] Warnock, Fowler told Thatcher, was 'an experienced and capable Chairman of proven ability and we think that that is an important attribute where such a breadth of interests and complexity of issues will need to be involved' in an inquiry 'which will be important and intellectually difficult, raising social and moral issues'.[46] Thatcher simply wrote 'Yes'.

Warnock had attended Oxford (studying Classics) at the same time as Margaret Thatcher,[47] but they did not meet until the 1970s.[48] In 1974, when Warnock had returned to Oxford to teach philosophy, Thatcher invited her to chair an inquiry into teaching children with special educational needs. In 1977, although Warnock had switched from Labour to Conservative supporter in the mid-1960s, the Labour government had asked her to join an advisory committee on one of the key areas of biomedical ethical debate, animal experimentation.[49] 'Applied ethics', a relatively new philosophical field, was split at the time between those, such as Richard Hare, who argued that philosophy could reconcile different groups by identifying good and bad arguments, and others, such as A. J. Ayer and Warnock, who thought such reconciliation could not be forced by reason.[50] She brought this experience to the inquiry, which sat for two years.

The *Report of the Committee of Inquiry into Human Fertilisation and Embryology* (the Warnock Report) finally arrived in midsummer 1984.[51] With chapters on artificial insemination, IVF, egg and embryo donation, surrogacy, sex selection, the freezing and storage of eggs, embryo and semen and the regulation of scientific research, it contained 63 recommendations. The most general recommendation was that the government should establish a new organisation that could both advise and have the executive authority to grant licences, including for research. Such a body should have 'substantial lay representation' and a lay chairperson. Warnock managed the problem of arriving at recommendations despite fiercely conflicting views first, by championing philosophical pluralism and second, by allowing the inclusion of three expressions of dissent. Wilson links the philosophical pluralism to a rejection of the overriding authority of moral experts and thence on to Thatcherite individualism.

But pluralism was also perhaps the only pragmatic position that was likely to lead to a useable report and recommendations.

Following the report was six years of public debate. The statutory regulating and licensing body, the Human Fertility and Embryology Authority (HFEA), was finally established by an Act of Parliament in 1990, coming into effect in 1991. I will comment on three aspects of this period: the role of Parliament, the role of Thatcher's science advisers and, finally, the role of Thatcher herself.

First, the Parliamentary debates were unusually intense, unprecedentedly so for a science-related issue in the 1980s. The main debate, on 23 November 1984, was opened by Norman Fowler, who acknowledged 'the extremely strong feelings which the issues covered by the [Warnock] report both here and among the public'.[52] The Speaker of the House of Commons limited Members' contributions to just 10 minutes each from 11.30 am to 1 pm, and urged those 'fortunate enough to be called before 11.30 to bear in mind that many other colleagues are waiting to take part in this very important debate'. The debate itself ranged across abortion (not covered by the Warnock report), Jesus' virgin birth, Down's syndrome, the status and beginning of personhood in the embryo, the blight of miscarriage, commercial surrogacy, the regulation of experiments on embryos, matters of conscience and religion, infertility, Galileo, Sigmund Freud, congenital diseases in Wales, interspecies fertilisation, the freezing of embryos and the right to have children. 'We must settle those matters,' argued Sir Gerard Vaughan (MP for Reading, East), upholding the role and capability of Parliament, adding: 'We must not wait for the medical profession or public opinion to move one way or another'.[53] Frank Field (MP for Birkenhead), inveterate contrarian, thought that as the debate 'progressed I have become more, not less, confused', although that was 'not because of contributions by hon. Members, but because it is difficult to think coherently about the issues'.[54] There were statements from inflexible moral positions and calls for the need to have minds open to reasoned debate. If there was a consensus it was around an insistence that this issue should be subject to Parliamentary free discussion and decision, and that, as Kenneth Clarke, Minister for Health, said on closing the debate:

> One has to decide to what extent it is legitimate to carry out research upon [the embryo] while treating it with respect, and whether the bounds can be set within which research can properly be tolerated. Those bounds cannot just be left to the medical and scientific establishment.[55]

Further intense Parliamentary debates occurred, notably when Enoch Powell attempted to pre-empt Warnock legislation by introducing an Unborn Children (Protection) Bill in February 1985, during which the Speaker again invoked a 10-minute speech restriction.[56] The Bill would have made *in vitro* creation of embryos illegal for any purpose, notably research, other than to enable a named woman to bear a child.[57] Heavily supported by pro-life MPs, the Bill dismayed scientists – including Anne McLaren, the developmental biologist who had been most influential on the Warnock committee.[58]

Second, embryo research was an issue, again unusually, where Thatcher's central advisors – the Chief Scientific Adviser and the Number 10 Policy Unit – played a secondary role compared to the Parliamentary process. Nicholson advised that the Powell bill would create 'untenable' problems for researchers, and told Thatcher that the government should 'stress the importance to medical research of allowing embryo experimentation to continue'; on the question of the form of a statutory body he supported a 'Standing Royal Commission on Bioethics', although his first preference was 'professional self-regulation'.[59] Fairclough, Nicholson's successor, proposed a purely advisory 'National Bioethics Commission', independent of government, in November 1987, involving professionals and others.[60] Nicholson and Fairclough also advised on the minor issue of which experts to send to international bioethics conventions – convened by the Prime Minister of Japan in 1984, the President of France in 1985, the Chancellor of West Germany in 1986, and the Prime Minister of Canada in 1987.[61] Members of the Number 10 Unit worked on the commercial surrogacy issue (David Willetts preferred a 'total ban') and generally supported the pro-life side.[62]

Third, Thatcher herself, while expressing her moral views, did not play the scientific expert card on the issue of embryo research. She objected to 28-week abortions ('strenuous efforts are made to save such premature children')[63]. She supported the Powell bill, although not in Parliamentary vote.[64] She met with a delegation of pro-life MPs while simultaneously responding sympathetically to the anti-abortion and anti-embryo research campaigning group LIFE.[65] She agreed that her Chief Scientific Adviser's arguments supporting embryological experimentation displayed 'a touch of casuistry' when he had argued that 'all that can be said is that fertilisation brings into existence a genetically novel kind of cell, and that this cell has the potential … for becoming a human individual'.[66] She hoped 'we can prohibit the storage of frozen embryos', while she also intervened to insist that 'genetic engineering' of embryos (an issue that arose post-Warnock), 'should be forbidden'.[67] On

Warnock's central recommendation, on the question of the need for the statutory authority, she rejected her Chief Scientific Adviser's preferred choice of professional self-regulation. Interestingly this insistence, while in line with Wilson's argument that Thatcherites rejected professional self-regulation (although Nicholson therefore was an important exception), was *not* justified by her ideologically, but rather by the specific, unusually emotionally fraught aspects of the issue: the same reasons that made it unusually an issue for authentic Parliamentary deliberation and decision-making:

> Yes [let the Cabinet H committee consider self-regulation as an alternative to a statutory body] – but it will be difficult to leave such emotionally important matters – such fundamental matters to self-regulation.[68]

Thatcher also rejected Fairclough's idea of an independent, advisory commission, declaring that 'the assumption he makes that the problems – all of them – can be resolved by the setting of an Advisory Committee ... is not our experience'.[69] The reason we are 'finding it difficult to legislate' after Warnock, was not lack of such advice but 'because of genuine differences of view on quite fundamental things'. Again, interestingly, her reasons are not quite what one would expect – for example, she dismissed Fairclough's argument that a mixed-membership commission would possess the 'ability to look out for new problems on the horizon': 'No', she wrote, 'this comes from the professionals who know about it'.[70]

Public consultation ended in July 1987, and a white paper, proposing a statutory licensing authority, was issued in November 1987.[71] Royal Society pressure (Thatcher held a meeting there in February 1988, meeting the President, George Porter, along with Anne McLaren and the embryologists Christopher Graham and Richard Gardner) stabilised the chosen course rather than steered it.[72] With Cabinet and Parliament split, and the post-Warnock Report legislation regarded as 'nothing but trouble',[73] Thatcher's government edged slowly from 1986 to 1990 towards the establishment of the Human Fertilisation and Embryology Authority, guided on one hand by the Warnock recommendations and on the other by fine judgements on what legislation would pass a free vote in Parliament. These twin poles – both external to executive government – were more important than Thatcher's own position as a moral agent or ex-scientist, and more important than her central advisers' advice.

I have described the policy responses to AIDS and embryological bioethics as exceptional and atypical in comparison with other science-laden issues confronted by the Thatcher government. AIDS was an emergency with little precedent, at least in the postwar era. Embryology was subject to unusually intense Parliamentary debate, unrivalled by any other science-related controversy. Thatcher, unlike the cases of nuclear projects, missile defence, climate change or acid rain, did not invoke the privilege of responding to either AIDS or embryology as science questions. She intervened, certainly and occasionally, but on moral grounds, such as to prevent what she saw as intrusive and distasteful inquisitiveness by publicly funded researchers into sexual behaviour (in the case of AIDS) and to listen to anti-abortion campaigners and to forbid the unlikely step of genetically engineering embryos (in the bioethics debate). These were not science-laden issues where Thatcher took command.

Notes

1. Graham Mooney, 'Public health versus private practice: the contested development of compulsory infectious disease notification in late-nineteenth-century Britain', *Bulletin of the History of Medicine* 73 (1999): 238–67.
2. PREM 19/1863. McKessack to Addison, 28 August 1985.
3. PREM 19/1863. Acheson, 'HTLV3 infection, the AIDS epidemic and the control of its spread in the UK', June 1985.
4. PREM 19/1863. Draft letter to minister of state, Home Office, undated. Fowler to Thatcher, 25 September 1985.
5. PREM 19/1863. 'AIDS: measures taken', 28 August 1985. The CMO letter was sent on 15 May 1985. The CNO letter was sent on 15 July 1985.
6. Pamphlet, Health Education Council, *Some Facts about AIDS. Acquired Immune Deficiency Syndrome*, 1985. It was followed a year later by Health Education Council, *AIDS. What Everybody Needs to Know*, November 1986.
7. Pamphlet, National Blood Transfusion Service, *AIDS. Important New Advice for Blood Donors*, January 1985.
8. PREM 19/1863. Phillips to Joseph, 29 November 1985.
9. PREM 19/2775. Major to Thatcher, 24 September 1987.
10. An example was Robert Yarchoan et al, 'Administration of 3'-azido-3'-deoxythymidine, an inhibitor of HTLV-III/LAV replication, to patients with AIDS or AIDS-related complex', *The Lancet* 327 (15 March 1986): 575–80. The copy in PREM 19/1863 has Thatcher's underlining throughout. The CMO briefing took place on 27 November 1986.
11. PREM 19/1863. Armstrong to Wicks, 1 October 1985. Thatcher initials her agreement of the rejection of Armstrong's suggested Cabinet coordination on this document.
12. PREM 19/1863. Press release, 'The fight against AIDS – more government money', September 1985.
13. PREM 19/1863. Willetts to Thatcher, 25 September 1985.
14. PREM 19/1863. Addison to Ryder, 26 September 1985.
15. PREM 19/1863. Willetts to Thatcher, 24 February 1986.
16. PREM 19/1863. Thatcher, handwritten comments on Willetts to Thatcher, 24 February 1986. The 'risky sex' part of the advert contained a mention of anal intercourse.
17. PREM 19/1863. Whitelaw to Thatcher, 26 February 1986.
18. PREM 19/1863. Thatcher, handwritten comments on Laurance to Addison, 5 March 1986.
19. PREM 19/1863. Thatcher, handwritten comments on Addison to Thatcher, 22 August 1986. Fowler to Thatcher, 30 December 1986.
20. Peter Scott, *National AIDS Manual.* London: NAM Publications Ltd, 1st edition November 1988, 1st update April 1989. PREM 19/2775. Griffiths to Thatcher, 26 July 1989.

21. PREM 19/2775. Gray to Walters, 28 July 1989.
22. PREM 19/2775. Mellor to Moore, 24 July 1989.
23. PREM 19/2775. 'Proposed national survey of sexual behaviour: note on scientific issues by the Chief Medical Officer, Department of Health', undated (July 1989).
24. PREM 19/2775. Mellor to Moore, 24 July 1989.
25. PREM 19/2775. Thatcher, handwritten comment on Morris to Thatcher, undated (February 1989).
26. PREM 19/2775. Thatcher, handwritten comment on Slocock to Thatcher, 25 July 1989.
27. PREM 19/2775. Goldhill to Slocock, 28 July 1989.
28. Stephen Ward, 'Charity set to fund sex survey vetoed by PM', *Independent* (13 October 1989). The Wellcome Trust made their decision on Thursday 12 October, phoned the PM's office and said that a press release would go out on Monday 16 October. The PM was on the way to Conservative Party Conference in Blackpool and her speech was on the afternoon of 13 October. The *Independent*'s scoop therefore came out on the morning of the speech. PREM 19/2775. Gray to Thatcher, 13 October 1989.
29. Wilkie, *op. cit.*, p.33.
30. Duncan Wilson, *The Making of British Bioethics*. Manchester: Manchester University Press, 2014, p.2.
31. Wilson, *op. cit.*, pp.8–9.
32. Wilson, *op. cit.*, p.12, see also p.106.
33. Wilson, *op, cit.*, pp.118–19.
34. Wilson, *op. cit.*, p.13.
35. Wilson, *op. cit.*, p.141. The Thatcher quotation comes from John Campbell, *Margaret Thatcher, Volume One: The Grocer's Daughter*. London: Vintage, 2007, p.377.
36. The list of government concerns was: 'a) Surrogate mothers … b) Cloning … c) Fertilisation of human eggs when there is not intention of allowing them to develop. The purpose here would be to use the developing embryos for research purposes. d) The freezing of human early embryos … e) Trans-species fertilisation experiments … f) The ultimate development of a human embryo in the laboratory, just to see how far development can proceed. It is suggested that some scientists will try to do this to satisfy their own curiosity'. PREM 19/1855. 'The techniques associated with test tube babies that are now being called into question'.
37. Wilson, *op. cit.*, p.154.
38. PREM 19/1855. Pattison to Clark, 12 February 1982.
39. PREM 19/1855. Williams to Thatcher, 10 February 1982.
40. Specifically, a proposed private member's Motion that asked for an 'interdepartmental interdisciplinary advisory committee' to report on social, medical, legal and ethical issues involved in the creation of *in vitro* embryos. Half the membership, thought Abse, should be women.
41. PREM 19/1855. Fowler to Thatcher, 20 February 1982.
42. PREM 19/1855. Fowler to Whitelaw, 4 April 1982.
43. PREM 19/1855. Thatcher, handwritten comments on Fowler to Whitelaw, 4 April 1982.
44. PREM 19/1855. Young to Fowler, 21 April 1982.
45. PREM 19/1855. Stowe to Armstrong, 20 May 1982. The other names were: James Sutherland (solicitor, former vice-chair of Glasgow Eastern Health Council), Lady Wagner (chair, Barnardo's), and Sir Norman Lindop (academic, ex-chair of the Council for Professions Supplementary to Medicine and the Data Protection Committee).
46. PREM 19/1855. Fowler to Thatcher, 3 June 1982.
47. Wilson, *op. cit.*, p.142.
48. Mary Warnock, *A Memoir: People and Places*. London: Duckworth, 2000, p.169. Warnock bemoans Thatcher's 'deep philistinism', 'appalling rudeness', 'crudity' and 'aggression'.
49. Warnock, *op. cit.*, p.181. Wilson, *op. cit.*, p.146.
50. Wilson, *op. cit.*, p.150.
51. DHSS, *Report of the Committee of Inquiry into Human Fertilisation and Embryology*, Cmnd. 9314, London: HMSO, 1984.
52. 'Human Fertilisation and Embryology (Warnock Report)', *Hansard* (23 November 1984), p.526.
53. 'Human Fertilisation and Embryology (Warnock Report)', *Hansard* (23 November 1984), p.550.
54. 'Human Fertilisation and Embryology (Warnock Report)', *Hansard* (23 November 1984), p.567.

55. 'Human Fertilisation and Embryology (Warnock Report)', *Hansard* (23 November 1984), p.590.
56. 'Unborn Children (Protection) Bill', *Hansard* (15 February 1985), pp.642–706.
57. PREM 19/1855. Booth to Redwood, 20 March 1985.
58. H. John Evans and Anne McLaren, 'Unborn Children (Protection) Bill', *Nature* 314 (14 March 1985): 127–8.
59. PREM 19/1855. Nicholson to Thatcher, 23 October 1985. Nicholson to Thatcher, 11 February 1985. Robert Armstrong found the idea of a Royal Commission 'attractive'. Armstrong to Wicks, 4 October 1985. Nicholson (probable author), 'Comments on the Report of the Government inquiry into human fertilisation and embryology', undated (July 1984) for comment that 'professional self-regulation would be preferable'.
60. PREM 19/2345. Fairclough to Thatcher, 2 November 1987.
61. PREM 19/1855. Nicholson to Thatcher, 3 February 1986. PREM 19/2345. Fairclough to Powell, 12 December 1986.
62. PREM 19/1855. Booth to Redwood, 20 March 1985, on support for Powell bill.
63. PREM 19/1855. Thatcher, handwritten comments on Addison to Thatcher, 25 October 1985.
64. PREM 19/1855. Booth to Redwood, 20 March 1985. 'The Prime Minister is in favour'. The Division in Parliament is recorded in 'Unborn Children (Protection) Bill', *Hansard* (15 February 1985), pp.702–6.
65. PREM 19/2345. Scarisbrick and Davies to Thatcher, 24 February 1987, containing list of 79 MPs who signed LIFE's open letter to Thatcher.
66. PREM 19/1855. Thatcher, 'I agree' on Barclay's comment, on Nicholson to Thatcher, 11 February 1985.
67. PREM 19/2345. Thatcher, handwritten comments on Newton to Whitelaw, undated (April 1987).
68. PREM 19/1855. Thatcher, handwritten comments on Barclay to Thatcher, 6 July 1984.
69. PREM 19/2345. Thatcher, handwritten comments on Bearpark to Thatcher, 6 November 1987. Thatcher's decision not pursue Fairclough's proposal is recorded on Fairclough to Thatcher, 14 December 1987.
70. PREM 19/2345. Thatcher, handwritten comments on Fairclough to Thatcher, 14 December 1987.
71. DHSS, *Human Fertilisation and Embryology: a Framework for Legislation*, Cm. 259. London: HMSO, 1987.
72. PREM 19/2345. Fairclough to Thatcher, 19 February 1988. See also: Collins, *op. cit.*, pp.129–30.
73. PREM 19/2345. Thatcher, handwritten comments on Newton to Whitelaw, undated (April 1987).

4
Power/leaks

The combined civil and military nuclear projects were perhaps the most consequential of all postwar British scientific and technological endeavours. However, while the science was a mixture of basic, applied and, as privatisation was considered as a goal, what would begin to be called 'near-market' research, the deep commitment to nuclear for reasons of deterrence and national status meant that nuclear policy could never be just science policy. Margaret Thatcher was a committed supporter of nuclear power and the nuclear deterrent. 'Nuclear power,' notes Dieter Helm, 'held a fascination for her: as a scientist, for its technical achievements; as an advocate for a strong defence policy; and, as an opponent of the miners, in the form of an insurance policy'[1]

In Chapter 6 I will examine aspects of the military nuclear project through an examination of Thatcher's engagement with Ronald Reagan's plans for missile defence. Here I begin a study of the civil nuclear side. In particular, I trace Thatcher's publicised visits to UK nuclear establishments, especially Dounreay in Scotland and Sellafield in Cumbria. I will argue that these visits were productive and reciprocally supporting encounters between two forms, at the highest levels in postwar Britain, of contained power: one political and one technical. The 'Prime Minister' is an office invested with extraordinary political power, especially when the individual in place commands confidence and authority. Such authority was 'lent', by choreographed shows of association and support, to shore up struggling nuclear projects. In return Thatcher could also benefit – although not in an unproblematic way, given heightened anxieties about the nuclear in the 1980s – from imagery that presented her in a white coat against a backdrop of iconic modern technology. However, as I explore in the second half to this chapter, such power was also undermined by 'leaks' and other instabilities of various kinds.

Britain's nuclear landscape

Britain's nuclear projects were developed behind barbed wire in laboratories, factories and field testing establishments dotted across the land. At the foot of the North Wessex Downs, due south from Oxford, was the Atomic Energy Research Establishment at Harwell. The first reactors on British soil, GLEEP and BEPO, were built here in the late 1940s. Fifteen miles further in the same direction, across the Downs, lay the Atomic Weapons Research Establishment, Aldermaston. These two were the major United Kingdom Atomic Energy Authority (UKAEA) laboratories for civil and military nuclear research respectively through much of the Cold War.[2]

Britain had detonated its first atomic device, on the Monte Bello Islands, off the coast of Western Australia, in 1952. Its manufacture had required the rapid construction of a network of facilities, including reactors and plutonium processing at Windscale in Cumbria, a factory working with uranium at Springfields in Lancashire, with headquarters at Risley, further south in the same county, a gaseous diffusion plant for extracting fissionable material at Capenhurst in Cheshire and huts for final assembly at Foulness on the Essex coast, prior to shipping to the other side of the world. Britain's full-scale nuclear weapon production required two further major facilities in the UK, built at the Royal Ordnance Factories at Burghfield (near Aldermaston) and Cardiff, Wales, as well as test sites in Australia. Thermonuclear weapons were detonated, too, from 1957, beginning with the Grapple tests conducted on remote islands in the mid-Pacific.

Next to the Windscale factory was Britain's first nuclear power plant, Calder Hall, which opened in 1956. It produced electricity for the civil grid and plutonium for atomic weapons. Calder Hall possessed four Magnox reactors, a design that was developed and deployed a further 24 times at power stations commissioned from the late 1950s to the early 1970s and located on the coasts of England, Scotland and Wales. In 1965 the Wilson government ordered a second wave of nuclear power stations, choosing the Advanced Gas-cooled Reactor (AGR) design based on a prototype built at Windscale. Seven twin AGRs were installed, starting with Dungeness B, which produced power from 1983, and ending with Torness 2, online from 1989.

Meanwhile the government had invested in further experimental nuclear concepts. First, the fast-breeder reactor promised greater efficiency in fuel use at a time when supplies of fissionable material seemed likely to become scarce.[3] The fast-breeder reactors consumed uranium (including depleted uranium from Capenhurst), produced plutonium and were cooled by the circulation of tons of liquid sodium. Hopes for this UK-led technology were sky high: 'once fast-breeder reactors have been successfully developed to the commercial stage, we can be assured of cheap power for hundreds of years with no fear of electricity costs rising due to shortage of fuel,' the Prime Minister had been told in 1969.[4] UKAEA began its experimental Dounreay Fast Reactor, with its iconic steel sphere, in the mid-1950s. The Prototype Fast Reactor was built at the same site on the Scottish north coast from the 1960s, producing electricity from the mid-1970s. Second, the idea of producing energy from nuclear fusion, as opposed to fission, was pursued. In 1958 hopes were raised and then dashed as the Zero Energy Toroidal Assembly (ZETA) experiment seemed to be producing unexpected energy. Nevertheless, the government acquired a new site, Culham, not far from Harwell, where ZETA had been built. It became the home of the largest European science project on British soil, the Joint European Torus (JET), funded through the European nuclear energy community, Euratom.

In the mid-1970s, dissatisfaction with the AGR design led to a debate about choosing a third design for civil nuclear power stations in Britain.[5] Initially the government favoured its steam-generating heavy water reactor (SGHWR) design, prototyped at Winfrith in Dorset, over the American company Westinghouse's pressurised water reactor (PWR) design. But in the economic troubles of the mid-1970s the SGHWR contracts were terminated. Instead, the Callaghan Labour government announced that it would greenlight the final AGRs (at Torness and Heysham) and support the construction of future PWRs. In late 1979 the new Thatcher administration decided that a PWR should be built.[6] However, with the uncertainties provoked by a new design, the PWR at on the Suffolk coast at Sizewell, alongside ageing Magnox reactors, would only begin after a public inquiry, which opened for evidence and testimony in 1983. The Sizewell decision would be one of the major nuclear policy moments of the Thatcher years.

Throughout the late 1970s into the mid-1980s, however, the two major customers for nuclear power plants continued to articulate and lobby for different designs. The Central Electricity Generating Board (CEGB) wanted a 'minimum family' of four PWRs. While not ruling out further AGR systems, it considered PWR to have 'substantially lower capital costs per unit of electricity than AGR' and to be 'capable of generating cheaper electricity'; they also perceived PWR to be 'proven mainstream worldwide technology with some 150 reactors in operation worldwide, compared to 5 AGR stations exclusively in the UK'.[7] The South of Scotland Electricity Board, on the other hand, argued that AGR performance was improving and disputed the CEGB's economic assessments.

Britain's nuclear project had begun under a government department, the Ministry of Supply. As it expanded, new agencies, notably the UK Atomic Energy Authority, had been set up – along with company-like entities, such as the consortia that merged in 1971 to produce the National Nuclear Corporation – to oversee construction of power stations and negotiate with its customers, the Central Electricity Generating Board (CEGB) and South of Scotland Electricity Board (SSEB). In the same year, the production side of UKAEA had been split off to form further company-like entities: British Nuclear Fuels Ltd (BNFL), to handle and manage fuel and waste, and the Radiochemical Centre at Amersham, the subject of the privatisation story told in the Chapter 5. At Windscale the creation of BNFL split the site. Parts remained with UKAEA and other parts, notably the Calder Hall power station and the Magnox reprocessing plant, came under BNFL. To the latter was soon added, after a public inquiry, a Thermal Oxide Reprocessing Plant (THORP), envisaged to take oxide fuel from home AGR and foreign reactors.[8] In 1981 the BNFL side was renamed 'Sellafield', partly to escape association with the infamous 1957 Windscale fire. In 1984 BNFL became BNFL plc, although its stock was wholly owned by the government. By 1979, therefore, the nuclear landscape of Britain, both in terms of organisations and sites, was dispersed, complex and expensively maintained.

Throughout the history of the British nuclear project there have been opposing anti-nuclear voices. In the late 1950s, drawing on older traditions of anti-militarism in the Labour movement and the church, anti-nuclear protest became organised on a considerable scale, exemplified by the establishment of the Campaign for Nuclear Disarmament (CND) and its protest marches directed at Aldermaston.[9] Its momentum dwindled from the mid-1960s. However, anti-nuclear movements gathered force globally from 1975, partly as the end of the Vietnam War redirected activists' energy and partly as environmental concerns became stronger.[10]

As Cold War tensions increased, and NATO decided in 1979 to deploy Pershing and cruise missiles in Western Europe, including the UK, so anti-nuclear protest expanded. CND, led by the charismatic Monsignor Bruce Kent, campaigned against the neutron bomb, circulated leaflets and petitions, and screened the terrifying nuclear war docudrama *The War Game* (1965, but banned by the BBC) for students. The British branch of Friends of the Earth led an articulate critique of nuclear power, not least via the 1977 Windscale inquiry. Friends of the Earth also acted to publicise leaks of radioactive material. In 1981 the Greenham Common Peace Camp began to protest against the arrival of US cruise missiles. The women developed an extraordinary feminist culture of protest.[11] In April 1983 the Greenham Common movement attracted 70,000 supporters to form a human chain from Greenham to Aldermaston and Burghfield, an effective way of making the British nuclear landscape visible. In general, the anti-nuclear movement offered the public a critical view against the British nuclear projects' claims for modernity, safety and peaceful intent.

Contained and constrained: powers meet

In contrast with other chapters, where I trace Thatcher's engagement with science mediated by the memoranda, files and the other paper technologies of bureaucracy, here I will follow the Prime Minister as she encountered the UK's greatest postwar science project – that of nuclear power – face to face.

Thatcher visited Dounreay in 1979 and Sellafield in 1985. These were events that combined ceremony, publicity, information gathering, witnessing, discreet lobbying and displays of contained power and authority. Furthermore, as a cultural contact between powers, they invite and deserve a thick description; small details had larger resonances and meanings that together can be reconstructed, almost anthropologically, to understand how both political and technical power intertwined and operated. However, I will also argue that the containment and display of power had limits, and these will be explored through the metaphor and the actualities of 'leaks'.

Thatcher was not the first prime minister to visit Dounreay. Edward Heath had visited briefly in 1973, but a more substantial visit was made by his predecessor.[12] In July 1969 Harold Wilson had flown to Wick airport and then been driven to the nuclear site. Even before he arrived he had been contacted by staff representatives anxious about recession in Caithness triggered by a run-down in staff at the region's largest employer.[13] At its peak 2,500 people had worked at Dounreay, while the nearby town of Thurso had been transformed with 1,500 new houses, new schools, a technical college and upgraded amenities. The 'stimulus of new people with new ideas has given the area a new confidence and hope for continued growth,' Wilson had been briefed.[14] The local MP, Robert Maclennan, had warned him that 'your every word about the future of Dounreay will be hung upon, analysed and interpreted with the closeness of a medieval theologian by my constituents and by the press'.[15] Following a lunch of salmon steaks, cheese and beer, and an hour's tour of Dounreay's two fast-breeder reactors, journalists at the Inverness press conference had indeed probed Wilson on employment questions.

Ten years later, Thatcher's visit was not framed primarily by a single issue. Avowedly pro-nuclear, she had in her in-tray a decision to take about new nuclear power stations and their designs. Visiting Dounreay was an opportunity to associate herself with what was still being presented as a world-leading and distinctively modern technology, but it was also conducted in the context of UKAEA proposing 'the next stage in fast reactor development': a full-size, 1250 MWe commercial demonstration fast reactor (CDFR). The focus of this piece of public ritual was a single moment: Margaret Thatcher would press the button that would start the reprocessing unit at Dounreay in front of the gathered press. 'This is a very important stage in the closing of the fast reactor fuel cycle,' briefed the press release. It explained that this was to be an international 'first' – one 'in which fuel is "burnt" in the reactor to generate electricity, taken out of the reactor, reprocessed to separate the plutonium, which is sent to Windscale to be refabricated by British Nuclear Fuels Limited, into fresh fuel for the reactor'.[16] The closed cycle, instigated by the Prime Minister, was a demonstration of contained power and control, both technological and political. Yet neither cycle was hermetically sealed. The risks of a breach in the uranium–plutonium cycle of processing were the subject of debate. The political cycle of credit was opened when the press witnessed the pressing of the button. Neither cycle could be fully controlled. Power generated by the cycles could leak.

On the evening of 5 September 1979 a Queen's Flight Andover aircraft landed at Dounreay Aerodrome. Mr Beaumont flicked a switch and temporary floodlights illuminated the party as it stepped down from the plane and walked towards a convoy of two UKAEA and three Northern Constabulary cars. Security was on edge, not least because of the assassination of Lord Mountbatten a week before. Margaret Thatcher's party of seven – comprising herself, her husband Denis, the MP Ian Gow (her Parliamentary Private Secretary), her press officer Henry James, her civil servant private secretary T. Lankester, a personal secretary Mrs J. Humphris and two personal detectives – were welcomed on the tarmac by Clifford Blumfield, director of Dounreay. The convoy then whisked the party to the Royal Hotel, Thurso, where they spent the night.

After breakfast the next morning, the Prime Minister's party were issued passes and the convoy set off on the 20-minute ride to the nuclear site, accompanied by a police escort.[17] With passes checked at the main gate, AEA police signalled for the inner gate to be opened. The convoy proceeded to the Prototype Fast Reactor building. Here they were greeted by Sir John Hill, chairman of UKAEA, alongside Dr Tom Marsham, head of the Nuclear Directorate of UKAEA, and Blumfield – the trio of nuclear managers who would accompany Thatcher for the rest of the day. Taking the stairs to the PFR Exhibition Room, the Prime Minister, her husband and Gow signed the visitors' book. Hill then spoke on the fast reactor concept, while Marsham and Blumfield followed with a review of Dounreay projects. Now issued with even higher-level passes and donning white coats (including one which was the Prime Minister's 'special size'), the party passed the security turnstile, and began a tour of the PFR Control Room, the reactor top and then, through an air-lock, to a lift, which ascended to the fourth floor and enabled a view of the Turbine Hall.

Retracing their steps, the party was taken by car to the D1200 building – the fuel-reprocessing plant. After another tour, and the unveiling of a plaque by the Prime Minister, she then pressed the button initiating reprocessing. There had been some concern that the press photographers, admitted to the small control room, might have to take turns, meaning that the button would have to be pressed multiple times. Henry James, the press officer, explained to local organisers that 'although the Prime Minister did not like "faking" events of this kind she would probably agree to pressing the switch more than once'.[18]

With the images of Thatcher, scientist and prime minister, taken, the party discarded the white coats and overshoes, washed, had their hands and feet monitored for radiation and then walked across to the Director's Dining Room for a lunch of Scottish fayre: cock-a-leekie soup, local Strathy salmon salad, strawberry shortcake and Scottish cheese and biscuits. While Mr Wilson had been given beer, Mrs Thatcher had the choice of chilled orange juice or apple Shloer. She sat between Sir John Hill on her right hand and Blumfield on her left. On the dining table fresh flowers replaced plastic plants for the day.

After lunch, more senior Dounreay management joined the group, along with representatives of the Staff Side and trade union, for 'informal discussions'. Behind the scenes, much of the work and organisation of Dounreay had been rearranged for the day, as the site went into presentation and high security mode. Police were dotted around the route from Thurso to nuclear plant; on site, telephones were disconnected, tannoys suspended and even the regular PFR tea run was re-routed. A special female toilet was designated in the reprocessing building, complete with female attendant. The visit was to be smooth, discreet in the right places and professional. The publicity was carefully corralled and, despite worries that the press might baulk at a trip to a remote location, photographers duly snapped the expected pictures. At the end of the visit the Prime Minister fielded questions at a 20-minute press conference. She was asked about the future of Dounreay, whether she was in favour of the fast-breeder reactor and the British nuclear programme more generally, and the question of local jobs. Thatcher said that, while not pre-judging the results of an inevitable enquiry, she

> personally would like to see it go ahead. … You know my personal view – I have always been interested in more R&D and for Britain to keep ahead. The French are building Super Phenix – at one time we were ahead in the early stages of the fast breeder. I do not share the fears that some people have of nuclear power.[19]

The *Daily Express* (but not the official notes) records her as saying that Dounreay was 'absolutely safe'.[20] Thatcher spoke about the Magnoxes, AGRs and PWR to meet the demand for electricity, as well as of the need for a strategy in which the 'alternative supply' (to oil and coal) was 'continuous, which will not run out like fossil fuel' – the 'obvious' one being 'the fast breeder'.[21] She cast the closed cycle as a prudent use of resources:

> I pressed the button on the fuel reprocessing plant today … No-one else in the world has one of these. You generate fuel for the next round. Not like Windscale. They are taking fuel from Magnox reactors and extracting plutonium, which cannot be used again in Magnox reactors. The worst thing that you can do with plutonium is to leave it lying around. The best thing is to burn it – the safest thing is to burn it in a fast-breeder reactor. Oil, gas and coal will not last forever – some of these things should be conserved as a source material for chemicals. The chemical industry depends on coal, gas and oil. If you can find a different source of fuel you can put oil and gas to better uses in the future. I happen to be a conservationist of natural materials.

The containment of plutonium at Dounreay as part of a closed system of 'burning' and reprocessing was therefore an important justification for the fast-breeder programme. Publicly the politics of nuclear power depended on the extent to which this claim of control was believed. Thatcher had been briefed on environmental aspects, including the controlled release of gases, but the parallel control over the encounter with the press ensured that she was not pressed on the issue.[22]

Such choreographed meetings with the press are, of course, part of the routine presentational management of politics. Within a few weeks

Thatcher would be opening Milton Keynes shopping centre and answering press questions at the John Lewis store. But for all the meticulous planning, the attempt to control the news cycle could be interrupted. A case of this eventuality happened after Dounreay. The white-coated Thatcher had posed for various photographs, not least the button-pressing centrepiece, but one image caught her on her knees examining a cavity in the reactor top. *New Scientist* repurposed the image for satire: 'And this is the drain the microprocessor industry went down'.[23]

Control – of image and isotopes – was also the theme of Thatcher's visits to other nuclear sites, especially Sellafield. In October 1984 Con (Coningsby) Allday, chairman and CEO of British Nuclear Fuels plc (as it had just been made a public limited company), invited Thatcher to open two new and expensive facilities: Pond 5, a plant for receiving, storing and reprocessing Magnox fuel that had cost £315m and SIXEP, an ion exchange plant for effluents that had cost £126m.[24] Thatcher had not visited Sellafield since 1978, when she had been leader of the opposition, and had stalled a previous request to open a new building at Risley in the election summer of 1983. Since then, however, Sellafield, despite the change of name from 'Windscale', had become the focus of considerable public concern. In October 1983 the Yorkshire Television documentary *Windscale, the Nuclear Laundry* had reported increased local leukaemia cases and plutonium dust in nearby homes. In November radioactive ruthenium and rhodium had been found on Cumbrian beaches, which were subsequently closed to the public. BNFL would be prosecuted and fined on six counts and, with public and private expressions of reluctance, required to invest in SIXEP to clean discharged liquid waste.[25]

So the invitation of 1984 presented a quandary: Sellafield was in disgrace, but it was also being moved towards privatisation and in need of a demonstration of political support from the top. The advice of the Minister of Agriculture to Number 10 was that while a visit would help 'rehabilitate Sellafield following the damaging aftermath of last November's incident', the Prime Minister should at least wait until the court case against BNFL had concluded.[26] But Robin Nicholson, her Chief Scientific Adviser, suggested that she accept the invitation, in part because 'she would find it of very great scientific interest', but also because

> the successful operation of reprocessing activities at Sellafield, and the public acceptance of them, is an essential component of our nuclear power programme. The morale of BNFL has dropped considerably over the last year or so, following management changes, operational failures leading to accidental discharges, the

> uncertainty about elevated incidences of childhood leukaemia in parts of Cumbria and the Government decision to introduce tighter controls on discharges. A visit by the Prime Minister would be taken as an indication of Government confidence in the Company and in the nuclear power programme.[27]

The visit was therefore delayed until November 1985, but when it happened it was to be an expanded, large-scale and well-publicised series of events. Rather than just visit Sellafield, Thatcher would spend a day, ferried by helicopter, tracing the nuclear fuel cycle, with each stage emphasising the control of isotope and image.

She started at Capenhurst, the site near Ellesmere Port where uranium was enriched. It was also the location of URENCO, a UK–Dutch–West German collaboration that in the 1960s had developed a commercial gas centrifuge technique, in contrast with French and US diffusion methods. The project had been a diplomatic manoeuvre for Harold Wilson, undermining French leadership in Europe while straining relations with the United States, and this demonstration of an alternative path for Europe might have appealed to Thatcher.[28] There was also talk of the Chinese buying licences for the gas centrifuge technique.[29]

The improbably named Neville Chamberlain, BNFL's director of enrichment, guided Thatcher around the site, donned in the regulation white coat (with blue BNFL badge) and cloth overshoes, with press photographers present on arrival. She saw 'three generations of centrifuge': the first entirely British, the second having some Dutch and German components, and the third, still in its early stages, representing a completely collaborative project between the URENCO partners. In Capenhurst's Cascade Hall Thatcher saw the 'serried ranks of centrifuges spinning away silently at about three times the speed of sound and without maintenance for at least 10 years'. She was invited 'to roll out a 2-tonne holder of enriched fuel, which represents £½m worth of Capenhurst product' ('on a hover', her press officer reassured her, 'not heavy'). With more press photographs taken it was a 'good opportunity, early in the day, to get on lunchtime bulletins and into evening papers'.

Then it was back in the helicopter to fly north to the next stage of the cycle: Springfields, to see the fabrication of Capenhurst-enriched uranium into fuel rods, for use in Magnox stations, and fuel assemblies for AGR, Dounreay PFR and the old SGHWR at Winfrith. Springfields also made the uranium hexafluoride (HEX) that went back to Capenhurst. Again Thatcher pulled on a white coat and posed for photographs. There was space too for a heart-warming visit to the Apprentice Training

School, which not only turned 55 school-leavers a year into fitters, turners, machinists, electricians and even a blacksmith for the nuclear sector, but also 'did up' cycles for handicapped children, wheelchairs for ex-servicemen and 'multi-activity toys'. 'Press photographers here,' her press officer, Jean Caines, advised.

The next stop was Heysham ('pronounced Heesham by the locals') II, Britain's fifth AGR, the power station south of Morecambe Bay where some of Springfields' fuel assemblies would be burnt. In 1985 it was a massive construction site. 'I managed to walk the course in high heels, but I felt precarious and was clucked at by the male population,' noted Caines, adding that a 'hard hat and anorak are the order of the day'.[30] Here Thatcher would be joined by the CEGB chairman Lord Marshall, who was expected to take the opportunity to lobby her on the Sizewell question.[31] Indeed, while on the press tour of nuclear sites, with the Sizewell inquiry ongoing, Thatcher was lobbied by both sides in the AGR vs PWR debate. She was briefed on the arguments of both sides, but was advised to 'avoid being drawn into discussion as the matter is under consideration by the Sizewell Inspector'.[32]

And finally on to Sellafield. The continued growth of the UK's nuclear power capacity was, as Thatcher's Chief Scientific Adviser had emphasised, 'desirable for reasons of both energy production (in particular, the projected decline in fossil fuels from the North Sea) and environmental protection (the contribution of fossil fuels to air pollution [at this stage, acid rain] damage'.[33] He also warned that Sellafield was 'the "Achilles heel" of the industry … [because of] uncontrolled escapes of radioactive materials'. The anti-nuclear environmental groups publicised these leaks and made them the focus of campaigns. Jilly Perry, a Friends of the Earth supporter, teacher and daughter of a local farmer, recalls that when the 1983 radioactive pollution showed up in lobsters 'we had a big lobster costume that we launched on the beach at Sellafield. We took our petition to 10 Downing Street, with our lobster'.[34]

Thatcher's words to the press were made this time on arrival 'in order to be sure to make the evening news'. After the unveiling of a plaque she visited the Fuel Handling Plant, admiring the 'enormous pond in which spent fuel rods are stored', the politically important environmental fix of SIXEP, the work in progress on the £1.3 billion Thermal Oxide Reprocessing Plant (THORP) and the exhibition centre. She then attended a attending reception for industrialists and local VIPs, at which she also spoke about each of sites to the editor of BNFL's house journal, *BNFL News*.

Altogether there were 10 separate, planned photo calls and three opportunities to speak to the national press, including television.[35] Throughout the whole day a special press helicopter, paid for by BNFL, shadowed the Prime Minister's own. Bernard Ingham prepared draft responses to possible press questions.[36] The charge of overly cautious environmental protection was to be defused by a parallel drawn between the nuclear industry and early rail:

> Your question reminds me so much of the days when the railways were being developed and men were required to walk before the engine with a red flag. Contrast that with the recent record breaking run between London and Newcastle at an average speed, I believe, of over 125 mph! Railways are undoubtedly one of the safest forms of travel but people were naturally very, very careful about the first iron horses. It is therefore only natural – and indeed sensible – that we should develop nuclear power's future with care and due caution.

Likewise, anti-nuclear campaigners were dismissed: 'I think the opposition is understandable and perhaps inevitable but at the same time irrational and misguided … [The] kind of opposition we see to nuclear power is certainly irrational when, for example, you think of the appalling death toll over the years in the coal mines'. (The miners' strike had ended only months before.) She was briefed to add: 'I think we resist change and scientific progress at our peril'.

There were plenty of subsidiary messages – on European collaboration (of a selective and anti-French type), BNFL's corporate responsibility and commercial trade opportunities – as well as many moments for nuclear managers to press informal political points, such as Marshall on Sizewell, or BNFL on the modernity of its technology. However, the main public message was to be that the nuclear industry was safe, efficient, competitive and successful. In Ingham's draft wording suggested for Thatcher: 'Britain needs a thriving nuclear industry and is fortunate in the one that it has already got. As a scientist, I profoundly believe that to be the case'.[37] The identification of Thatcher, scientist, with the UK nuclear project was designed into the whole day, with the multiple publicity images of her in a white coat and her helicopter flight tracking the nuclear fuel cycle (although not right to the end – she would have had to be either buried or dumped at sea).

These highly public visits to Dounreay and to the UKAEA sites culminating in Sellafield were highly constrained and contained events.

Although involving months of preparation they were curiously static: over in a day, ritualistic and designed to project singular messages. They were also moments of gift exchange. The gift of public prime ministerial attention strengthened a nuclear industry that was under attack from environmental and anti-nuclear campaigners. The gift of photo opportunities in nuclear sites nevertheless enabled the Prime Minister to project a scientific image, and to associate with a project that was still a symbol, for some, of postwar progress, energy and modernity: it was also, significantly, 'not-coal'. The visits were powerful in proportion to their containment of power: the figure of the powerful Prime Minister travelling by helicopter, posing for photographers, pushing buttons to start controlled processes.

The dynamics of nuclear policy-making, however, took place elsewhere. Aside from the military nuclear issues (discussed in Chapter 6), I will show that three intersecting civil nuclear matters for decisions requiring Thatcher's attention in the 1980s all illustrate the limits of constraint, containment and certainty. Each of them served to complicate the message projected via the public visits: the questions of waste and the possibility of a link between radioactive leaks and clusters of leukaemia cases, and, pre- and post-Chernobyl, the possibility of privatisation.

Waste, leaks and leukaemia

'Radioactive wastes arise at all stages of the nuclear fuel cycle,' noted the Department of the Environment in 1979.[38] Waste – 'material of no commercial interest' – was produced by reactors as spent fuel elements, gaseous discharges and, in the future, through decommissioning. Spent fuel rods would be placed in a cooling pond, which in turn had two outputs. Liquid effluent would be treated and then either authorised for 'discharge' or turned to solid waste and stored for eventual 'disposal'. The cooled fuel elements would be passed on for reprocessing. Gaseous discharge from reactors would be filtered, with some products treated for disposal and others discharged to the atmosphere.

Reprocessing, as we have seen at Sellafield and Dounreay, was another complex industrial process. It involved further cooling in ponds (producing more liquid waste for turning to 'sludge' for storing or discharge at sea), decladding of the fuel elements, recladding of 'hulls' and processing the spent fuel producing a range of outputs – including useful recovered isotopes, gases discharged to the atmosphere, low active liquid waste discharged to the sea and highly active liquid waste that was stored prior to a future process of vitrification, storage again, and then 'disposal'. In 1976 there were 1,530 cubic metres of highly active liquid waste at Windscale (Sellafield) and Dounreay, as well as a further 7,000 cubic metres of highly active solid waste in the form of fuel cladding.[39] On top of this were 11,000 cubic metres of sludges, 3,000 cubic metres of plutonium contaminated wastes and 20,000 cubic metres of wastes stored at power stations. These figures were expected roughly to treble by the year 2000. The discourse was of containment and management – the latter 'used as a broad term to describe all or part of the process of minimising the creation of waste and the subsequent sequence of its conditioning, storage and disposal'.[40]

In 1979 the low active waste was disposed of by pipelines to seas and rivers (such as from Aldermaston to the Thames) or by annual dumping at sea. The highly active waste, with the liquid material locked in glass through vitrification, awaited a decision on the form of final disposal. The Royal Commission on Environmental Pollution had, in Lord Flowers' report of 1976, called, first, for a review of the adequacy of the research into disposal options and, second, for the establishment of a 'Nuclear Waste Disposal Corporation'.[41] The Callaghan government had, in response, granted the first request in its White Paper *Nuclear Power and the Environment*.[42] For example, Michael Heseltine, then Minister for Environment, had announced (via a reply to a Parliamentary Question) a series of test drillings in Scotland, Wales and England to investigate the geological suitability of certain areas for radioactive waste disposal.[43] (Thatcher read the reply in draft and judged it 'very interesting – but I doubt whether it will allay fears!')[44]

Yet by December 1981 the research drilling was suspended, as the government rethought its waste management policy.[45] Specifically, in explicit rejection of the Royal Commission's second request, for waste to be managed by a 'Corporation' – in other words, a public interest body akin to the BBC or the NRDC – the government announced, via a new white paper, that it was endorsing a 'Nuclear Industry Radioactive Waste Executive' – NIREX (the 'W' for waste, perhaps subconsciously, was suppressed) – formed by the nuclear industry and the generating boards. 'Definite role for the private sector. NIREX preferred to the Royal Commission's Nuclear Waste Disposal Corporation', as a summary of the main points of the white paper that was shown to the Prime Minister noted.[46]

NIREX faced considerable public opposition in the two areas it named as being possible locations: Billingham, in Cleveland, where there was a history of large-scale ICI chemical engineering, for deep storage, and Elstow, in Bedfordshire, for shallow storage.[47] With a commitment made that the Sizewell Inquiry should have an up-to-date statement of waste management strategy, the Minister for Environment sought Thatcher's approval, in July 1984, for a procedure that involved the identification of at least six sites followed by a single major inquiry.[48]

Like the siting of underground waste storage, the dumping of waste at sea was, in the government's eyes, to be underpinned by scientific evidence and rationality. Britain dumped low-level radioactive solid waste 500 miles southwest of Land's End, a site recommended by the OECD. Greenpeace, which had already protested at Windscale and Capenhurst, and occupied the parts of the Torness power station construction site, made the annual dump, now the responsibility of NIREX, a target of environmental campaigning. In 1983 they won an effective ally to their cause: the National Union of Seamen, with support from other unions, instructed its members not to sign on for the dump ship, MV *Atlantic Fisher*. The drums of waste, containing mostly concrete and steel packaging as well as radio-isotopes from medical uses, power stations, civil and military establishments, sat on trains parked in Bicester (near Harwell), Thatcham (near Aldermaston) and Winfrith (the site of UKAEA reactors). 'If the dump is prevented this year, it will be almost impossible to resume next year,' warned one civil servant to another, adding that 'Greenpeace will consolidate a victory won on non-scientific grounds'.[49] Yet, with strong support for the action shown at the Trade Union Congress, the block of the 1983 dump was indeed successful.

The government's attempt to contain the issue was hampered severely by leaks. Greenpeace received a document from an interdepartmental meeting concerning dumping at sea of plutonium waste, and immediately publicised it at a press conference on 1 September 1983. This was the start of a mode of challenging government authority by the means of leaks: the following month the FCO civil servant Sarah Tisdall leaked the arrival dates and Heseltine's publicity handling tactics of the US cruise missiles at Greenham Common (she would be sentenced to six months in jail in 1984), while in March 1984 Clive Ponting passed to Tam Dalyell two documents on the sinking of the *Belgrano* (and successfully pleaded public interest – the jury did not agree with the judge's instruction that 'the public interest is what the government of the day says it is'). With the support of Robert Armstrong, Cabinet Secretary, investigation of the leak of the sea dumping documents was handed over to the police and the Director of Public Prosecutions.[50]

Also, as the Minister for the Environment complained, the appeal to science was faltering:

> There is some prospect of changing the TUC's attitude and I am meeting them for that purpose ... But this could not be achieved before next year at the earliest and might have to await completion of the further review of scientific evidence within the London Dumping Convention [the international framework under which dumping was permitted]. Of course, the unions' view is completely unsupported by scientific evidence.[51]

There was a faint hope that conceding dumping at sea in favour of disposal on land would 'give a fairer wind' to NIREX's land-based proposals, 'which are vital to the continuing credibility of nuclear power in this country'.

This situation of leaks, controversy over sea dumping and concern about public credibility was the context in which news of the unauthorised discharge from Sellafield in November 1983 became public. Ministers fielded questions in Parliament with four investigations ongoing: the Sizewell inquiry, one from the Radiochemical Inspectorate of the Ministry of Environment and one (unpublicised) from the Nuclear Installations Inspectorate of the Health and Safety Executive, both into Sellafield discharges, plus a fourth one on leukaemia clusters to which I will turn shortly. The first two concluded with the Sellafield management being criticised and the case being turned over to the Director of Public Prosecution – the body now pursuing two Sellafield cases, one relating to the leak of documents and the other against the management that had allowed up to 4,500 curies of radioactive material to be discharged down the pipeline to the Irish Sea and the beaches of Cumbria.[52]

The public contradictions drew press comment. A particularly amusing one came from John Twidell, a physicist at the University of Strathclyde, who wrote to *The Times* pointing out that in consecutive sentences Patrick Jenkin, Minister of Environment, had told Parliament that the radioactivity of the beach effluent was below the level that would constitute 'any hazard to the general population in the area' and that the handling of such substances 'could exceed the annual dose limit for the skin'.[53] 'Obviously 1984 has arrived,' said Twidell, invoking George Orwell; such 'statements are a clear case of 2 + 2 = 5'. Remarkably, Jenkin replied with his own letter to *The Times* editor, saying that he had meant it: the radioactive material was safe to the general public, unless the public actually handled it.[54]

The fourth ongoing investigation was by Sir Douglas Black into clusters of childhood leukaemia. Black was a prominent physician and previously had authored a 1980 report into health inequalities that had not been warmly welcomed by Thatcher's government. Following the airing on 1 November 1983 of the Yorkshire Television programme *Windscale – the Nuclear Laundry*, which suggested the 'possibility of a link' between an apparent cluster of cancer cases, particularly five cases of leukaemia in children, to Sellafield,[55] Black was appointed to lead a working group investigating the issue. Their task was to examine the evidence of clustering of cancer cases, consider the need for further research and make recommendations.

The group collected and read existing studies (of which there were nine, from the 1950s to the 1980s) that rested on data from various sources, including an Office of Population Censuses and Surveys examination of death certificates in the Copeland District (the southwest administrative area of Cumbria around Sellafield), and both the Manchester and Northern Children's Tumour Registries. In July 1984 government received Black's report.[56] The result was consternation. Black had concluded that there was unusual, statistically significant incidence of leukaemia in the village of Seascale, but also that the incidence could not be explained by the combination of background radiation and known discharges from Sellafield. Furthermore, as the Secretary of State for Energy, Peter Walker, told his Prime Minister: 'Black's conclusions imply the possibility of some kind of link for which there is no evidence (an <u>unplanned</u> emission which was <u>undetected</u> by the monitors and affected the population by an <u>unsuspected</u> route)'.[57] (Indeed the committee set up to investigate the clusters further soon received evidence of discharges that had not been known to Black's group.)[58]

Walker was concerned that publication 'could provide a propaganda success for opponents of nuclear power'; Number 10 civil servants described the report as 'imprecise and not at all reassuring' and a 'an unsatisfactory outcome which will not reassure public opinion'.[59] Three observations can be made of the government's initial response to the Black report. First, while seen as imprecise and unhelpful, no attempt was made to suggest that Black, whose politics ran counter to Thatcher's, had politicised his science. Second, the appeal to scientific evidence did not settle the issue and the significance of clusters would remain controversial into the 1990s.[60] Third, the empirically justified uncertainty in Black's results could not only be seen as potentially undermining public confidence, but also deployed to perform an opposite action. Thatcher, for example, was briefed before her Sellafield visit that an 'independent inquiry by Sir Douglas Black into claims about an increased incidence of cancer in the vicinity of Sellafield found no evidence of any general risk to health for children or adults living near the plant when compared to the rest of Cumbria and gave a "qualified reassurance" to the local people'.[61] The science was Janus-faced.

Margaret Thatcher chaired the meeting at Downing Street on 24 July 1984 to discuss the interconnecting issues of Sellafield, health and waste strategy. Present were all the relevant ministers, as well as her Chief Scientific Adviser, Nicholson. Rather than poor management, it was now the age of the Sellafield plant that was seen as the root cause of 'current levels of discharge … higher than those from any other reprocessing plant in the world'.[62] They noted Black's recommendation of a critical review of the need for discharges at their present level, but also, responding to their constituents' concerns, the pressure from backbenchers for action. Yet there was also frustration expressed at the mismatch between scientific advice and public concern:

> On the one hand, it was argued that there was no scientific case for any further reduction, since discharges were already comfortably within the agreed international standards, which were themselves very low. There were many alternative uses for the money which would make a higher contribution to the nation's health. On the other hand, it was argued that international standards were virtually certain to reduce further … ; that questions of public confidence were just as important as scientific facts; …

Thatcher paid great attention to Nicholson's advice, and subsequently also asked and received from him a separate briefing on the 'distinction between alpha, beta and gamma radiation … [as] this is fundamental to the strategy for reducing discharges'.[63] However, with public confidence judged to be a more important factor than science, the meeting endorsed, said Thatcher, the introduction of the relatively modest, although still expensive, remedial technologies – the ones she would see and open at her Sellafield visit, described above.[64] They would be announced as part of a deliberately low-key response to the Black report. On the broader question of the storage of waste the meeting lost its nerve, accepting the point that 'a major change in the Government's approach in these matters', such as the multi-site inquiry that NIREX had proposed and Jenkin supported, 'required further and deeper consideration. There was a danger that by multiplying the number of sites the Government would simply multiply the opposition to any land disposal of waste: an alternative approach would be store waste only at existing nuclear sites' (Lord Marshall promoted this option).[65]

Decision was pushed back, again, by months. In December 1985 further ministerial exchanges on the selection of four sites for further investigation took place, with a possible view of proceeding with one or, given defence interests in a coastal site, two. Sites mentioned were Elstow (Bedfordshire) again, Fulbeck (Lincolnshire, not far from Thatcher's birthplace of Grantham), Bradwell (on the Essex coast), the privately owned Woburn Estate near Ridgemont (also Bedfordshire) and later South Killingholme (also coastal, in Lincolnshire).[66] Investigation of all except Woburn proceeded. In March 1986, the Prime Minister was fending off local MPs with binders of angry letters from constituents complaining specifically about NIREX's high-handed attitude, but more generally that nuclear waste might be stored in their backyards.[67] However, within days a much more serious challenge to the nuclear project erupted.

Chernobyl and Sizewell

On 26 April 1986, reactor no.4 at the Chernobyl nuclear power plant ruptured during a system test, explosively starting a fire in its graphite core.[68] Plumes of radioactive material rose high and drifted westwards. Outside the Soviet Union, Swedish nuclear scientists first raised the alert on 28 April when their monitors detected the radioactivity. In London, Whitehall began to buzz with hurried intelligence briefings and telegrams sent back and forth detailing evacuation plans for British nationals. On 2 May 1986 Thatcher, who was in Seoul, read two lengthy documents of advice, both sent by emergency telegram, the first from Lord Marshall, the leading figure in the UK nuclear industry, and the other from her new Chief Scientific Adviser, John Fairclough.[69]

Marshall compared the Chernobyl design to those of reactors in Britain. 'The reactor it least resembles is the PWR which we are proposing to build at Sizewell,' he noted, of his favoured project.[70] The one most similar was the SGHWR – like Chenobyl it was a boiling water, pressure tube design, but it used heavy water rather than graphite to moderate the chain reaction – that had been 'passionately advocated' by Marshall's rival, Frank Tombs, as well as the South of Scotland Electricity Board. Marshall reminded Thatcher that the SGHWR had been abandoned by the Labour government, on Marshall and John Hill's advice, because it 'failed to pass British safety rules'. In other words: a 'very much better reactor concept failed to get safety approval in the United Kingdom, but the poorer Russian design got safety approval in Russia and 27 reactors of that type are now operating in Russia'. Marshall suspected that the devastating but secret nuclear accident in Kyshtym in 1957, almost coinciding with the Windscale fire, was of a similar type.

'I am sorry to tell you,' Marshall also informed Thatcher, 'that, this morning, for the first time, we detected fall-out from the Chernobyl reactor with our monitoring instruments in Kent.' While he thought the 'levels of contamination are, of course, very low and do not pose a health hazard to the population', their 'psychological effect will, however, be very large'. Indeed, it was public perceptions that commanded Marshall's attention:

> Clearly this is a big setback for nuclear power. In my public speeches I am stressing the difference between our safety rules and that of the Russians and I am using the SGHWR story … to demonstrate that my arguments are not based simply on assertion but are based on historical fact. I have been pleased by the way people have received my arguments. I believe informed commentators and opinion formers think it is intrinsically plausible that the Russians have different and lower standards than ourselves. I am therefore hopeful that a massive public presentation campaign with the support of Government will retain the overall tolerance of the British public. However, we must expect greater local resistance to the siting of power stations …

Fairclough, on the same day, told Thatcher that the 'Chernobyl accident provides an opportunity to test existing models of reactor safety, and the effects of a major accident, against real data', so long as the Soviets could be persuaded to release technical information.[71] He could add to Marshall's Kentish evidence the news that 'monitoring in East Anglia has today picked up increased levels of radioactivity in the grass and there may be further fall-out with some contamination of food crops and water supplies'. Fairclough, like Marshall, also noted that there was 'certain to be an adverse effect on public attitude to all issues of nuclear power, including radioactive waste management', and the Government would want to 'reassure the public'. But, he strongly urged, ministers must 'resist the temptation to make any categoric statements such as "a similar accident could not possibly happen in the UK"', while the data was missing.

But the British areas of contamination from Chernobyl followed the contingencies of where rainfall, carrying the radioactivity down to earth, happened to be heaviest as the plumes passed. By chance, one such major area was Cumbria, around Sellafield. News of the high Cumbrian measurements travelled from the Ministry of Agriculture, Fisheries and Food

Incident Room to Thatcher on 5 May: 'This is not to do with Sellafield. It is the result of the Chernobyl incident'.[72] 'But who will believe this?' Thatcher has written by this comment. Restrictions on the movements of milk, water and sheep seemed imminent.

The first days of the government's response to fall-out were, as a hurried review admitted, uncoordinated and unedifying.[73] Senior figures, including Thatcher and her press secretary Bernard Ingham, were in the Far East. Whitehall had plans for an emergency arising from a UK nuclear installation, but not for one arising from a foreign source. 'Anxious telephone callers inundated MAFF and seriously hampered communications' while ministers squabbled; one of them, William Waldegrave, gave out the number for the Department of the Environment car pool on Radio 4, mistaking it for the emergency information hotline. A Number 10 Policy Unit adviser concluded that the 'ill-coordinated nature of the information and advice aroused rather than calmed public anxiety'. While the 'nuclear professionals' (perhaps thinking of Lord Marshall) 'performed satisfactorily', it was noted that 'without careful translation [their] language can be confusing and sometimes alarming to the public'.[74] One particular pair of contrasting statements exemplified what was seen as a failure of public understanding of science:

> Many people don't understand statistical probability, especially of very low order. For example, on the day that Kenneth Baker assured the public that the risks from the Chernobyl fall-out were insignificant, John Dunster, Head of the National Radiological Protection Board, was saying that the death toll in the UK would run to tens of people. Both conclusions derived from the same assumptions and analysis. Dunster was quantifying what he regarded as an insignificant risk. The next day he had to explain that the tens of deaths would arise over the next 30–40 years, during which time millions would die from cancer wholly unconnected with the Chernobyl incident.[75]

'Now that the initial pressure over Chernobyl is beginning to die away,' wrote a Number 10 official to the Department of Energy in early June 1986, barely a month since the accident, 'the Prime Minister has asked if your Secretary of State could consider how best to counteract the mistaken impressions which the affair has left about nuclear power'.[76] 'Timing is of great importance,' Peter Walker, Minister for Energy, told Thatcher, 'and one of the major impacts upon timing is our receipt of the Sizewell Inquiry Report within the next three months', while 'campaigns mounted without knowing the contents of that report might have problems'.[77] Nevertheless, he promised coordination and a 'major and effective longer term campaign' featuring a 'major think-piece speech which will put these matters into the correct historic and long term perspective' and mobilised through a gathering of 'all the nuclear interests into a major group to discuss ... public relations and advertising'. The Number 10 Policy Unit argued that evidence showed the most effective, practical way of 'winning public confidence' was 'to take people literally around the nuclear power stations'.[78]

Bernard Ingham, Thatcher's trusted adviser on public presentation, set out the general problems that were 'intensified' but 'not changed' by Chernobyl:

> 1.– winning and maintaining confidence in a fuel with, by definition, uncertain long term effects on people – but effects which could produce a painful, lingering death and/or, it is believed, deformities in the unborn;
> 2.– coping with well-organised pressure groups whose aim is not to improve the safety of the industry but to close it down.

He noted that these were 'complicated in this country' by, among other factors, 'the dominance in the nuclear power and reprocessing industries of scientists who have a passionate belief and confidence in their fuel and an insensitivity to public concern'.[79] The charge of insensitivity was unfair. Within UKAEA the 'need to consider the effect on the public image of the Authority as a consequence of (a) the Chernobyl accident, (b) concerted attempts by some groups to denigrate the Nuclear Industry and (c) the media's occasional one-sided and exaggerated view of minor incidents often because of leaked comments ... ' was a topic of urgent and anxious discussion.[80]

Extensive past initiatives of public relations were reviewed[81] and new actions proposed. Both in terms of the media and in terms of reactor safety, nuclear industry staff, from Lord Marshall through to the UKAEA press officers, believed the key word was *containment*. Not only 'Chernobyl', but also the 'apparent reduction in status of politicians, scientists, technologists ... are all part of the problem of convincing the media and through them the public that nuclear power is essential for the wellbeing of the country,' argued the UKAEA. 'This note compares the containment features of different types of reactor systems,' wrote Marshall, 'We hope that this will reassure the public about the safety of our own reactors.'[82]

This intuition – that the problem was a combination of the falling status of scientists and the solution was clearer public communication – was the core of the new public understanding of science (PUS) movement. However, it is an open question whether this political and technical crisis encouraged it at all. A committee under Walter Bodmer had reported on PUS in 1985, having begun deliberations as early as 1983, while in 1986 the UK's key institutions – the British Association for the Advancement of Science, the Royal Institution and the Royal Society – were beginning to discuss PUS initiatives.[83] It is striking that this activity was completely below the political horizon viewed from Number 10 or the commanding heights of the nuclear industry, despite apparently similar deep concerns about public confidence.

Ingham had thought that it did not make sense to plan a major public confidence campaign until the Sizewell Inquiry had reported. Interestingly, he added 'I assume that post-Chernobyl we shall not feel able to go ahead with PWR'.[84]

Thatcher received the draft Sizewell Inquiry report in December 1986. While she herself did not read the 13 volumes and 109 chapters her staff did, and they summarised them for her. Frank Layfield, the main author and chair of the Inquiry, had concluded that there were no serious doubts about safety; the economic benefits and security of energy supply justified the irreducible risks of building Sizewell B and that they outweighed environmental detriments. 'An accident at Sizewell B would almost certainly have tolerable consequences, at worst requiring measures such as the banning of milk near the station,' wrote Layfield in the draft report. He added 'Theoretically possible accidents which could cause hundreds or thousands of deaths would almost certainly not occur' and 'It can be inferred that Sir Frank would endose the CEGB's contention that a chain of shortcomings akin to that which led to Chernobyl would never happen here'.[85] 'The Sizewell Inquiry Report', one of 'record-breaking duration and detail', as Thatcher was informed, 'justifies the wait'.[86]

Overall, the report was read with relief (the historical judgement has not been so kind).[87] The nuclear project was important for many reasons, and intersected with other issues, such as the action on acid rain discussed in Chapter 7.[88] The only gloomy note was that Layfield suggested that the construction of any further PWR station would have to be preceded by a further expensive inquiry. Indeed in 1988, when the question of a new PWR programme was being considered, the question of inquiries, their cost and public attitudes following Chernobyl were interlinked. 'Had Chernobyl not happened', then cheap, local inquiries, starting with Hinkley Point C, might have proceeded, considered

Norman Fowler. However, because of 'Chernobyl, it appears that the nuclear safety issue must have another public airing'.[89] With regards to the Sizewell report, publication was set for January 1987. Peter Walker was 'anxious to prevent leaks before publication'; he was 'maintaining the tightest security in his Department and proposes similar highly-restricted circulation' elsewhere.[90] Government approval for Sizewell B was given in 1987 and the reactor started in 1994.

Conclusion

This chapter has had two contrasting halves. The first tracked the visits of Margaret Thatcher to UK nuclear sites, one to the Scottish fast-breeder project at Dounreay and the other a multi-site encounter transported by helicopter. The main point is that these were displays of contained power: the technical, engineering control of the nuclear cycle and the controlled public representation of a prime minister at the peak of her political power. Each gave something to the other: the pro-nuclear Thatcher gave support to a controversial project, while the nuclear industry provided the backdrop for press photography of Thatcher in a white coat. The identification between the two was tight: the helicopter flight tracked the cycle of nuclear material, while the photograph of her pressing the button to start reprocessing combined engineering and political control in one image. But we might wonder just how much autonomy does even the prime minister have within these tightly closed cycles of public representation and nuclear engineering?

In the second half of the chapter, I have stressed the limits to this control. Leaks undermined the message of safety, photographs were repurposed as satire, issues of waste disposal and disease were not settled by appeals to science and campaigning groups and unexpected external events, not least Chernobyl, challenged the containment of messages about safety and policies for future nuclear expansion.

For Anthony Giddens, writing in 1985 in *The Nation-State and Violence*, perhaps the most influential political sociology text written in Thatcher's Britain, power, metaphorically, was something contained. Furthermore, the containment of power was linked to 'locales': 'the settings of interaction, including the physical aspects of setting – their 'architecture' – within which systemic aspects of interaction and social relations are concentrated'.[91] So while, for example, castles, manorial estates and cities had performed the role in earlier societies, now business firms, schools, universities, hospitals, prisons and, above all, the nation-state were the 'power containers' of the modern world. What was distinctive about the nation-state, said Giddens, was that the nation-state power container was a bounded and unified one, in which the 'administrative purview corresponds exactly to its territorial delimitation'.[92] Within its boundaries power was 'generated' by a number of distinctive mechanisms.[93]

The overriding image is of the nation-state as a battery, driven by cells of power – or even, when, in his magpie-like way Giddens picks up Talcott Parsons' notion of 'power deflation', perhaps a gasometer that is powerful when at full capacity but becomes less governable when dissipated.[94] Parsons developed a particular theory of power: one that attempted to show that power could be both consensual and coercive, could be produced anew rather than merely being a 'zero sum game' and was a symbolic medium whose structural role was to enable effective collective action.[95] As a systems theory it was a product of twentieth-century engineering, with a dash of 1960s cybernetics. But Giddens also pointed to some significant features of power that will interest us here: that it is a symbolic system and that political leaders were the creators of new power in a manner analogous to a sound bank issuing new credit. It was symbolic power that was dispensed, and to some extent earned, in the visits to nuclear sites by Thatcher.

The identification of Thatcher with nuclear power stands in a much longer tradition of political machinery. I have traced elsewhere how the nineteenth-century journalist Walter Bagehot, in his 1867 classic *The English Constitution*, likened the political system to a steam engine, with 'regulators' and 'safety valves' ultimately powered by the 'potential energy' of Queen Victoria.[96] Before Bagehot, checks and balances and automata provided mechanical metaphors for different forms of political power. Energy, power and waste were concepts that developed both as political and technical languages. Containment, whether mundane as in a battery cell or large-scale and modern as in a nuclear plant, was a twentieth-century variant, particularly prominent during the Cold War, as historian Paul Edwards has argued.[97] When containment at Chernobyl failed it was read as a metaphor for a crumbling Cold War superpower.

Commenting on Giddens's notion of contained power, the geographer Peter Taylor noted that the container in the 1980s was leaking.[98] He was thinking of the nation-state, and how it was being challenged from without by supranational entities such as the European Community and from within by the withering of state power through Thatcherite policies such as privatisation. But we have seen in this chapter how, despite the efforts to shore up and contain both political power and the nuclear project, they were seen to be undermined by leaks.

A touchstone belief of the protagonists of the UK nuclear project was that the technology was controllable, under expert command, and therefore safe to deliver power to the nation. 'The principal problems of nuclear power,' declared UKAEA chief Sir John Hill in 1977, 'are not now engineering or technology but problems of political will and public acceptability.'[99] He would later guide Thatcher around Sellafield.

Thatcher's visits to nuclear sites were choreographed displays of political will conducted in the name of public acceptability. Sociologist Brian Wynne, in his ground-breaking study, began with this quotation from Hill; he went on to argue that public inquiries, such as those into Windscale and Sizewell, were rituals that sought to delimit public debate to matters of fact rather than allow expression of the full range of social and emotional responses.[100] Inquiries were intended, says Wynne, as tools to control and contain public discourse in the face of anti-nuclear opposition. But control had its limits. Neither Windscale nor Sizewell inquiries settled the nuclear question. On a smaller scale, I have shown how Black's inquiry into the evidence for leukaemia clusters, the geological surveys into potential waste disposal sites or the attempts to save dumping at sea, while all appealing to the rationality of science, did not serve to close these controversies.

Despite all the efforts to exert control, the public acceptability of the nuclear remained volatile. The public understanding of science movement, whose growth in the 1980s I have noted, conceived the general issue as one of a deficit of knowledge among the public. Lord Marshall, after Chernobyl, thought that public acceptance could still be won by a controlled, expert delivery of facts. I will give one more example. In 1987, and again in 1988, when 'the subject certainly has increased pertinancy following the PM's recent Royal Society address', a Dr Eric Voice, resident of Thurso, wrote to the UKAEA with a proposal.[101] 'What single factor operates against the harmonious adoption of nuclear electricity generation?' Voice asked, rhetorically, before giving his answer: 'It is obvious that this factor is neither technical nor economic … but rather the world-wide public perception of risk'. Voice suggested an experiment:

> There is a large literature … on the possibility that radiation around 'background' is not only harmless but positively beneficial, and even essential, for healthy life. If such a fact could be established with wide media publicity, the greater part of public opposition to nuclear power would vanish. To this end, take a large and well-characterised group of experimental animals (several thousand mice?), let half the group live and breed for generations in a

'normal background' environment, let the other half live and breed … within a deep cavern in limestone or chalk, and on a diet containing only isotopically-separated ^{39}K.

With the fact of healthy, long-living radiated mice established and publicised, Voice argued that the informed public would accept nuclear power. It might cost £500,000, but, as Voice put it: 'Half-a-million to start the swing of public opinion in Britain and world-wide in favour of nuclear power and waste disposal? It would be the most worthwhile sum that the UKAEA could ever spend'. Voice received a rejection letter, but a polite one – not least because he was so evidently pro-nuclear.

Perhaps Voice's suggestion was not so far-fetched. Plenty of money was expended in the 1980s to produce matters of fact about nuclear safety, and there was a widespread belief that an absence of public knowledge was a cause of the problem. But ultimately it was not to be facts about safety, nor the stemming of leaks, that settled the future nuclear project, but rather the corrosive effect of a different kind of calculative rationality: the economic assessment of costs that were disclosed during moves towards privatisation, the subject of the next chapter.

Notes

1. Dieter Helm, *Energy, the State, and the Market: British Energy Policy since 1979*. Oxford: Oxford University Press, 2010 (revised edition), p.130.
2. Margaret Gowing, *Britain and Atomic energy, 1939–1945*. London: Macmillan, 1964. Margaret Gowing assisted by Lorna Arnold, *Independence and Deterrence: Britain and Atomic Energy, 1945–1952*. Vol.1: *Policy Making* and Vol.2: *Policy Execution*. London: Macmillan, 1974. Walter C. Patterson, *Going Critical: an Unofficial History of British Nuclear Power*. London: Paladin Books, 1985.
3. A. M. Judd and K. F. Ainsworth, 'Fast reactors in the U.K. 1946–1996', *Energy* 23(7–8) (1998): 609–17.
4. AB 48/1640. 'UKAEA. Dounreay Experimental Reactor Establishment', undated (1969).
5. Helm, *op. cit.*, p.92.
6. Mike Parker and John Surrey, 'Contrasting British policies for coal and nuclear power, 1979–1992', *Energy Policy* 23(9) (1995) 821–50.
7. PREM 19/1982. 'AGR v PWR debate', undated (October 1985).
8. For the inquiry, see: Brian Wynne, *Rationality and Ritual: the Windscale Inquiry and Nuclear Decisions in Britain*. Chalfont St Giles: British Society for the History of Science, 1982.
9. John Minnion and Philip Bolsover, eds, *The CND Story*. London: Allison & Busby, 1983. Holger Nehring, 'Cold War, apocalypse and peaceful atoms. Interpretations of nuclear energy in the British and West German anti-nuclear weapons movements, 1955–1964', *Historical Social Research* 29 (2004): 150–70.
10. Lawrence S. Wittner, 'The forgotten years of the world nuclear disarmament movement, 1975–78', *Journal of Peace Research* 40 (2003): 435–56.
11. 'Hear the women's voices', in Barbara Harford and Sarah Hopkins, eds, *Greenham Common: Women at the Wire*. London: Women's Press, 1984.
12. AB 48/1640. UKAEA press release, 1979.
13. AB 48/1640. 'Memorandum from Dounreay Staff Side', undated (1969).

14. AB 48/1640. 'UKAEA. Dounreay Experimental Reactor Establishment', undated (1969).
15. AB 48/1640. Maclennan to Wilson, 3 July 1969.
16. AB 48/1640. UKAEA press release, 'The Prime Minister's visit to Dounreay – 6 September 1979'.
17. AB 48/1640. Dounreay Nuclear Power Development Establishment, 'Programme for the visit of the Prime Minister, the Rt Hon Mrs Margaret Thatcher, PC, MP, on Thursday 6 September 1979'.
18. AB 48/1640. Faxed message, Vey to Blumfield, 3 September 1979.
19. AB 48/1640. 'Note of meeting with the Press', 10 September 1979.
20. *Daily Express*, 7 September 1979. https://www.margaretthatcher.org/document/103900.
21. AB 48/1640. 'Note of meeting with the Press', 10 September 1979.
22. AB 48/1640. UKAEA, 'Background notes for the visit of the Prime Minister to Dounreay Nuclear Power Development Establishment', 30 August 1979. Solid waste was segregated 'by specially developed instruments' and then placed into silos prior to probable burial. Low-level liquid waste was discharged to sea by a 600m pipeline. The plan for high-level waste was storage followed by vitrification and then 'permanent disposal'. 'A small amount of gaseous activity is released to the atmosphere through filters via stacks which are monitored,' the Prime Minister was briefed, 'The gases are mainly short life Argon, a natural constituent of air, which surrounds the reactor vessel, and Krypton from the reprocessing of fuel. These discharges are controlled within limits agreed with the Scottish Development Department'.
23. Doctored photograph at the base of the 'Ariadne' column, *New Scientist* (25 September 1980), p.976. My thanks to Dominic Berry for drawing my attention to this image.
24. PREM 19/1892. Allday to Thatcher, 29 October 1984.
25. Hansard, Nuclear Waste (Sellafield), HC Deb 21 November 1983, vol.49 cc19–22, for Waldegrave (USS Environment) answering on behalf of Patrick Jenkin, confirming SIXEP. PREM 19/1892. Nicholson to Thatcher, 30 October 1985: 'Senior BNFL management, both privately and publicly, have occasionally expressed doubt about the need for such investment in expensive plant'.
26. PREM 19/1892. Harborne to Barclay, 8 November 1984.
27. PREM 19/1892. Nicholson to Barclay, 9 November 1984.
28. Susanna Schrafstetter and Stephen Twigge, 'Spinning into Europe: Britain, West Germany and the Netherlands – Uranium Enrichment and the Development of the Gas Centrifuge 1964–1970', *Contemporary European History* 11(2) (2002): 253–72. Stephen Twigge, 'A baffling experience: technology transfer, Anglo-American nuclear relations and the development of the gas centrifuge 1964–70', *History and Technology* 19(2) (2010): 151–63.
29. PREM 19/1892. Caines to Thatcher, 14 October 1985.
30. The underlining here is Thatcher's.
31. PREM 19/1982. Wybrew to Addison, 29 October 1985.
32. PREM 19/1982. 'AGR v PWR debate', undated (October 1985).
33. PREM 19/1982. Nicholson to Thatcher, 30 October 1985.
34. Hunter Davies, ed., *Sellafield Stories: Life with Britain's First Nuclear Plant*. London: Constable, 2012, p.292. For local response to Greenpeace's Sellafield campaign, see also p.38.
35. PREM 19/1892. Caines to Thatcher, 14 October 1985.
36. PREM 19/1982. Ingham to Thatcher, 29 October 1985.
37. PREM 19/1982. Ingham to Thatcher, 29 October 1985. This line was used, for example in the interview in *BNFL News*.
38. Departments of the Environment and Transport, *Review of Research on Radioactive Waste Management and Radioactivity in the Environment*, Research Report 32, 1979, p.11.
39. Departments of the Environment and Transport, *Review of Research on Radioactive Waste Management and Radioactivity in the Environment*, Research Report 32, 1979, p.11.
40. Departments of the Environment and Transport, *Review of Research on Radioactive Waste Management and Radioactivity in the Environment*, Research Report 32, 1979, p.3.
41. Royal Commission on Environmental Protection, Sixth Report, *Nuclear Power and the Environment*, Cmnd 6618. London: HMSO, September 1976.
42. *Nuclear Power and the Environment. The Government Response to the Sixth Report of the Royal Commission on Environmental Protection*, Cmnd 6820. London: HMSO, May 1977.
43. PREM 19/1791. 'Draft answer', July 1979. Desk research by NERC's Institute of Geological Sciences had indicated that 'potentially suitable rocks lie under about 16% of the land area of the country'. The provisional list of areas suitable for further investigation included land

in Cheshire, Cumbria, Grampian, Gwynedd/Powys, the Highlands and Western Isles of Scotland, the Leicestershire/Nottinghamshire border, Northumberland, Somerset, Strathclyde and Worcestershire.

44. PREM 19/1791. Thatcher's handwritten comments on Bristow to Pattison, 23 July 1979.
45. PREM 19/1791. King to Secretary of State, 25 June 1982.
46. PREM 19/1791. 'Radioactive waste management. Main points in the white paper', undated (July 1982).
47. A good source for the range of views in opposition to Billingham and Elstow is the House of Commons debate, Hansard, 'Nuclear Waste (Disposal), 8 March 1985.
48. PREM 19/1791. Jenkin to Thatcher, 23 July 1984. Jenkin also proposed using a Special Development Order so that investigative work could begin; this itself would allow for Parliamentary debate.
49. PREM 19/1791. Ballard to Scholar, 28 July 1983. 'There will be repercussions on AWRE, Harwell and Amersham International in particular, who rely on prompt disposal in managing their wastes'. Another example of market forces disrupting the old Radiochemical Centre.
50. PREM 19/1791. Jenkins to Dowling, 5 October 1983.
51. PREM 19/1791. Jenkin to Walker, 19 October 1983.
52. PREM 19/1791. Jenkin to Thatcher, 19 December 1983.
53. Letter to the editor, *The Times* (30 December 1983).
54. PREM 19/1791. Jenkin to editor of *The Times*, 30 December 1983.
55. Statement by James Cutler (Producer of *Windscale – the Nuclear Laundry*, Yorkshire Television) to the Black Advisory Group, 12 December 1983. Cutler had begun the project as an investigation into workers' health at Windscale before having his attention drawn to the incidence of cancers in children around Seascale.
56. Douglas Black (chair), *Investigation of the Possible Increased Incidence of Cancer in West Cumbria. Report of the Independent Advisory Group*. London: HMSO, 1984. For an investigation into the media representation, local public responses and 'languages of risk', see: S. M. Macgill, *The Politics of Anxiety: Sellafield's Cancer-link Controversy*. London: Pion, 1987.
57. PREM 19/1741. Walker to Thatcher, 17 July 1984.
58. The Committee on Medical Aspects of Radiation in the Environment (COMARE) was set up on the recommendation of the Black report. 'The main new evidence [reaching COMARE] comes from Dr Jakeman, and stems from his detection of high radiation levels in his garden in 1954/5'. PREM 19/1741. Addison to Thatcher, 18 July 1986. While COMARE were able to satisfy themselves that the newly known discharges did not account for the increased incidence of cancer, the evidence came 'at an unfortunate time, after Chernobyl. It will be helpful to those wishing to scare people about nuclear power and about Sellafield in particular'. Jakeman is an interesting case of semi-citizen science.
59. PREM 19/1741. Barclay to Thatcher, 16 July 1984. Pascall to Barclay, 16 July 184.
60. AB 45/323 contains some of the conflicting claims, with more sophisticated methods for varying cluster size (the 'Geographical Analysis Machine') being deployed to undermine cluster claims, while Gardner, a member of COMARE, published results in the *British Medical Journal* in 1990 that he argued showed linkage.
61. PREM 19/1892. 'Sellafield. Points to make', undated (October 1985).
62. PREM 19/1791. Barclay to Ballard, 25 July 1984.
63. PREM 19/1741. Nicholson to Thatcher, 20 July 1984, on options. Nicholson to Thatcher, 27 July 1984, on alpha, beta and gamma radiation, both with many underlinings.
64. PREM 19/1741 'Options for reducing Sellafield discharges. Paper by officials', July 1984, sets out the options. Option 1 (£280 million) was 'relatively low-cost' and involved a new floc precipitation plant; option 2 (£290 million) was the same but accelerated; option 3 (£525 million) added the removal of beta/gamma activity from all effluents and was the form of treatment used in the French standard comparison of Cap de la Hague; option 4 (£2,510 million) was considerably more expensive, used evaporation and solidification rather than floc precipitation and would exceed any other reprocessing plant then built in terms of reducing discharges. They are summarised for Thatcher in Pascall to Barclay, 20 July 1984. The meeting of 24 July 1984 endorsed option 1, with an exploration of the possibility of acceleration.
65. PREM 19/1741. Barclay to Ballard, 25 July 1984, records Jenkin's support.
66. PREM 19/1791. Baker to Thatcher, 4 December 1985, followed by other ministers' comments. Booth to Addison, 17 December 1985. H Committee decided on 17 December 1985 to proceed with geological exploration of Elstow, Fulbeck, Bradwell and South Killingholme. Before a

Special Development Order was issued, the Ministry of Defence was to 'make urgent inquiries about using a coastal site on MOD land for the disposal of MOD waste'. The MOD did indeed decide that neither coastal site met its needs. By this point the Environment Minister had changed his mind again, and decided that coastal sites were unsuitable.

67. PREM 19/1739. Contains details of meetings between Douglas Hogg MP as well as Hogg's collection of constituents' letters, including one about a confrontation in the Hare and Hounds between NIREX's Deputy Chairman, Mr Ginniff, and locals: 'Trust is earned and Mr Ginliff certainly managed to destroy all trust we could possibly have had'. Wicks to Young, 21 March 1986: 'The prime minister asked … whether his constituents were aware that much of the waste concerned had relatively low levels of radioactivity … Mr Hogg confirmed that this was well known, but local reaction was emotional and impervious to logical argument'.
68. There is now a considerable literature on Chenobyl, most recently Serhii Plokhy, *Chernobyl: History of a Tragedy*. London: Allen Lane, 2018. Other major works include the garlanded 1997 oral history by Svetlana Alexievich, *Chernobyl Prayer: a Chronicle of the Future*, translated by Anna Gunin and Arch Tait. London: Penguin Modern Classics, 2016.
69. Thatcher also later read *New Scientist* articles to find out more about the Chernobyl design. Tom Wilkie and Roger Milne, 'Chernobyl: sorting fact from fiction', and Geoffrey Greenhalgh, 'The Soviet drive to nuclear power' (both 8 May 1986), were the articles read.
70. PREM 19/3656. Marshall to Thatcher, 2 May 1986. 'Passed to PM party in Seoul on 2 May by telegram'.
71. PREM 19/3656. Fairclough to Thatcher, 2 May 1986. Fairclough also asked for and was granted resources to scope the 'longer-term scientific and technical implications' of Chernobyl. PREM 19/3656. Fairclough to Addison, 7 May 1986, with signed approval from Thatcher, 14 May 1986.
72. PREM 19/3656. 'Radioactive milk', 5 May 1986.
73. PREM 19/3656. Wybrew to Thatcher, 16 May 1986.
74. AB 38/1972. 'Second meeting of the Chernobyl Issue Management Review Group', 12 May 1986, shows that Marshall was 'very angry' at not being 'able to speak for a united industry at a time of difficulty'. The quotations come from PREM 19/3656. Wybrew to Thatcher, 16 May 1986.
75. PREM 19/3656. Wybrew to Thatcher, 16 May 1986.
76. PREM 19/2134. Norgrove to Dart, 2 June 1986.
77. PREM 19/2134. Walker to Thatcher, 5 June 1986.
78. PREM 19/2134. Wybrew to Thatcher, 6 June 1986. Wybrew was quoting CEGB evidence provided for the Environment Select Committee Report on Radioactive Waste.
79. PREM 19/2134. Ingham to Norgrove, 11 June 1986. The role of the scientists was Ingham's first listed complicating factor; the others were the failure of the nuclear industry to grasp the issue of public confidence, the different interests of Government departments and a system of incident reporting that 'plays into the hands of active effective and highly political pressure groups'.
80. AB 38/1972. C. W. Blumfield. 'Possible Authority actions to improve the public image of nuclear power', 16 May 1986.
81. Past initiatives had included lectures, using the widely circulating house journal *Atom News*, 'mobile, local and permanent exhibitions', 'Assistance … given to the Science Museum', providing consultants for television programmes and preparing 'large amounts' of 'presentational literature for visitors, schools and interested organisations'.
82. AB 38/1972. Marshall, 'The accident at Chernobyl – the containment issue', 12 May 1986.
83. Simon J. Lock, 'Lost in translations: discourses, boundaries and legitimacy in the public understanding of science in the UK', PhD thesis, University College London, 2008. CAB 134/4757. Minutes, STO(83)6th,13 September 1983, shows that civil servants responsible for science had begun to pay attention to 'public understanding of science' as an agenda item.
84. PREM 19/2134. Ingham to Norgrove, 11 June 1986.
85. Quoted in PREM 19/2134. Wybrew, 'Synopsis of the Sizewell Inquiry Report', undated (November/December 1986).
86. PREM 19/2134. Wybrew to Thatcher, undated (November/December 1986).

87. Helm, *op. cit.*, p.104, gives a damning summary of the Sizewell report's 'mistakes': 'the limiting of the choice of technology to that between coal and PWRs, and the associated failure to anticipate the move to gas ... The report lacked serious consideration of environmental problems and had little by way of a detailed examination of the waste issues and the related THORP and MOX consequences, relying instead on the assumption that THORP would provide a viable waste-management option. The significant first-of-a-kind costs were not fully appreciated, despite the evidence of Magnox and AGRs. This was in part due to the further assumption that Sizewell B would be followed by other PWRs. Finally, and crucially, ... the assumptions about fuel prices were flawed, despite the fact that, by the time Layfield had finished, the oil price had collapsed'.
88. 'If CEGB (perhaps following a rough ride at Sizewell) get out and dust off their standard design for a coal-fired station, they will have put something in it to reduce sulphur. Even if this adds another 15% to capital cost'. AT 82/203. Wedd to Harrop, 23 July 1982.
89. PREM 19/2136. Fowler to Parkinson, 14 April 1988.
90. PREM 19/2134. Wybrew to Thatcher, undated (November/December 1986).
91. Anthony Giddens, *The Nation-State and Violence*. Cambridge: Polity Press, pp.12–13.
92. Giddens, *op. cit.*, p.172.
93. These would include: 1) those that increased the scope of 'time-space distanciation', such as the mechanisation of transportation, the severance of communication (such as telegraphy) from transportation (together enabling time–space convergence, essentially an increased ability to coordinate and timetable), an upsurge in the collection and use of administrative information, which in turn, reflexively, constituted new social objects, ie objects of social study and knowledge; 2) those that increased internal pacification, including not only Foucauldian supervision, but also direct supervisory relations, and the reservation of the prerogative of violence to the state.
94. Giddens, *op. cit.*, p.202. Parsons' deflation (his example was McCarthyism) was in fact developed as part of an extended comparison of money and power as generalised symbolic media.
95. Talcott Parsons, 'On the concept of political power', *Proceedings of the American Philosophical Society* 107(3) (1963): 232–62. 'Power then is generalised capacity to secure the performance of binding obligations by units in a system of collective organisation when the obligations are legitimised with reference to their bearing on collective goals and where in case of recalcitrance there is a presumption of enforcement by negative situational sanctions.' For other theories of power, see: Steven Lukes, ed., *Power*. Oxford: Basil Blackwell, 1986.
96. Jon Agar, *The Government Machine*. Cambridge, MA: MIT Press, 2003, pp.31–2.
97. Paul Edwards, *The Closed World: Computers and the Politics of Discourse in Cold War America*. Cambridge, MA: MIT Press, 1996, p.10.
98. Peter Taylor, 'The state as container: territoriality in the modern world-system', *Progress in Human Geography* 18(2) (1994): 151–62, p.158.
99. John Hill, 'The politics of nuclear power', *Atom* 243 (1977): 1–9, p.1.
100. Brian Wynne, *Rationality and Ritual: the Windscale Inquiry and Nuclear Decisions in Britain*. Chalfont St Giles: British Society for the History of Science, 1982. The Hill quotation is on p.2.
101. AB 45/323. Voice to Collier, 27 May 1987. Voice to Collier, 8 October 1988.

5
Radioactive privatisation

Privatisation

In 1979 nationalised industries contributed 10 per cent of the Gross Domestic Product. Privatisation, for Thatcher, was justified both economically and ideologically, making industry more efficient by increasing exposure to the market and rolling back socialism in the name of freedom. Major privatisations include the return into private hands of the recently nationalised British Petroleum (reducing government-held shares by 5 per cent to 46 per cent in October 1979), British Aerospace (51 per cent of government shares sold in February 1981, remaining shares sold in 1985, retaining a single 'Golden Share'), Cable & Wireless (nationalised in 1947, half of the government's share was sold in November 1981), the Radiochemical Centre (sold as Amersham International in February 1982), the British Transport Docks Board (sold as Associated British Ports in February 1983), Jaguar cars (August 1984), British Telecommunications (the telecoms side of the Post Office had been split off as BT in 1981, and over half of the government's shares sold in December 1984), British Gas (December 1986), British Airways (February 1987), Rolls-Royce (May 1987), the British Airports Authority (July 1987), British Steel (December 1988), the many regional water companies (December 1989) and regional electricity boards (a month after Thatcher's resignation, December 1990. Non-nuclear electricity generation followed in March 1991, with PowerGen and National Power being carved from the Central Electricity Generating Board).[1] Research and development was integral to many of these companies.

Privatisation was an issue for very small as well as medium and large science-based organisations. For example, the National Collection of Industrial and Marine Bacteria (NCIMB) was one of 15 national collections of microorganisms. With origins in the 1950s, the NCIMB was held at the Torry Research Station in Aberdeen. The marine side of the collection was closely connected to Torry's work, including bacteria that spoiled fish or caused fish diseases, as well as marine microbes of general interest. The industrial side contained bacteria relevant to the rest of the food sector. With the Ministry of Agriculture, Fisheries and Food being asked to identify staff cuts, a proposal emerged to make the NCIMB a private company owned by the University of Aberdeen. Tam Dalyell, responding to an alarm raised by the United Kingdom Federation of Culture Collections, who were 'concerned about a transfer to a private company of national assets worth about £1 million', wrote to Thatcher directly, citing her statements, discussed in Chapter 2, of her being 'responsible for science'.[2] In this case, the government's response was delegated back to the ministry, but it is a good example of how her reservation of responsibility shaped the political response in practice. The privatisation was completed in 1982 when the NCIMB was transferred to the University of Aberdeen, and was much later spun out as an independent company in 2000.[3]

In this chapter I will explore how the policies of privatisation and those for science and innovation intersected through two case studies. The first takes a long view of the Radiochemical Centre. This example stands out because it was in many ways a trial of the policy of privatisation as a whole: it was early, it was completed smoothly (to produce Amersham International Ltd), it was a first full privatisation (although the government retained a sole 'special share') and lessons were learned for the much larger, flagship privatisations that followed. I show that, while the Radiochemical Centre had already taken steps towards the market in preceding decades, not least by operating a trading fund, privatisation involved a market economy disrupting other economies of exchange. Amersham is also intriguing, in the context of this volume, as an experimental privatisation because it was of a science-based body.

The second case study examines the faltering steps towards privatisation of other, larger parts of the British nuclear project, including the UK Atomic Energy Authority and the nuclear power bodies. I will suggest a possible influence of the shift in science policy, identified in Chapter 3, on the course taken. Privatisation of the nuclear also clashed with some of the environmental aspirations discussed in Chapter 7. In particular, following the vision set out in Thatcher's 1988 Royal Society speech, the desire to make commitments to stabilise carbon dioxide emissions at 1990 levels conflicted with the planned privatisation of the electricity industry, since it would add costs and make the shares less attractive to investors.[4] Privatisation was not a science policy *per se*, but consequences flowed in both directions.

Amersham

Radium was discovered by Marie and Pierre Curie in 1898. A powerfully radioactive element, which on decay produces radon, which is also radioactive, radium was soon used in a wide range of medical applications, from tumour treatments to more dubious quack medicines. It was, however, a very expensive substance. In 1929 a National Radium Trust and a National Radium Commission were established to build and coordinate radiotherapy centres across Britain.[5] With the establishment of the National Health Service, these functions would be taken over by the Ministry of Health. Typical prewar radiotherapy took the form of either the insertion of 'radium needles' carrying small quantities (1–5 mg of radium) directly into the tissue) or the use of 'mass radiation units', in which large quantities (5–10g of radium) generated wide beams of radiation in a fashion similar to x-ray tubes. Radon for tumour therapy was produced from radium in the form of capsules (known as 'seeds') that could be implanted in patients.

Before the Second World War radon was made at a local level, within hospitals, or by the Radon Centre at Barton-le-Clay in Bedfordshire. Radium sources later manufactured at Amersham for industrial radiography were typically mid-sized (250–500 mg). Finally, scientific research created a demand for more specialised radium products: as an ionising agent, as a radiation standard (1 μg and 500 mg) and as a source of neutrons (radium mixed with beryllium). These complemented the supply of other research sources – polonium, mesothorium and radiothorium.

Medical radiotherapy was practically suspended during the Second World War. Indeed, as one official wrote in 1940, 'the requirements of public safety have made it necessary to bury a large proportion of the radium stocks in this country'.[6] Instead a new demand for radium came from its use in luminescent instrument dials in military aircraft. But radium supplies from abroad could not be relied upon. It was in this context that the Ministry of Aircraft Production contracted Thorium Ltd to start a small radium refinery.[7]

In 1940 Dr Pat Grove, a chemist, had founded a private company called Thorium Ltd in the Buckinghamshire village of Amersham.[8] The company refined radium and produced for the Ministry of Supply luminescent paint for tank and aircraft instrument panels. By 1943 the laboratory was producing 'mesothorium, radiothorium, radium D, polonium, etc', in addition to 'large quantities of radioactive luminous compound'.[9] The wartime rate of radium refining was about 10 g per annum. After the war 'several grammes of radium were recovered from surplus radioactive luminous compound and old aircraft instruments'.

In 1946 the Ministry of Supply purchased the assets of Thorium Ltd, appointing the company as managing agents, and the Radiochemical Centre was established 'as a commercial enterprise under public ownership'.[10] Expansion and rebuilding of the site, using the Mobile Labour Force of the Ministry of Works, took place between March 1947 and March 1949. This expansion included the installation of radon extraction equipment, which used four grammes of radium, in operation by October 1948. Initially it was hoped to move to bigger site, but increases of cost (additional building space, equipment, accommodation and conformity to health standards)[11] forced a decision to stay and adapt at Amersham. The Council, concerned for its leafy, suburban location and fearing a factory, objected to any 'large scale production'; the government promised only 'laboratory scale production' would take place.[12]

In 1948 medical provision in England and Wales was centralised under the umbrella of the National Health Service. Under the National Health Service Amersham, charged to 'acquire and hire out the radium already in the UK', supplied radon to many hospitals (48 in 1949), where previously the radon had been made in-house.[13] The larger radium beam units were supplied to a more select list, mostly specialist cancer hospitals. A consequence of the transfer of control of hospitals from local to a more centralised form of authority was that the organisation of a centralised supply of key materials also needed to be rearranged. The following year, therefore, the Radiochemical Centre became the central node in the network of radium and radon supply – part of a material, but not necessarily market economy. By 1949 the number of scientific and administrative staff employed stood at 70 people, up from 12.[14] Amersham also produced sources for industrial radiography, used for inspecting metal castings and forgings.

The transfer of radium from the Ministry of Health to the Ministry of Supply (ie to Amersham) raised two financial questions: what should happen to the National Radium Trust funds and what should be charged by Amersham to resupply the hospitals of the National Health Service? In winding up the National Radium Trust it was capitalised at £180,000.[15] The rate for hire for radium from Amersham in 1949 was £225 per gram per annum. Radon (presumably as 'seeds') was available at between 6 and 18 shillings (under the prewar Medical Research Council arrangements it had been 1 to 2 shillings, the former for hospitals and the latter for private patients, per millicurie)[16]. When supplied by the Radon Centre at Barton-le-Clay, the 'receipts covered the cost of Radon made for sale, but the receipts did not cover the whole cost of the Centre'.[17] Radon for research was supplied free of charge. In other words, radon supply (like radium) was subsidised.

The arguments made by the Treasury were that the supply of Amersham radium and radon should be financially self-supporting, while the supply of isotopes should either be charged to the Ministry of Supply or 'put on an economic basis as they emerge from the experimental stage'.[18] Subsidies to the National Health Service, furthermore, should not be concealed.[19] The economy of radioisotopes was brought to the surface – made visible – in a move that was, ironically, triggered by the arrival of the NHS. Indeed there had been nominal charging even before the NHS – there was always money in radioactivity. But what was new was the move to charge to cover costs, essential to later conceptions of Amersham as a potential trading company or even potential private entity.

As the factories and laboratories of the atomic state came into operation, so Amersham also deepened its role in the processing of radioisotopes. The Atomic Energy Research Establishment, Harwell built atomic piles that substantially increased the supply of radioisotopes; it also trained scientists in radioisotope techniques.[20] In response to increasing medical and industrial demand, Amersham did much of the work preparing and distributing sources. These included carbon-14, radiophosphorus and radioiodine.

Over the years products have also included, in addition to radium for the NHS and other radioactive sources for medical uses, highly radioactive substances such as cobalt-60 for irradiation and neutron triggers for starting nuclear reactions. Amersham acted as the distributing agent for these materials on behalf of the Medical Research Council. The products of Amersham were also sold abroad, especially to the United States, bringing much needed dollars back in return. In the 1950s Amersham's materials were even exported to the Congo – a radioactive equivalent of coals to Newcastle. In 1954 Henry Dale opened new, remote-controlled facilities for the production of radioisotopes (carbon-14 and refined fission products from reactors). These were described by the *Financial Times* as a vast improvement on the previous mere 'enlarged laboratory equipment', meriting the label of being a full 'chemical production plant'.[21] Also in 1954 the United Kingdom Atomic Energy Authority (UKAEA) was set up. It took over formal responsibility of the Radiochemical Centre, alongside its much larger military and civil nuclear programmes. A further £250,000 expansion of Amersham was requested in 1957.[22]

In 1964 the United Kingdom Atomic Energy Authority put the Radiochemical Centre under its trading fund, along with what would become British Nuclear Fuels Ltd (BNFL). The director (Grove) operated under a Board of Management, responsible to UKAEA and chaired by a UKAEA Member. By then the Radiochemical Centre catalogue contained over 2,000 items, 'mainly based on artificial isotopes produced using irradiation in reactors and cyclotrons'.[23] Since many other countries also had such facilities by the mid-1960s, work at Amersham increasingly focused on isotopes and compounds that were more sophisticated and more difficult to produce. Orders were greater than 50,000 and income was £2m per annum, half of which came from exports (mostly to the United States via an entity called Nuclear Chicago). Total staff in 1965 was 450, of whom 120 were professionally qualified scientists. Following the Atomic Energy Authority Act of 1971 UKAEA was split. UKAEA continued research at Harwell and Aldermaston; the Radiochemical Centre Ltd (TRC) and BNFL became separate entities, while remaining wholly owned subsidiaries. UKAEA, noted a civil servant, 'will expect the company [TRC] to act commercially and to have as a principal objective the earning of an adequate commercial return on its capital employed'.[24]

While the Radiochemical Centre charged for its products – and indeed, as I have noted, the advent of the National Health Service made this accounting more visible – monetary payments were not the only forms of exchange that shaped the TRC under public ownership. We can see this clearly in the case of blood products. The principle of freely donating blood was essential to the British system. In 1967 the LSE social policy researcher Richard Titmuss had surveyed nearly 4,000 blood donors, asking after their motivations. The results, published in *The Gift Relationship* (1970), showed that altruism was the leading reason to donate.[25] Titmuss's aim was to counter the free market ideology emanating from the gathering force of the Institute for Economic Affairs.[26] Blood moved through a gift relation, freely given between strangers and with consequent expectations of free use – no charging for blood products. Such a system, argued Titmuss, worked far more effectively than the market systems for blood supply found in the United States.

The Radiochemical Centre was at the edge of this free, gift exchange system. From the early 1950s it received specially purified proteins (albumin and fibrinogen) from the publicly owned Blood Products Laboratory and labelled them with radioisotopes to produce a diagnostic tool. The Radiochemical Centre had agreed that the donors' gifted origin of the proteins required it to label the tool clearly with the phrase '"Human albumin [or fibrinogen] provided free of charge by DHSS" [in order] to avoid any misunderstandings by donors that their blood was being sold'.[27] TRC was allowed to sell the excess product overseas.

However, even though the arrangement created overheads for both the Blood Products Laboratory and the Radiochemical Centre, this gift exchange principle of supply free of charge was confirmed in the late 1970s.[28] Requests coming from for-profit, private companies for supplies from the Blood Products Laboratory were denied, on the same reasoning.[29] Such altruistic relationships would be ended by privatisation. The standard work, Parker's two-volume official history of privatisation, misses this dimension of moral economy.

One reason why the Radiochemical Centre seemed to Thatcher's administration a good candidate for an early trial of privatisation was that not only was the Centre small and had been instructed to act commercially, but also that increased exposure to the market had been discussed for a decade. The Heath government considered selling shares in TRC in 1971 and invited N. M. Rothschild & Sons to advise.[30] Tie-ups to big commercial companies, singly or as a consortium – British Oxygen, Wellcome Burroughs, Beecham, ICI, Glaxo and Fisons were all mentioned – were also considered.[31] The Radiochemical Centre, under Grove, resisted such moves, feeling that 'participation by private industry was not in the company's best interests'; he believed it would 'damage the independent and impartial image which they enjoyed among customers' and have 'harmful effect on their overseas sales since their distributors might be competitors' of the firms involved.[32] In 1972 Rothschilds nevertheless suggested a sale of 49 per cent of the government-owned equity, combined with 'a firm declaration of the Government's readiness to undertake a public flotation within 2–3 years'.[33] The government announced its preference, a public flotation with sale of a minority of shares, in 1974, to be carried out at a time dependent on 'general business and market conditions'. This would have made TRC a publicly owned company, but one with a substantial minority private stake.

In July 1979, two months after Thatcher's general election victory, David Howell, charged with reviewing which nationalised industries to 'dispose of' within the financial year, wrote to Nigel Lawson. He picked out the Radiochemical Centre Ltd, by then a company with a turnover of £39 million, as a leading candidate for privatisation:

> The position is different with TRC, a small but rapidly expanding company which has a first-rate record in its business producing radioactive materials for use in medicine, industry and research.[34]

TRC and UKAEA had yet to be consulted (Sir John Hill, of UKAEA, objected and TRC's long-standing director, Grove, promptly retired). Ministerial approval to legislate on disposal of shares in TRC was given by a Cabinet subcommittee within the month.[35] Some issues, such as pension rights, needed resolution. It was not until late 1981, after an initially preferred option to sell to a single corporate buyer failed,[36] that the sell-off of a rebranded TRC, Amersham International, through public flotation on the Stock Exchange, was announced. The sale took place in February 1982. It was massively oversubscribed, as was internally admitted, and Labour accused the government of underpricing its assets – indeed the official historian has called it 'the largest mis-pricing of a privatisation issue during the first Thatcher government'.[37] Nevertheless, the Conservative government regarded Amersham as a successful experimental trial of privatisation, with added significance here for being a privatisation of a science-based venture under a science-trained prime minister.

In one sense, the sale of the Radiochemical Centre Ltd to become the fully privatised Amersham International was simply the final step into the marketplace of an already commercial outfit. But a more nuanced view notices the disruption of other non-marketbased relationships. One concerned the gift relationship of blood products. While a minor product for the Radiochemical Centre, isotope-labelled albumin and fibrinogen diagnostic tools were made from gifted material. In 1980, with privatisation looming, the Blood Products Laboratory sought approval to charge for the production of fibrinogen. This proposal was accepted by 1982, although it also sparked a request for back payments and a court case was threatened.[38] One consequence of the privatisation of Amersham, therefore, was that it encouraged the blood donation sector – the exemplar of an altruistic, even socialistic gift culture within the health service – also to think and act in more marketised terms. Another non-market relationship was that of the circulation of scientific information and knowledge, which would certainly have been threatened if a single corporate buyer had been found. As was noted in the *Financial Times*:

> Amersham has established close links with the research centres of all major drug companies and regards these links as vital to a company operating at the frontiers of medical science. If one [corporate buyer] gained control, access to the others would cease. 'The day we cut off our flow of information, we're dead,' says Dr Stuart Burgess, managing director.[39]

The government was obliged by market rules to disclose information that was price-sensitive. There was also an obligation to mine opposition statements for information about any proposed re-nationalisation. This kind of knowledge had previously been held discreetly. Now it too had a price. There was therefore a changing information economy in addition to exchanges in the marketplace or by gift. Furthermore, privatisation itself raised the prospect of commercial secrecy as a more general obstacle to the flow of information. This became an issue in the run-up to privatisation, as the government had to decide what and how much price-sensitive information to release.[40] After privatisation, commercial secrecy became the norm.

I think it is useful to picture Amersham as a centre of interconnecting flows. One flow is of material: radioactive outputs from reactors at Harwell and elsewhere entering Amersham, along with blood sera and proteins, plastics for packaging, paper for documentation and so on. At Amersham these materials were transformed into products, which were shipped nationally and internationally through distribution networks. Radioactive effluent and waste also flowed outwards, for disposal on land, in rivers and sea – for comparison, Amersham created more radioactive waste than the Dungeness power stations.[41] The Thames was the conduit for Amersham's radioactive tritium.[42] A second set of flows was of information: accounts to government, scientific knowledge in and out. Only some of these movements were regulated by the market, even when the Radiochemical Centre was instructed to act commercially. Others – notably waste – proved resistant. Products might be sold at profit or at cost. Inputs might be bought or gifted. Privatisation decisively shifted these relationships toward a market form.

The constraint of the market: the faltering privatisations of the nuclear

Privatisation was an ideological aim of the Thatcher government, although the justifications shifted over time.[43] While, as we have seen, the isotope production arm of UKAEA, as Amersham International plc, had been a successful, path-breaking privatisation, the other, substantially larger components of UK nuclear industry – UKAEA, BNFL and the nuclear power stations of the Central Electricity Generating Board and the South of Scotland Electricity Board – would provide far more difficult challenges. While there exist excellent historical analyses of the privatisation of nuclear power, not least Dieter Helm's *Energy, the State and the Market* and Parker's official histories, the proposed privatisation of research has not received close attention.[44]

In March 1984 Peter Walker, Secretary of State for Energy, launched a 'comprehensive review of the activities and role' of UKAEA, with the 'objective of establishing a long-term framework for the Authority's activities and for public expenditure in this area'.[45] The review would involve consultation with a large number of bodies, each with different interests, notably industry (CEGB, the Nuclear Installations Inspectorate, BNFL), the Treasury, Ministry of Defence (which had 'sizable repayment contracts'), the Department of the Environment, the Department of Education and Science and the Research Councils, in addition to UKAEA itself. As Andrew Turnbull, Thatcher's private secretary responsible for economics advice, noted, seeking her consent for Walker's plan:

> A thorough review of UKAEA must be overdue. We have employed 15,000 people for 30 years but has the nation got value for money? Or is it a producer dominated organisation? Does it need to be financed so much (50 per cent) by the taxpayer or could customers (the electricity utilities and BNFL) contribute more? Has one branch of science absorbed too much of our research effort? Content?[46]

'Yes' was Thatcher's immediate response. Keith Joseph wanted the inquiry to ask whether more research to support the nuclear project should be done by universities and polytechnics (an issue from the debates I discussed in Chapter 3).[47] The review was completed in August 1984. Chaired by an insider, Ivor Manley, a civil servant at the Department of Energy and a UKAEA Board member, the report listed reasons why privatisation of UKAEA 'while possible in principle' was 'not in practice'. These included the fact that its 'monopoly status in the core nuclear programmes, particularly the fast reactor ... means there is no early prospect of competition and the operation of market forces', the unlikelihood of a sale coming close to recouping the book value of the net assets of £154 million, the Authority's 'major continuing liabilities, for example in radioactive wastes and decommissioning' and a 'public concern over nuclear issues which might make it inappropriate to seek to launch this key element of the nuclear industry at this time'.[48] The review did, however, recommend that a step towards a customer/contractor approach would be to establish the UKAEA as a trading fund, which would require all work to be accounted for on a 'fully commercial basis, impose additional discipline through the requirement to meet financial objectives, create financial flexibility between years, highlight major issues which need to be dealt with on commercial terms, and facilitate eventual privatisation'.

Robin Nicholson was scathing. In its 30 years' existence, he told Thatcher, the UK had slid from 'first rank to second rank position in civil atomic energy', CEGB, UKAEA's 'main customer', had been 'forced to buy reactor types other than those it would have chosen on technical, operational and commercial grounds'. In addition, UKAEA had contributed 'negligible export of reactors' while consuming £5 billion of taxpayers' money.[49] Furthermore, there had been an opportunity cost in research:

> During most of this period the UKAEA has, by virtue of the interest of its research and its employment conditions, creamed off a significant slice of the UK's R & D talent and created a substantial technological asset. Because of the failure of the civil nuclear programme and the small amount of technology transfer to non-nuclear activities, the contribution of this technological asset to the wealth-producing sector of the economy has been negligible.

Nicholson argued that the review had failed to address this 'dismal record' – largely because, with its terms of reference and working group composition, 'the voice of the status quo dominated'. The Number 10 Policy Unit agreed with the Chief Scientific Adviser's assessment. Nicholas Owen said he had

> reminded us that some of the best scientific brains of a generation have been squandered on misdirected work on civil nuclear power. Since our scientists are among the ablest in the world, the finger points to political misjudgements and intervention. … Can we do better over the next 30 years?
>
> Our priority should be develop a framework for nuclear research which allows a maximum role for the market and little at all for political interference. The best solution would be to privatise the AEA.[50]

Both Nicholson and Owen supported the trading fund idea, as a step towards the private sector.

Some ministers pushed back. Patrick Jenkin, speaking for Environment, supported minor changes in the funding sources of research, but warned that 'furthering commercial exploitation of waste management' must 'avoid creating barriers to the free interchange of new ideas and technologies in a field where public safety is of such paramount importance'. Addressing the subtext of the review, he said that 'it would be unrealistic to go for privatisation at the present time, not least because of current public sensitivities in this field' (a year after the Sellafield leak and during public outcry, led by environmental groups, against dumping at sea).[51] Nevertheless, two main points were carried forward, and confirmed in Cabinet committee.[52] First, the issue of nuclear research should be framed by the customer–contractor principle and second, that this framing was seen as another step towards privatisation that would complement the transition to trading fund status.[53] Hopes were delayed in 1985, however, when a short Atomic Energy Authority Bill, necessary to implement the trading fund policy, was crowded out of the 1985/86 Parliamentary programme. The situation was further complicated by the question of whether Nicholson should be appointed the new chair of UKAEA, as I have discussed in Chapter 2.

In April 1986 the Secretary of State for Energy sounded out Lord Weinstock (GEC) and British Nuclear Associates about the possible sale of the UKAEA shareholding in the National Nuclear Corporation. This semi-privatisation was already regarded as a matter presenting 'formidable practical difficulties' before the Sizewell report; following Chernobyl it became 'even more difficult'.[54] Any sale had to be further delayed.[55]

Nuclear power was getting ever more expensive. This increase was due to a combination of factors that went beyond mere rising costs, but also included changing regulation frameworks, attitudes to next generation and experimental projects, choices made between paths to privatisation and, in a connected way, the increasingly hard-headed use of cost–benefit accounting that began to dislodge other 'strategic' commitments. First, even relatively small regulatory changes, such as the new unwillingness to dump waste at sea, added extra costs.[56] Second, ministers increasingly baulked at the rising expense of the more experimental, research-laden, next generation nuclear projects. The fast-breeder projects were assessed and found wanting in 1988. Likewise, early in the same year the future of the fusion project was in doubt, both as an extension to JET and a possible Next European Torus (NET), and, beyond that, the International Thermonuclear Experimental Reactor (ITER) were questioned by ministers and advisers.[57]

The issues were both local and continental: at one level, South Oxford District Council might refuse permission (since decommissioning to a green field would be delayed by 200 years), while at a European level the UK wanted Europe to pay more, by cancelling the 'host country premium' and by contributing to spiralling decommissioning costs.[58] 'The economics have changed radically since the project was agreed,' George Guise informed Thatcher, adding that the 'decommissioning costs are an order of magnitude higher than originally proposed whereas the economic case for a fusion reactor is weaker'.[59] With a review of fast-breeder policy also underway, the question for Thatcher was to 'ask whether we need expenditure on both the fast breeder and fusion', or indeed on either?[60]

Third, the cost of future inquiries was rising – not least with the realisation, post-Chenobyl, that a second (or more) PWR inquiry could not be restricted to just the local specific issues not covered by the Sizewell B inquiry. Fourth, it was not only the costs of inquiries, but also what might be included in inquiries that added financial uncertainties. For example, when, in 1987 ACOST delivered a review on the industrial impact of Sizewell, while there was some relief that it was 'better than it might have been, given Sir Francis Tombs's opposition to the PWR and Lord Flowers' stance on environmental aspects of nuclear power', there was also consternation that the draft contained:

> two difficult issues which will be important at the next PWR inquiry (Hinkley Point). The issues are first, how far cost benefit analysis can be applied to safety assessment criteria; and second, how a more prescriptive approach by the NII [Nuclear Installations Inspectorate] could be introduced into safety assessment, and whether it is an essential prerequisite of privatisation.[61]

The Chief Scientific Adviser, Fairclough, helped to negotiate the removal of the offending passages.[62] But the significance was that measures to ensure safety might add costs which conflicted with a desire to move nuclear power into the private sector.

The fifth factor was therefore that choices had to be made between paths to privatisation. In late 1988 Thatcher's government announced that the CEGB would be split to form three new utilities: National Power and Powergen, both responsible for power stations, and the National Grid Company, responsible for electricity transmission and owned by the 12 regional electricity distribution companies. The South of Scotland Electricity

Board, with its nuclear and non-nuclear power stations, remained, for the moment, intact. But investors were deeply unwilling to buy into the privatisation while there was unresolved, and indeed unresolvable, uncertainty over the costs of nuclear waste disposal and decommissioning.

In 1990, in what was to be the last year of Thatcher's prime ministership, the nuclear parts of National Power and the SSEB were thus separated, forming the corporations Nuclear Electric and Scottish Nuclear respectively (they would combine under a holding company, British Energy). The Secretary of State for Energy, John Wakeham, had announced this move in November 1989, and one implication was a moratorium on the construction of new PWRs. Lord Marshall, who had moved from CEGB to be chair of National Power, promptly resigned, attacking the government for being driven by short-term market considerations.[63] Another implication was the diminished support for the fast-breeder programme, ultimately leading to the end of Dounreay.[64] As I traced in Chapter 4, Thatcher had visited Dounreay in 1979; now in 1988 she was considering closing the plant. 'Closure of [the Prototype Fast Reactor] and the reprocessing plant would mean the loss of 1,500 out of 2,000 AEA jobs at Dounreay,' Cecil Parkinson warned her. 'The total effect on the Caithness economy would be very severe.'[65] In 1991 the government sold its majority stake in National Power (now free of the nuclear costs) and in Powergen. In 1995 a final examination of the options for privatisation of nuclear power concluded that although the most modern nuclear power stations might be put into private hands (with the establishment of British Energy, responsible for the eight AGR stations and the Sizewell B PWR), the public purse would not support a new PWR programme and plans for Hinkley Point C and Sizewell C were stopped.[66] Meanwhile the old Magnox stations, which had been removed from privatisation plans in 1989, were folded into BNFL. UKAEA, which had eventually become a trading fund in 1986, was made a plc in 1996.

In following this tortuous path to privatisation, one effect was the light that was shed on some of the obscurities of 'nuclear economics'.[67] The secrecy over the cost of the nuclear programme could not be hidden from the market. Highly critical economic analyses were conducted by the Comptroller and Auditor General, and subsequently by the Public Accounts Committee and the Select Committee on Energy in 1984. In another example, in 1988 the Non-Fossil-Fuel Obligation – essentially a requirement that the newly privatised electricity companies should be required to buy the output of nuclear power stations, balanced by a Fossil Fuel Levy on all electricity bills – has been described as 'a tacit recognition by the government of the extra cost of nuclear power'; the minister, Cecil Parkinson, observed that 'for the first time, as a result of our proposals, the public is being told what nuclear costs are'.[68] Likewise, National Power could only be privatised once its hidden nuclear costs had been removed. And again, the ways in which BNFL's costs were calculated were changed, moving from costs-plus to fixed price, so the risks moved away from private sector to the BNFL rump, which in turn became even less attractive to private investors. And again, British Energy's decision in the mid-1990s not to build PWRs 'reflected the judgement that private investors would not be keen to invest in a company that would build new nuclear plants'.[69]

Finally, there are suggestions of a connection between the radical shift in science policy identified and described in Chapter 3 and the withdrawal of direct government support for nuclear projects. The Government 'still recognised the long-term need for fast reactors, [but] it believed that the technology had been proved,' noted BNFL energy policy analysts Judd and Ainsworth of the 1988 decision drastically to cut back R&D expenditure at Dounreay, and 'Consequently the responsibility for further development should be taken by private industry on a strictly commercial basis, and there was no longer a place for nationally funded activity'.[70] This argument is precisely that of George Guise's persuasive (at least, to Thatcher) attack on near-market research that I have uncovered in detail. The research-laden parts of the nuclear project were especially vulnerable. If this connection holds – and it is not the reasoning offered by the Secretary of State for Energy,[71] so the connection, if present at all, would have to be in Thatcher's own mind, as part of her new understanding of the place and role of government-funded R&D – then it is an example of nuclear policy being shaped in the late 1980s not only by market reasoning, or deepened critical economic analysis, but by a science policy that had been derived largely by anecdote and ideology. Amusingly, at the time of the fast-breeder reactor research decision, Guise was writing to Thatcher that 'cold fusion could even become the greatest economic benefit from particle physics!'[72]

Notes

1. David Parker, *The Official History of Privatisation. Volume I: the Formative Years 1970–1987*. London: Routledge, 2009. David Parker, *The Official History of Privatisation. Volume II: Popular Capitalism, 1987–1997*. London: Routledge, 2012.
2. MAF 453/6. Melling to Dalyell, 3 March 1982. Professor Joseph Melling (Centre for Applied Microbiology and Research (CAMR), Porton Down) wrote as honorary secretary of the British Co-ordinating Committee on Biotechnology, relaying the UKFCC's concerns. Dalyell to Thatcher, 9 March 1982.
3. http://www.ncimb.com/defaultinfo.aspx?page=about-ncimb.

4. PREM 19/2967. Wakeham to Thatcher, 5 March 1990.
5. The National Radium Trust was financed through four routes: a public subscription of £150,000, an initial Exchequer grant of £100,000, a later loan from the Ministry of Health under section 3 of the Cancer Act, 1939, of £155,000 and a mixture of bequests, donations and interest on investments. T 225/518. PDP to Hale and Bligh, 13 May 1949.
6. T 225/518. Usher to Chatfield, 27 June 1940.
7. T 225/518. Ministry of Supply, 'The Radiochemical Centre. Amersham', pamphlet for press visit, 11 May 1949.
8. David Fishlock, 'Amersham International will come to the market shortly', *Financial Times* (11 January 1982).
9. T 225/518. Ministry of Supply, 'The Radiochemical Centre. Amersham', pamphlet for press visit, 11 May 1949.
10. Fishlock, *op. cit*. The move was approved by the Lord President's Committee on 24 May 1946. HLG 79/10. Peirson to Phillips, 21 May 1947.
11. 'In planning the extensions, it has been necessary to provide for safeguards to protect the health of staff and to ensure that no danger to the public arises from the disposal of effluent from the Centre. The health standards laid down by the Medical Research Council are very high, and have recently been made more stringent; their observation, particularly in the provision of ventilation and high quality internal surface finish, has been the principal reason for a considerable increase in the estimated capital cost'. HLG 79/10. 'The Radiochemical Centre – Amersham', undated (1947).
12. HLG 79/10. Notes of meeting, 23 July 1947.
13. T 225/518. 'Radium', Oram to Proctor, 11 May 1949. Ministry of Supply, 'The Radiochemical Centre. Amersham', pamphlet for press visit, 11 May 1949.
14. T 225/518. Ministry of Supply, 'The Radiochemical Centre. Amersham', pamphlet for press visit, 11 May 1949.
15. T 225/518. Bligh to Hale, 13 July 1949.
16. T 225/518. Fairie to Bligh, 9 June 1949. 'Amersham's charge of 6/- covers from .5 to 2.0 millicuries and in <u>gold</u> container'. Oram to Bligh, handwritten on Bligh to Oram, 11 July 1949. Platinum on radon seeds required an additional charge of 1/s, with credit given on return seeds.
17. T 225/518. Fairie to Bligh, 9 June 1949.
18. T 225/518. Bligh to Hale, 13 July 1949.
19. T 225/518. Millet (?) to Smith, 12 November 1949.
20. Néstor Herran, 'Spreading nucleonics: the Isotope School at the Atomic Energy Research Establishment, 1951–67', *British Journal for the History of Science* 39 (2006): 569–86.
21. 'Radio-active materials. New plant', *Financial Times*, 8 April 1954.
22. T 225/518. Ridley to Leeming, 31 January 1957.
23. AB 48/576. 'The Radiochemical Centre, Amersham', 26 April 1965.
24. EG 9/35. Doggett to Cunningham, 4 May 1972.
25. Richard Titmuss, *The Gift Relationship: from Human Blood to Social Policy*. London: Allen & Unwin, 1970.
26. Philippe Fontaine, 'Blood, politics, and social science. Richard Titmuss and the Institute of Economic Affairs, 1957–1973', *Isis* 93 (2002): 401–34.
27. MH 160/621. Maycock to Parrott, 11 November 1977.
28. MH 160/621. Parrott to Maycock, 28 November 1977.
29. MH 160/621. Thompson (Duphar Medical Ltd, part of the Philips multinational) to Fenner, 4 May 1971.
30. EG 9/27. Eden to Hill, 16 December 1970.
31. EG 9/27. Davies to Thynne, 14 September 1972.
32. EG 9/27. Minutes of meeting at DTI, 14 October 1971.
33. EG 9/27. Hearne to Herzig, 11 February 1972.
34. EG 9/54. Howell to Lawson, 4 July 1979. Parker, *op. cit*., vol.I, p.136, records that the 'disposal of TRC had not been mentioned in the Conservative's Election Manifesto and the possibility of selling it surfaced during the Chancellor's urgent pursuit of state assts to sell during the summer and autumn of 1979'.
35. EG 9/55. Minutes, E(DL)(79), 5th meeting, 26 July 1979
36. T 465/29. Gieve to Wicks and Willetts, 26 November 1981.

37. EG 9/83. Henderson to Moore, 18 February 1982. Merlyn Rees in *Hansard* (22 February 1982). Parker, *op. cit.*, Vol.I, p.136.
38. MH 160/621. Godfrey to Harris, 30 November 1982.
39. Fishlock, *op. cit.*
40. EG 9/82. Wheldon to Harris, 16 October 1981, in which the Treasury Solicitor identified the problem in the case of the sale of shares in Cable & Wireless. Gieve to Weldon, 11 December 1981, on its occurrence in the Amersham sale.
41. POWE 74/204. 'Power reactors. Dungeness. Radiation protection. health physics returns. Environmental monitoring. 1966–1983'.
42. CAB 124/3044.' Radioactive Substances Advisory Committee. Panel on the disposal into the Thames of tritium waste from the Radiochemical Centre at Amersham 1964–1965'. AB 6/2684. Grove, 'UKAEA. The Radiochemical Centre Board of Management. A note on the problem of tritium', 1965.
43. Broadly speaking from reducing public spending and increasing industrial efficiency through competition to arguments for the benefits of 'popular capitalism' centred on a share-owning public. Parker, *op. cit.*, vols. I and II.
44. Dieter Helm, *Energy, the State, and the Market: British Energy Policy since 1979*. Oxford: Oxford University Press, revised edition, 2010, footnote 6 on p.190, is the extent of discussion of UKAEA privatisation. Parker, *op. cit.*, vol.II, pp. 276–92, likewise focuses on the much larger issue of nuclear energy privatisation.
45. PREM 19/1740. Walker to Thatcher, 2 March 1984.
46. PREM 19/1740. Turnbull to Thatcher, 5 March 1984. Thatcher's underlining. The 50 per cent referred to Department of Environment nuclear research spending. Much of the rest of the funding of UKAEA research came from public funds, albeit indirectly, including 11 per cent from BNFL, 10 per cent from the generating boards CEGB and SSEB, as well as more from Environment (6 per cent non-nuclear), Defence (3 per cent), DTI (2 per cent). Europe (mostly through JET) contributed 6 per cent, while 'private sector/overseas' paid for 5 per cent. Department of Energy, *Review of the UK Atomic Energy Authority. Report by a Departmental Working Group*, August 1984.
47. PREM 19/1740. Joseph to Thatcher, 12 March 1984.
48. PREM 19/1740. Department of Energy, *Review of the UK Atomic Energy Authority. Report by a Departmental Working Group*, August 1984. Other reasons given were an only notionally funded statutory pension scheme (£500 million), the need for significant working capital and security and policing implications.
49. PREM 19/1740. Nicholson to Thatcher, 12 October 1984.
50. PREM 19/1740. Owen to Thatcher, 12 October 1984. Owen dismissed each of the reasons given in the report against privatisation: public sensitivities were increased, not decreased, by public ownership ('Maintaining the AEA in the public sector sustains the belief that nuclear power is a uniquely dangerous beast, rather than a clean, proven energy source with demanding, but manageable, safety requirements'; the AEA may be a monopoly but it would have to deal with equally powerful customers (the generating boards), while other assets have been sold for less than book value (he cited Sealink).
51. PREM 19/1740. Jenkin to Thatcher, 2 October 1984.
52. The committee was E(NI).
53. For Nicholson the situation to avoid was that of the failed application of Rothschild to medical research. 'The Rothschild transfer of MRC funds to the DHSS failed because DHSS did not take steps to become an informed customer and eventually this transfer had to be reversed. … My own view is that the Departmental experts will need to be backed by independent technical consultants, perhaps drawn from the international arena. This might cost up to £1m pa, but would be money well spent if it enabled DEn to choose the right R & D programmes in their £175 spend'. PREM 19/1740. Nicholson to Thatcher, 12 October 1984.
54. PREM 19/2134. Walker to MacGregor, 9 June 1986.
55. PREM 19/2136. Parkinson to Rifkind, 23 December 1987, noted that Weinstock wanted GEC to purchase 35 per cent of the National Nuclear Corporation.
56. In 1985 the London Dumping Convention extended the moratorium on dumping nuclear waste at sea indefinitely. House of Lords Select Committee on Science and Technology, 'Management of nuclear waste', Third Report, 10 March 1999.
57. The ministers involved were Cecil Parkinson, Nigel Lawson and John Major.
58. PREM 19/2136. Monger to Thatcher, 22 January 1988.

59. PREM 19/2136. Guise to Thatcher, 22 January 1988. Guise's fire was aimed at NET and he approved the extension of JET (essentially to start injecting tritium as a step towards higher power): 'To stop the JET project before tritium is introduced would be like discontinuing work on a bridge when it had almost reached the other side of the river. The scientific case for continuing is sound, and should be supported'.
60. PREM 19/2136. Monger to Thatcher, 22 January 1988.
61. PREM 19/2136. Wilson to Norgrove, 2 December 1987. There was also concern that the draft contained a 'number of passages [that] might cause damage to the commercial interests of British industry, eg by implying that they could not build a safe pressure vessel, or that their prices for PWR components might not be internationally competitive'.
62. PREM 19/2136. Fairclough to Gray, 25 January 1988.
63. Markku Lehtonen and Jenny Lieu, 'The rise and fall of the fast-breeder reactor technology in theUK: between engineering "dreams" and economic "realities"?', SPRU Sussex Energy Group report, 2011, p.24. Lord Wakeham's oral history interview, conducted under the British Library's Oral History of the Electricity Supply in the UK project, reveals some of the complexities of this resignation. https://sounds.bl.uk/Oral-history/Industry-water-steel-and-energy/021M-C1495X0022XX-0001V0.
64. Lehtonen and Lieu, *op. cit.*, p.58, note that following the February 1988 white paper on electricity sector privatisation, 'the CEGB withdrew its modest share in fast breeder R&D'. Then, a few months later, Parkinson 'declared that the government would stop paying the £50 million a year running costs of the PFR and that the associated annual R&D expenditure would be cut from £51 million in 1988–90 to £10 million a year from 1990–91' – just enough to sustain international collaboration. See also Mike Parker and John Surrey, 'Contrasting British policies for coal and nuclear power, 1979–1992', *Energy Policy* 23(9) (1995): 821–50.
65. PREM 19/2136. 'Fast reactor R&D: the need for review', 22 April 1988.
66. DTI, *The Prospects for Nuclear Power in the UK*, Cm 2860. London: HMSO, May 1995. Grahame Danby, 'Nuclear privatisation', Research Paper 96/3, House of Commons Library, 15 January 1996. Parker, *op. cit.*, vol,II, pp.286–92.
67. Critics of the UK nuclear project, such as Walt Patterson, co-founder of UK Friends of the Earth and an important documenter of changing nuclear policy, noted the 'abrupt' change of plans as the markets thwarted privatisation plans and ministers only found out the 'true scale of nuclear costs' late in the day. Yet such critical accounting of costs was hardly new in 1988. Walt Patterson, 'Thatcher's failed romance with nuclear power', *Bulletin of the Atomic Scientists* 46 (3) (1990): 9–12, p.9, p.12.
68. Mark Winskel, 'Autonomy's end: nuclear power and the privatization of the British electricity supply industry', *Social Studies of Science* 32(3) (2002): 439–67, quotes Parkinson. Lehtonen and Lieu, *op. cit.*, p.23.
69. Lehtonen and Lieu, *op. cit.* p.27.
70. A. M. Judd and K. F. Ainsworth, 'Fast reactors in the UK, 1946–1996', *Energy* 23(7–8) (1998): 609–17, p.615.
71. PREM 19/2136. 'Fast reactor R&D: the need for review', 22 April 1988, by Cecil Parkinson, argues that it was low uranium prices that undercut the strategic justification for FBRs (which would re-use uranium and plutonium, and so make the UK less dependent on potentially expensive uranium imports), which in turn meant that they were not attractive to privatised industry bodies. 'HMG is therefore likely to have to finance most of any of the continuing programme, should there be one'. Of course, it is this willingness for the government to fund research that was at issue and that, in turn, would depend on what 'near-market' meant.
72. PREM 19/2136. Guise to Thatcher, 22 January 1988.

6
The Strategic Defense Initiative and the politics of research

America does possess – now – the technologies to attain very significant improvements in the effectiveness of our conventional, nonnuclear forces. Proceeding boldly with these new technologies, we can significantly reduce any incentive that the Soviet Union may have to threaten attack against the United States or its allies.

As we pursue our goal of defensive technologies, we recognize that our allies rely upon our strategic offensive power to deter attacks against them. Their vital interests and ours are inextricably linked. Their safety and ours are one. And no change in technology can or will alter that reality. We must and shall continue to honor our commitments.

I clearly recognize that defensive systems have limitations and raise certain problems and ambiguities. If paired with offensive systems, they can be viewed as fostering an aggressive policy, and no one wants that. But with these considerations firmly in mind, I call upon the scientific community in our country, those who gave us nuclear weapons, to turn their great talents now to the cause of mankind and world peace, to give us the means of rendering these nuclear weapons impotent and obsolete.

Tonight, consistent with our obligations of the ABM treaty and recognizing the need for closer consultation with our allies, I'm taking an important first step. I am directing a comprehensive and intensive effort to define a long-term research and development program to begin to achieve our ultimate goal of eliminating the threat posed by strategic nuclear missiles. This could pave the way for arms control measures to eliminate the weapons themselves. We seek neither military superiority nor political advantage. Our

> only purpose – one all people share – is to search for ways to reduce the danger of nuclear war.
>
> My fellow Americans, tonight we're launching an effort which holds the promise of changing the course of human history. There will be risks, and results take time. But I believe we can do it. As we cross this threshold, I ask for your prayers and your support.[1]

Even before March 1983, when Ronald Reagan called upon the 'scientific community … those who gave us nuclear weapons' to turn their 'great talents' to the task of strategic defence, laser weapons were a topic on which Britain's new, science-trained Prime Minister took a personal interest.[2] 'This,' Margaret Thatcher would later recall, was 'one of those areas in which only a firm grasp of the scientific concepts involved allows the right decisions to be made,'[3] And 'As a scientist I shall understand this better than any of my ministers and therefore I am the one that the Americans should be talking to' was how her Private Secretary recollected her first instinct.[4] Indeed, the archives show that space-based missile defence – what the Americans called the Strategic Defense Initiative (SDI) – was indeed an area in which Thatcher 'ran' policy directly. While Thatcher's biographer, Charles Moore, concludes, on the basis of a later interview with her Cabinet Secretary, that Thatcher's scientific background contributed to her being sceptical about SDI, the primary source evidence, presented here, shows she was in fact distinctly more favourable to the project than her key ministers and advisers.[5]

In the first months of office, a document warning of Soviet developments was placed in Thatcher's box of working papers. 'Specifically, the Soviets are in the process of creating a prototype space laser capable of knocking out any land- or sea-launched ballistic missiles,' the author wrote, adding that 'American scientists consider the coming breakthrough in laser technology to be as important as were the development of atomic and nuclear weapons and of ballistic missiles in their day'.[6] Such laser weapons ('the death ray of science fiction') would challenge the ABM Treaty of 1972 and fall outside the proposed Article XI of the draft SALT II Treaty, in which signatories undertook not to place into Earth orbit nuclear weapons or any other kinds of weapons of mass destruction, as well as 'make the American concept of Mutual Assured Destruction … technologically obsolete'. Thatcher underlined much of this document and requested advice from the Ministry of Defence, specifically from its chief scientific adviser, Ronald Mason.[7]

The Ministry of Defence took a month to reflect on the request. The advice, when it came, was sceptical: while a laser powerful enough to damage a guided weapon was experimentally possible, there were 'major practical difficulties to be overcome before such a system could be developed as an effective weapon'.[8] Fog and cloud would reduce range significantly and immense energy was needed for operation: 'large quantities of heat, of the order of a megawatt or more, and waste gases would have to be dissipated'. This might be possible on board a ship or on land. In space, while the range might increase, any installation would be 'more complex and probably highly vulnerable'. The Russians, it was known from intelligence sources, were nevertheless committing $300 million a year to research and development of laser damage weapons, apparently 'with some success'; in a passage underlined in Thatcher's blue pen, the Ministry noted 'uncorroborated evidence that they have been examining the feasibility of locally heating part of the re-entry vehicle of a ballistic missile ... so as to make it unstable ... and to miss the intended target'. Finally, it was noted that any deployment would also contravene the 1967 Outer Space Treaty.

Prime ministerial interest rested for a year.[9] In January 1981 stories carried by the *New Scientist* and *The Times* caught her eye.[10] The latter story, by Reuter's Washington Correspondent, speculated that the new space shuttle might perform pointing and tracking tests for lasers.[11] The former story reported an MIT study that said that space laser weapons, while theoretically possible, were 'not technically feasible in the real world'; they could be jammed or triggered to fire at the wrong target.[12] Significantly, it was the 'theoretically possible' aspect that attracted her rather than the grounds for doubts. It is a feature of the SDI question that leading politicians, not least Reagan and Thatcher, were far more credulous than experts or journalists. The Ministry of Defence reported that, even in the eyes of the US Department of Defense's scientific expert who coordinated the programme, 'no application had yet been found for laser damage weapons'.[13] Yet the *Minister* of Defence would write to Thatcher, on top of the same advice, that

> Sooner or later there will be a breakthrough in research on these weapons and it is important that we keep track of what is happening. The implications for eg Trident could be considerable. For the moment, however, nothing startling appears to be imminent.[14]

Thatcher firmly underlined the first, alarming sentence, but not the keep-calm-and-carry-on sentiment of the last.

Then, in March 1983, Reagan made his dramatic speech. The key moment came at the end of the half-hour televisual address to the American people, in which he asked citizens to oppose cuts to the defence budget. Having castigated the previous administration for running down the armed services, alarmed his audience with the scale of Soviet military expansion and revealed satellite photographs of MiGs on Cuban airfields and an immense new runway built in Grenada with communist cash for a Caribbean state with no air force, Reagan began to speak on a different register. 'I've become more and more deeply convinced that the human spirit must be capable of rising above dealing with other nations and human beings by threatening their existence', he said, looking the camera in the eye. The President then added that even if arms control reductions should succeed:

> it will still be necessary to rely on the specter of retaliation, on mutual threat. And that's a sad commentary on the human condition. Wouldn't it be better to save lives than to avenge them? Are we not capable of demonstrating our peaceful intentions by applying all our abilities and our ingenuity to achieving a truly lasting stability? I think we are. Indeed, we must.[15]

Was there an alternative? Yes, according to the President. 'I believe there is a way. Let me share with you a vision of the future which offers hope. It is that we embark on a program to counter the awesome Soviet missile threat with measures that are defensive … [turning to] the very strengths in technology that spawned our great industrial base and that have given us the quality of life we enjoy today.' It might take years, with failures along the way, but 'tonight we're launching an effort which holds the promise of changing the course of human history'. And to do so, the President reached out to 'the scientific community … those who gave us nuclear weapons' – was there in that phrase a tone of invited redemption? – 'to turn their great talents now to the cause of mankind and world peace'.

Reagan's public speech, which was reaffirmed by a strong private message from the President to the Prime Minister, was immediately recognised in the Ministry of Defence as being of 'potentially crucial importance'.[16] What should the UK response be? In particular, given that this might be a research and development programme of immense scale and resources, what were the consequences and opportunities for scientists in Britain? For UK interests more broadly, four questions of concern were identified: 'the questionable validity of the technical proposals'; 'the possibility [given a shift of emphasis from deterrence to defence] that any nuclear exchange would be confined exclusively to Europe while both superpowers remained immune behind the shield of their improved ABM defence'; the 'credibility of the United States' commitment to arms control, particularly in relation to the 1972 ABM Treaty'; and, last but not least, 'the future of the British independent strategic nuclear deterrent'. The last alone would force a choice between confronting the United States on the feasibility of SDI or admitting that Trident would be redundant if SDI was successfully built and replicated on the Soviet side.[17] The questionable validity of the technical proposals, examined specifically by the Defence Scientific Staff, could likewise be broken down into areas of concern: deployment would be at least 30 years away, would require diverting the Shuttle programme, present 'considerable command and control difficulties' and be vulnerable to Soviet anti-satellite and other countermeasures.[18] Nevertheless, noted the Defence Scientific Staff, while there was 'nothing in the laws of science which says that a space based directed energy weapon system for ABM defence cannot be built ... it is clearly a greater challenge than the US project to land a man on the moon'.

In the United States, an inter-agency group spent the six months following Reagan's speech on reviewing possibilities.[19] Its conclusions, reported by diplomatic telegram, were that there was 'a good deal of momentum in favour of a substantially expanded research and development programme', one which would have 'implications ... for Britain and French systems'.[20] It was suggested that Thatcher take the 'opportunity during her talk with the president to emphasize that US exploration of this area of technology is of direct concern to us'.

The discussion in the UK took place in the context of a European interest in taking positive action towards arms control – of anti-satellite systems in the shorter term and under the influence of SDI in the long term. In February 1984 the French, backed up by the Germans, emphasised their analysis of military developments in outer space at a Western European Union ministerial meeting. They expressed the belief that there was a danger of a new arms race and that European allies should take a view on arms control options.[21] Geoffrey Howe, the Foreign Secretary, would have wanted to support this position, but felt constrained not to say anything while Ministry of Defence's views were yet to be circulated in Whitehall. When these were finally expressed in June, they were, thought senior diplomat Percy Cradock, especially critical of SDI:

> the longer term issue, SDI, is much more diffuse and worrying. In essence it involves US research into the possibility of a multi-layered system of ballistic missile defence (BMD) capable of destroying incoming missiles at various points in their trajectory. It is highly speculative, would be horrendously expensive and it is hard to see how a flawless system providing 100% cover could be devised. It could also be highly destabilising in terms of the super power [*sic*] balance. Research and testing would eventually contravene the 1972 Anti-Ballistic Missile Treaty.[22]

Yet a president would nevertheless be tempted to develop a defence that could 'at least reduce the prospect of a total holocaust'.

In September 1984 a wide-ranging set of consultations with Western European allies took place at the Pentagon and the US State Department. While no substantial new information about SDI was shared, UK observers nevertheless took home two messages. The first – 'very welcome' – was that the Americans were now more willing to think of SDI as concerning the 'defence of the alliance as a whole (not just continental United States)'. The second was the revelation that while there was 'scepticism among some senior US officials as to whether a comprehensive strategic defence system will ever be deployed, if only for resource reasons', there was equally a 'feeling that there may be valuable spin-off from the research along the way and that the scale and effort going into the President's initiative of strategic defence has at the very least made a strong impression on the Russians, which may be no bad thing'.[23] With a US election fast approaching, after which President Reagan might be expected to commit fully to SDI, the pressure was on for the UK to agree a firm line of response. Geoffrey Howe and Michael Heseltine, Foreign Secretary and Minister of Defence respectively, teamed up to persuade Thatcher. In a jointly signed letter the duo, highlighting both the technical problems and the doubts,[24] urged, on the basis of a substantial review of policy, that the 'key question is whether at this important juncture the Government should be willing to engage with the Americans in serious discussion of the underlying arguments for and against such a concept'.[25]

SDI, with its mixture of scientific research and drive from the highest levels of political leadership, was an area where the Prime Minister was indeed 'running ... policy directly'.[26] Thatcher, aided by points made by Charles Powell, her Private Secretary for Foreign Affairs and Defence, resisted Heseltine and Howe's assault. Crucially, the counter arguments rested on an assessment of scientific research.

In particular, a point from Powell received a big blue tick from Thatcher. It was that:

> not enough weight is given in [Howe and Heseltine's argument] to Soviet potential and capabilities ... [they are] ahead of the US in important areas of research. Given what we know the Soviet Union are up to, it seems to me that the Americans have no option but to push ahead in this area.[27]

Powell's next point is underlined:

> The paper underestimates the dynamics of scientific progress. You can't disinvent DEW or KEW [directed energy and kinetic energy weapons] technology. There's no question of choking BMD at birth as the paper seems to suggest. The goal should be to manage the new technology in as economical way as possible to add to the West's overall security.

In late 1984 Thatcher met successively the new Soviet general secretary Mikhail Gorbachev and the re-elected President Reagan. She assured the latter that the former would not be allowed to drive a wedge between the UK and the US via the issue of SDI. She also 'told the President' of her 'firm conviction that the SDI research programme should go ahead'.[28] At the same time the public debate on SDI and arms control heated up. Four veteran heavyweights of US foreign policy – Robert McNamara, McGeorge Bundy, Gerard Smith and George Kennan – authored a paper in *Foreign Affairs* that argued that SDI and arms control were incompatible.[29] This argument was rejected by Thatcher and the American Secretary of Defense, Caspar Weinberger.[30]

The winter of 1984–5 saw high-level diplomacy between the United States and the Soviet Union grappling, or perhaps rather posturing, over the arms control issue, including control of space weapons. At one stage, on the latter, the Soviets were accused of 'not taking yes for an answer'. Yet suddenly the ground shifted. Margaret Thatcher received a briefing on developments at Geneva, where George Schultz and his counterpart Andrei Gromyko met in January 1985. Robert 'Bud' McFarlane, Reagan's National Security Adviser and leading advocate of SDI, was accompanied by a team including the US ambassador when he briefed Thatcher; she in turn was accompanied by a top UK diplomatic team. While 'no-one knew for sure what had persuaded the Soviet Union to return to serious negotiations', McFarlane was convinced that 'above all it was the US decision to pursue research on the Strategic Defense Initiative'.[31] Elimination of the SDI research programme had been the constant demand, filling the first day of the Schultz–Gromyko talks. Thatcher wrote personally to Reagan to thank the President for keeping her, and the UK, informed, adding pointedly that the Alliance stood the test of 'serious problems over the past year in the arms control field' precisely because 'intensive consultations' encouraged cohesion.[32] Meanwhile the Soviet ambassador reiterated to Malcolm Rifkind, junior minister at the Foreign Office under Howe, that Soviet aims were the 'non-militarization of outer space', including a ban on the 'development, testing and deployment of "attack space weapons"', which in turn were a condition on further progress in other arms control areas.[33]

It was in this context that a particularly powerful reframing of SDI formed within the highest UK government circles, one that put the politics of *research* at its centre. On 20 February 1985 Margaret Thatcher was scheduled to hold a 'seminar' in Washington on SDI. A scheme for a *coup de theatre* was carefully plotted. It was noted that at an earlier meeting, held at Camp David in December 1984,[34] Thatcher had achieved considerable impact on the President when she spoke beyond the content of the pre-circulated papers and addressed Reagan with a fresh directness. The old President had perked up and listened attentively. With this lesson learned, the plan was to prepare for the Prime Minister 'something new and important to say when she meets the President, in order to make the maximum impact ... with ideas which have not been pre-digested by his advisers'. It was in this spirit that the British ambassador to the United States, Oliver Wright, made a clear and crucial suggestion.[35] Thatcher had, at Camp David, made a 'very helpful distinction' between SDI research and SDI development.[36] This had, said Wright, 'made possible the armistice': it drew support from other European governments and 'ensured that we were not on the side of the Russians in opposing SDI'. But now there was an opportunity to go further. Wright put the point as follows:

> So far so good. Except on one point. This concerns our unwillingness to involve ourselves in the research programme now picking up steam under the SDI rubric. My Defence Staff and my Chancery have received repeated overtures from General Abrahamson, the energetic Air Force General running the SDI programme, about the possibility of his visiting the UK to brief British industry. So far the decision has been not to pick up the offer.
>
> I think this is wrong. It is also inconsistent with our policy of approving research.

Furthermore, argued Wright, the 'SDI under the President's inspiration has fired the American imagination' with the result that its 'popularity now extends widely throughout the relevant branches of the Administration, the US armed services, and Congress, who will in my judgement vote at least the funds to whether it will work'. He observed that 'SDI is one of the things, perhaps the most important, that Ronald Reagan wants to be remembered by, like Jack Kennedy and putting a man on the moon'. As with Apollo, it represented the 'American "can-do" spirit, something to aim at'. Therefore, asked Wright, the key question was, given this momentum, 'are we going to exclude ourselves from the revolution in defence technology that the SDI research programme is likely to ignite?' 'If we continue to spurn US interest in involving us,' Wright warned, 'I see a real danger of our missing the bus.' Past and present examples were brought forward to back up the point that investment now would save money later:

> If we indeed miss the bus, as we missed the space bus,[37] we shall only have to instigate, in 10–20 years time, another Alvey catch-up operation in order to stay in the field of nations competent in the most advanced technologies. Isn't it better to get in on the ground floor? Isn't there high grade employment here for a lot of British brains? Isn't there work for eg Plessey, British Aerospace, Racal, GEC? And shouldn't our defence scientists be given a chance to remain up with the US front-runners?

The conviction here was that involvement in research would lead to spin-offs, both military and civilian, provide business opportunities for 'enterprising British companies' and be a window of influence on American 'policy decisions on testing and deployment issues if we know what we are talking about'. So, urged Wright, Thatcher should tell the President 'of our interest in the research programme and our wish to take up the US offer of a piece of the technological action'.

Here was the 'something new and important to say'. Powell included the proposal to 'volunteer to participate in SDI-related research' as the first point of discussion at the pre-meeting at Chequers, where main UK positions would be agreed prior to departure for Washington; he thought it 'could have considerable political impact, above all with the President'.[38] The drawback, what he saw as the likely MoD and FCO response, was that such a bold embrace of SDI research 'would identify us publicly with the SDI, which they would see as a mistake in itself and likely to cause problems in the

Alliance' (that is to say, with Western European states). Again a dividing line was being drawn with Reagan, pro-SDI US officials and generals, Thatcher and Powell aligned on one side, and the FCO and MoD (represented by Howe and Heseltine) and France, Germany and other Western European allies on the other.[39]

When Thatcher met Reagan on 20 February 1985 it is clear that she did indeed offer UK research for SDI. Direct evidence – the record of the meeting – is unfortunately still retained. But subsequent enthusiastic, even frantic, exploration of what level of research might be funded is supplementary evidence. Furthermore, that Wright's specific proposal was closely read by Thatcher is shown not only by the blue ink underlining parts of his letter to Howe, and the opinion of her adviser Charles Powell,[40] but also in the echoes of Wright's language in Thatcher's personal message to Caspar Weinberger: 'I was and remain very impressed by the vigour and ingenuity which is going into tackling the immense technical problems in the best American "can do" manner'.[41]

Before turning to the British attempt to secure considerable research funding for SDI from the United States, we need some sense of four aspects of the SDI issue: the visioneering of SDI technology; the quality and sources of knowledge about it; the relationship between scepticism and expertise; and the sometimes extraordinary back channels of diplomatic pressure to support SDI.

What was SDI? Knowledge, uncertainty and criticism

The SDI project had its joint origins in Cold War science and the science fiction imaginary.[42] Rooted partly in past and present research programmes in the US national laboratories but also in technically detailed visions of future technologies, what historian Patrick McCray has called 'visioneering', SDI also changed shape over time.[43] The Cold War US defence research establishment was home to both proponents and critics of anti-ballistic missile systems. In the late 1950s and 1960s the Army had wanted to install a massive system of ground-based interceptor missiles. As Rebecca Slayton has shown, it was largely physicist-advisers who had criticised the Army's proposal, on the grounds that it would not only be expensive and of doubtful effectiveness but would also disrupt deterrence.[44] She has also shown that this specific, physics-centred expertise did not grasp the other, more computational risks of such systems.

Nevertheless, research towards various means of ballistic missile defence continued, many encouraged by the influence of Edward Teller, the staunchest hawk among defence scientist-advisers.[45]

By 1981, two years before Reagan's announcement, various programmes of research were already underway.[46] The Lawrence Livermore laboratory in California was investigating accelerators for the Navy, funded by DARPA; this was the so-called 'Chair-Heritage' programme. Los Alamos had a research programme under a Dr Knapp, investigating a space-based system. The US Department of Energy had an interest in developing technologies suitable for particle beam weapons and the Department of Defense had convened a 'task force', led by a Dr Franken of the University of Arizona. The Department of Defense also had a Laser Damage Weapon Programme, coordinated in part by a Dr Richard Airey.

In response to Reagan's 1983 speech, two investigative surveys were commissioned, the second of which, known as the Fletcher report, offered a public vision of SDI without releasing sensitive, technical details.[47] The effort to organise, as well as establish and defend the feasibility of, SDI in the United States was led by General Abrahamson. Abrahamson, described by one British newspaper as a 'cherubic Luke Skywalker [who] positively exudes good intent,'[48] conceived of SDI as a 'system of systems', including satellites, ground radars and command and control infrastructure.[49] Likewise the UK Ministry of Defence's new Chief Scientific Adviser, the chemist Professor Richard Norman, in his personal briefing of Margaret Thatcher on the subject in January 1985, said that 'a review of all recent available evidence showed that the US conception was that of a layered system'.[50] Elements of these layers included: surveillance and identification of Soviet missile launches by eight satellites in distant, geosynchronous orbits (36,000 km); tracking of missiles by 20 satellites at 10,000 km, with infrared and long-wave radar, and battle station satellites, numbering at least 100, probably at low 1,000-km orbits. The attack would be optimally made during the boost phase of the ICBMs, when all the parts – missile, warhead, decoys – were together before deployment of countermeasures. Attack during the boost phase, while desirable, meant that the detection and the decision to engage had to be taken extraordinarily quickly.

Norman described the three possible types of weapon: kinetic, directed-energy and particle beam. Kinetic-energy weapons might take the form of small projectiles (5 kg masses, rocket-propelled, travelling at 6 km per second), smaller bullets (weighing 1 kg but launched by an electric rail-gun and travelling at 10 km per second), or even the form of a tiny missile, 10 g in mass, accelerated by an electric rail-gun to speeds of 200 km per second. Directed-energy weapons were predominantly envisaged to be space-based laser systems, although ground-based lasers, firing at space-based mirrors in geosynchronous orbit – that would in turn reflect the energy to high-altitude, ground-based 'flighting mirrors' 25 m in diameter – were also considered. The types of laser under investigation included hydrogen-fluoride infrared devices as well as Teller's favourite: an extraordinary, space-based, x-ray laser powered by a nuclear explosion. The giant pulse of x-rays would be split and redirected onto Soviet missiles by hundreds of smaller mirrors carried by the satellite. It would, of course, work only once. The particle beam option was considered least promising.

Tying these systems together would be the computer and communication networks necessary for 'battle management'. As historian Rebecca Slayton has emphasised, the difficulties regarding software – not only making it work, and work fast enough, but also ensuring that the programming did not have bugs that would lead to failure or disaster – emerged as SDI's Achilles heel.[51] Norman, too, said to Thatcher that the United States was:

> very advanced on individual components, but not on the management of the system as a whole. The software which would be needed was far beyond anything conceivable in the present state of the art. There were also major technical problems in the fields of optics and vast amounts of energy needed for laser weapons.

However, knowing what SDI might be was much more difficult than my summary of Norman's briefing to Thatcher has suggested. SDI existed as components under past and continuing research, system diagrams of the whole that would over-simplify and change over time and future projections of technologies that pushed up against, and possibly beyond, physical and computational limits. Furthermore, technical information about SDI was restricted, even for close allies of the United States. 'An assessment of US attainments and capabilities is, in some ways, more difficult to make [than that of Soviet systems]' ran one complaint. 'We have received a number of US briefings on the SDI programme, but most of these have concentrated on strategic and political issues and aims, rather than technical ones.'[52] The US could – and did – exclude British eyes and ears. At a meeting at NATO on SDI by Abrahamson's team in August 1984, for example, the UK 'was allowed to attend all the briefings at Secret level but was excluded from a session involving nuclear devices, eg X-ray lasers and high endoatmospheric discrimination'.[53] Technical details would, even when known, be either classified secret or, if published in technical journals, be of unknown reliability.[54] Likewise relevant intelligence reports, presumably only of the Soviet side, would also be secret. Thatcher, wary of slipping in public, ordered an audit and summary of what she could tell Parliament about SDI.[55] This audit found that the 'few hard facts available in the open literature' were 'largely derived from "leaks" which have not been confirmed by official sources'.

Knowledge of SDI was therefore incomplete for a number of reasons. This situation meant that all decision-makers, and those who would influence them, including critics, worked in conditions of incomplete knowledge and uncertainty. Debate, both in and out of government circles, relied on incomplete, private briefings and published, limited, self-interested documents such as the Fletcher report,[56] – and, increasingly, the critical accounts and analyses published, for example, by the activist Union of Concerned Scientists.[57] Furthermore, newspaper and popular science press coverage was also highly influential, not least because these were a main source of imagery that visually encouraged the view that SDI was feasible ('how it will work', captioned once such piece in the *International Herald Tribune*)[58] even when the text itself might be critical (see, for example, similar imagery of the SDI 'system of systems' in the *Scientific American* and the *Economist*)[59].

So in a curious way, the centre – at least the UK centre of government – was not much more knowledgeable about what SDI was than were the experts and media commentators outside the walls. In the United States, the counterattack of criticism against SDI was mounted most prominently by academic scientists such as Hans Bethe of Cornell, Kosta Tsipis of MIT, Herbert Lin, also of MIT, and the Union of Concerned Scientists.[60] British scientists who organised against SDI, such as David Caplin and Tom Kibble of Imperial College, were just as critical, but had little access to technical details or political decision-makers.[61] In Whitehall experts, as well as officials within the Foreign and Commonwealth Office and the Ministry of Defence, were also sceptical of SDI's feasibility. 'The FCO and MoD always want you to read the *Scientific American* article,' complained Thatcher's gatekeeper, Charles Powell, to his Prime Minister, 'because they agree with it'.[62] Such scepticism was echoed, albeit again for different reasons, in the public debate on Star Wars in Britain.[63]

Against such scepticism, one extreme response was to see widespread expert criticism as a positive indicator of radical innovation. Such, for example, was the rather significant opinion of Caspar Weinberger, US Secretary for Defense. It was expressed at a conference on the US/UK Relationship in the Field of Defence and Security, held at Ditchley Park in the Oxfordshire countryside, supposedly under the Chatham House Rule, reported by Michael Heseltine to Thatcher:

> He had 'rather complete' confidence that the SDI would work. Indeed he took encouragement from the consensus among scientists that the goal was not achievable; Einstein himself had been a lone voice.[64]

If we put aside the possibility that Heseltine, who, as we will see later, may not have been playing with a straight bat in the SDI research debate, was here misrepresenting or satirising the views that Weinberger actually expressed, then this seems a remarkable dismissal of expert opinion by the Secretary for Defence. Of course, back in Washington there was a small circle of powerful experts, notably around Edward Teller, who had promoted and justified extreme military technologies throughout the Cold War, from the hydrogen bomb to Star Wars.

Extraordinary back channels

As well as the direct, official US–UK contacts, exemplified at the highest level by that between Margaret Thatcher and Ronald Reagan, US diplomacy involved the circulation of views through back channels which, of course, were also monitored and weighed at the centre. A curious example of these back channels is a breakfast between Lord Thomas and Henry Kissinger that took place in February 1985. Hugh Thomas, a Foreign Office diplomat-turned-historian, was chair of the Centre for Policy Studies, the free-market, Thatcherite think-tank based in London. Kissinger, the distinguished, immensely well-connected foreign policy adviser had, under Reagan, no formal affiliation to the President, but continued to network, consult, comment and influence.

At breakfast, Kissinger spilled his views on SDI. It was, he thought, essential, 'the only way out of our nuclear dilemma'. In particular, he suspected that, 'in the long run', the conviction necessary to the success of the policy of deterrence through threatened massive nuclear retaliation would be eroded. Any such threat had to be credible. Would any president really authorise a nuclear attack on a Russian city in response to a conventional attack in Europe? 'Even now,' he doubted that

> anyone around President Reagan would advise, say, a nuclear attack on Kiev in response to a Russian takeover of Berlin: particularly not Nancy Reagan who, in his opinion, would be the determining voice if she were around at the time when the President had to take a decision of this nature (and she would make it her business to be around).[65]

Aside from this extraordinary glimpse into the imagined dynamics of nuclear decision-making, the point had consequences for strategic defence. SDI was therefore, for Kissinger, 'a way of avoiding' the otherwise eventual 'move towards unilateral nuclear disarmament', as this lack of credibility became widely recognised. Russian attitudes to SDI were seen as rational and their worries about SDI would drive them to the negotiating table. European – that is, French, German – concerns were dismissed as ordinary responses to shifts in the balance of power, Kissinger's favourite framework of analysis.

Thomas dutifully forwarded Kissinger's message to Thatcher.[66] There is no evidence that it played any particular role in shaping policies, and was probably only read for its amusing tittle-tattle. However, its timing – a fortnight before Thatcher's meeting with Reagan and in the moment of decision regarding the pitch for research – certainly did not detract from the momentum for support of SDI among Thatcher and her closest advisers.

Pitching for research

In the weeks before Thatcher's critical meeting with Reagan on 20 February 1985 there were signs that a similar change of heart was occurring in other European countries. In West Germany, Helmut Kohl and his defence minister Manfred Wörner, softened by the 'moral/philosophic' framing of SDI – that it was a step towards reducing civilian casualties in the instance of a nuclear war – signalled that West Germany was also open to European collaboration in SDI research, albeit under 'certain conditions', including the strong one of 'no secrets', or full sharing of fruits of research.[67] As we will see, one of the motivating factors behind the shift in the German position – that, in Kohl's reported words, the SDI programme 'would give the US a significant technological leap forward and the European allies should not become technologically dependent' – was equally at play in British discussions. It was also noted in the German press that 'US offers of research collaboration were designed in part to create pro-SDI lobbies in Europe', a wise cynicism that was not in evidence in London.

At a gathering of defence ministers on 26 March 1985 Caspar Weinberger, detecting 'unanimous support', announced that a formal invitation for European allies to collaborate on SDI research would soon be issued.[68] Indeed, just such a formal invitation landed on Michael Heseltine's desk that day.[69] Similar letters arrived in Bonn, Paris, Rome,

Tokyo and Jerusalem. An indication of interest was requested within three months. Heseltine's first response was warmly to endorse the suggestion of Lord Carrington, the ex-Foreign Secretary who was now Secretary-General of NATO, of putting up a 'concerted European effort'.[70] For Heseltine such a joint proposal from Britain, Germany, France and Italy would have two advantages. First, it would 'help the process of maintaining a shared European approach to the wider strategic issues raised by SDI', not least because it would discipline individual countries from over-indulgence – a problem, apparently, 'because of the lure of participation in the technologies of the future'. Yet Heseltine's second reason addressed precisely this gain: a joint response would mean that 'four major European countries working together could share the benefits of their collaboration'. Within this approach it is interesting that Heseltine ruled out the 'CERN model': some 'may argue for a European centre of excellence … as the mechanism for a joint European contribution', but such an institution 'would rapidly acquire a vested interest in the pursuit of SDI which might not be helpful to our own effort to think through the issues in political and strategic terms rather than allow technological and industrial factors to dominate'.

Again Thatcher was guided by Powell. 'Perhaps I am too suspicious,' her adviser wrote:

> but it seems to me that one reason why the Defence Secretary proposes a joint European project on SDI research is that he wants to build up a body of opinion sceptical of SDI. There doesn't seem any good industrial or scientific reason for a joint response: so it must be political.[71]

Clearly Powell thought Heseltine's long-term game was undermining SDI rather than building European collaboration. Thatcher agreed, rejecting the suggestion of a joint European proposal and instead offering mere 'coordinated rational responses', while reserving each country's 'national freedom of manouevre' [*sic*]. In communicating this decision, Powell added other reasons – which may originate from him or may come from a conversation with the Prime Minister:

> She suspects that we would be more likely to lose from it. She doubts whether the French would share scientific knowledge or technology with us in areas where they are ahead: and we should be net contributors of expertise in regard to other European countries. Politically she sees a risk that a joint response might place an

> undesirable restraint on the position which we take towards SDI. She also thinks that we can reasonably expect to cash in with the Americans the credit we have built up by giving a lead to European support for SDI research, and can best do so bilaterally.[72]

Perhaps taking the hint, and perhaps extending it deliberately too far, Heseltine's response would be to attempt to cash in this credit for as much as it could buy. With the Foreign and Commonwealth Office, on balance, in favour, and the Department of Trade and Industry sidelined, it was the Defence Secretary who led the negotiations.[73] In anticipation, much of the necessary bureaucratic machinery was set up, such as a Cabinet-level official committee to gather views and coordinate discussion, not least on how SDI research might fit more broadly with research and development, in policy and in industry.[74]

The European context for considering SDI research was also moving fast. The French very quickly rejected Weinberger's invitation. On 17 April 1985 President Mitterrand proposed 'Eureka', a European fund for research that would aim to provide the technological boost without the SDI strings.[75] He thereby 'presented himself as taking bold action to hasten the "technological renaissance of Europe" at a time when the SDI proposal was reminding Europeans of their technological weakness'.[76] In Germany a split opened up between the pro-SDI chancellor Helmut Kohl and pro-Eureka voices both within cabinet (such as Hans-Dietrich Genscher) and outside (such as Kohl's predecessor Helmut Schmidt). On 31 May Germany endorsed Eureka. The UK soon followed, but, by ambitiously also seeking SDI funds, kept two irons in the fire.

Nevertheless, following the rejection of the joint European plan, Heseltine seems to have dragged his heels. He met with Weinberger and grumbled about potential violations of SALT II, which the US Defense Secretary dismissed out of hand.[77] The two were said to be 'at loggerheads'.[78] When Heseltine reported to Thatcher in late May, two months after Weinberger's invitation, he could not hide the absence of progress. 'The truth is that very little progress has been done,' remarked Powell to Thatcher, who replied, derisively 'The Germans have got further than we have. This won't do'.[79] She demanded action, in the form of firm proposals, within three weeks. Geoffrey Pattie, the Minister of State for Industry and Information Technology, also complained that the Ministry of Defence were going too slowly, channelling UK industrialists' grumbles that the German government was being far more active and supportive.[80] Pattie's claim that the 'Americans regard SDI as a giant pull through of new technology, whether or not the eventual military aims are ever achieved' was supported by the intervention of Thatcher's Chief Scientific Adviser, Robin Nicholson. He, too, argued that 'whether or not SDI succeeds in its strategic aims, the very large US spend will produce technical advances in areas of importance to conventional defence and to civil industry'. Indeed, since even the United States did not have the resources – including the demand for 40,000 scientists and engineers alone – to pursue SDI alone, the UK, despite 'a real resource cost to the UK', had, Nicholson believed, 'no choice' but to contribute, willingly through negotiation or unwillingly through a new brain drain.[81] Yet Heseltine's proposals were so far 'very feeble'. The Chief Scientific Adviser's recommendation to his Prime Minister was therefore:

> We have a unique and hard won position of being the only country with a respected and trusted position on defence science and technology with both the US and Europe. We should exploit this position ruthlessly.[82]

Nicholson was therefore critical of Heseltine's revised position, as of July 1985, not only to negotiate bilaterally with the United States, but also to share a 'pool of information' with European partners, coupled with active Eureka participation. So were others.[83]

Heseltine's next, bold response would strain relations with Thatcher significantly. The Secretary of Defence met his US counterpart, Caspar Weinberger, at the Pentagon 'at short notice', on 22 July 1985. He was accompanied by his private secretary and the British ambassador. Weinberger brought along Major General Colin Powell. The official minute abbreviates the first item on the agenda, on SDI, and turned quickly to concerns about early warning systems and armaments sales to Argentina.[84] The private summary sent by Heseltine to Thatcher reveals that the SDI discussion was incendiary.[85] Heseltine, claiming the support of British industry, told Weinberger that 'to secure proper British participation, we needed to put on the table at the outset a bid for our share [of SDI research] expressed in money terms'. So

> Taking account of the advice from industrialists, I did, however, decide to put on the table a figure for the amount of US-funded SDI work to be placed in Britain ... I decided to pitch this at $1.5Bn out of the $26Bn which the US plans to spend over the period 1985–1989.[86]

Heseltine's stated reasoning was that a smaller commitment would carry all the political risks (of criticism from, say, the Labour Party, and of a brain drain of the top talent) but few of the benefits (in access to new technology, in balance of trade). 'The scale of our bid,' Heseltine noted with a straight face, 'clearly came as a surprise'.

The scale of Heseltine's bid deserves underlining and context. If successful, it would have made the US SDI programme a bigger funder of UK research than, for example, the Medical Research Council. 'The Defence Secretary slapped a high bid for "5% of SDI work or nothing" which rather rocked the Americans,' glossed Powell to Thatcher. "Bold and I hope not intended as a "wrecking" bid.'[87] That Powell had to ask suggests that the motives of Heseltine were obscure, even to those at the centre of the administration. What game was Heseltine playing? Was he bidding high because he wanted UK investment and was playing a high-risk strategy? Or was he bidding high because he knew it would be refused – in which case was this because he did not agree with the aim or approach of SDI, or for some other cause?[88] We know that his department, the Ministry of Defence, like the FCO but for different reasons, had reservations about SDI. Also, we can recall Heseltine's reasons for rejecting the 'CERN model' for SDI (because it would create a vested interest in the 'pursuit of SDI') as extra evidence that Heseltine was sceptical of SDI as an end. If he thought that the 5 per cent bid could have been successful then he was guilty of seeking to create just such an interest. Therefore, if he was not being inconsistent, Heseltine probably was indeed offering a wrecking bid.

In April 1985 the Chief Scientific Adviser of the Ministry of Defence (Norman) and the Director of AWRE Aldermaston, on a tour of US nuclear establishments, had taken 'the opportunity to have a long exploratory discussion with General Abrahamson and his staff on possible co-operation in the SDI research programme'.[89] The Ministry of Defence now began to work out in detail, responding to Thatcher's demand for such evidence, the areas in which the UK could do more than a billion dollars of SDI work.

Nearly one-third of this would be straight 'research', conducted in government labs, universities and industry. Specifically 18 areas were identified (see Table 6.1 in Appendix), many of which were in niches where UK research was particularly strong, from optical computing and lasers to software security and electronic sensor materials. These research areas would contribute £250–300 million. They were complemented by an smaller but more lucrative set of proposed areas for 'SDI research validation experiments' – essentially experimental demonstrations of the abilities of parts of the SDI 'system of systems' to work effectively (see

Table 6.2 in Appendix), conducted largely by private companies. This development work would bring in £840 million, making a total of over £1 billion (meeting Heseltine's $1.5 billion bid).[90]

From this point on, however, the SDI bid began slowly to unravel. Some of the problems were due to contradictions internal to UK priorities. The DTI expressed 'pessimism about the ability of UK industry to exploit SDI technology', while the Cabinet Office, at the same time as the MoD composed its ambitious list of research and development areas, noticed that it begged 'the fundamental question of whether it would be in the UK's overall economic interest to allocate resources to work of this scale and kind'. After all, as I show and discuss in Chapter 3, 'one of the main issues that Ministers are now tackling .... [was the] recommendation that defence R & D spending should be progressively *reduced* so that scarce industrial resources, particularly in electronics, can be re-allocated to more economically productive purposes'.[91] In other words, tying science policy to SDI had considerable, unanalysed opportunity costs, despite the lure of immense dollar funding. Nicholas Owen, of the Number 10 Policy Unit, wondered aloud whether the $1.5 billion was 'wishful thinking', adding 'We are pursuing contradictory objectives: straining to secure a substantial SDI workshare, which we cannot afford, and which we probably could not deliver, while at the same time, trying to contain or reduce defence expenditure, particularly on R&D'.[92] Keith Joseph argued that 'Key skills in both UK industry and UK universities and Research Councils would be at risk of brain drain', with the impact on university research teams 'disproportionately great'.[93] Furthermore, commitments to SDI and Eureka were being made at a time when, said Joseph, 'we continue to squeeze funding for our science base' (also a topic addressed in Chapter 3).

Another set of obstacles were in place across the Atlantic.[94] The American negotiators wanted the 'smallest acceptable British contribution covering technologies of greatest interest to the US, on terms favourable to the US and with no commitment beyond the next year or so,' complained a British official.[95] The rules on transferring defence technology, which was essential if demonstration experiments would be funded, were being interpreted inflexibly and in line with narrow US interests – both those of security and those of a 'Buy American' protectionist kind. A small 'Pathfinder' programme of work was offered – perhaps £150 million over five years and a substantial disappointment. Off the record, at least some on the US side thought the British were 'being greedy', and briefed that the Reagan administration had 'other fish to fry up on the Hill'.[96]

Margaret Thatcher had all these arguments before her when reviewing SDI research policy in October 1985. Her Chief Scientific Adviser,

Nicholson, considered the likelihood of spin-off, noting that while some of the technologies were 'likely to be of major importance to civil industry', SDI participation would 'naturally direct the technologies towards military applications, and UK industry has a poor record of extracting the potential civil benefits from military R&D'.[97] And her gatekeeper for all this advice, the diehard Charles Powell, concluded that if the $1.5 billion fixed figure was dropped it should be in return for 'satisfaction on all our other requests', while urging that, in view of the desire to find cuts in defence research, UK 'participation in SDI research should be an alternative to not additional to existing defence R&D'.[98]

Further disappointment, perhaps not unexpected, came when Caspar Weinberger informed Heseltine that, due to 'legal constraints and other difficulties', no fixed 'set aside' of SDI R&D funds could be allocated to the UK.[99] There would be no $1.5 billion investment. This decision, surely reflecting the Reagan administration's realistic assessment of Congressional politics, came despite the UK offering a 'comprehensive exchange' of 'all relevant information', relating to both government defence research establishments and British companies. (Such openness would have been modelled on the post-1958 sharing of nuclear secrets that had reversed the heavy postwar restrictions of the McMahon Act.) A much more vague but secret Memorandum of Understanding was nevertheless drafted – essentially a framework for further exploration of the R&D topics identified.[100]

Heseltine told the Prime Minister that he thought that 'the original approach of seeking a specific sum' had 'played an important part in achieving this outcome'. Howe, Heseltine's supporter in this matter, thought the Minister of Defence had done well. So, it seems, did Thatcher.

However, by November 1985 the prospect of substantial UK SDI funding was receding fast. Lord Carrington was told, off the record, by Robert McFarlane, Reagan's National Security Advisor, that 'any countries interested in getting some benefits from SDI research should sign on quickly because SDI was not going to last'.[101] McFarlane thought that doubts about the system's effectiveness, as well as a squeeze on funds, would strangle the project. That month Reagan told Thatcher, finally if apologetically, that no substantial funding would be available.[102]

SDI continued to be a matter of high diplomacy, as the endgame of the Cold War began. At the summit at Reykjavik in October 1986 the Soviet desire to confine SDI to the laboratory was the stumbling block that prevented a remarkable proposition to eliminate strategic nuclear weapons within ten years. Reagan's refusal – despite the fact that his vision of a nuclear-free world was within grasp – has been interpreted very differently: from intransigence to a step 'crucial to the victory over communism'.[103] Confinement to laboratories would be, in practice, precisely the fate of SDI in the 1990s and 2000s.

Conclusion

The academic analysis of Reagan's Strategic Defense Initiative has, perhaps understandably, been dominated by strategic issues. How did SDI feature in high diplomacy? Would SDI have upset the balance of deterrence? Was SDI a strategy, consciously or by accident, for undermining the Soviet Union, perhaps critical to the end of the Cold War? These are, of course, important questions. Thatcher considered Reagan's SDI decision 'the single most important of his presidency'.[104] The virulence of Soviet opposition to SDI, which culminated at Reykjavik, the causes of which were unclear at the time,[105] now seems to be a recognition, as Thatcher also later wrote, that 'they had lost the game … they could not hope to match the United States in the competition for military supremacy'.[106] The physicist Freeman Dyson once insightfully noted that Soviet leaders such as Khrushchev pursued grand defence by bluff, taking high technology not as functional but as psychological.[107] In the 1980s this is precisely what Reagan himself did – SDI worked in, and on, the imagination – and the Soviet leaders took it seriously because that was their mindset too.

While in this chapter I have shown that there existed a serious division of opinion on the wisdom of SDI, the significance for this book's argument is as a case study of an issue identified by Margaret Thatcher as a science-based, high-stakes issue: 'one of those areas in which only a firm grasp of the scientific concepts involved allows the right decisions to be made'. She talked directly to Reagan. But her Minister for Defence, Michael Heseltine, who had no science background – his education was a Philosophy, Politics and Economics degree from Oxford and his working experience was as an accountant, property developer and publisher – negotiated with his American counterpart, Caspar Weinberger. Within the context of this slight loosening of prime ministerial control, Heseltine pitched for an extraordinarily large sum of research and development funding. Science, and science policy, was thus a critical aspect of SDI as it related to the United Kingdom.

In 2003 the historian Holger Nehring posed several probing questions to a witness seminar held to discuss the British response to SDI, specifically on the science question.[108] Two can now be answered. First, Nehring asked: 'Was British support for the technological side of SDI motivated by the perception of Soviet competition in the arms race, or by the desire to boost the economy by establishing SDI-related R&D?' From the documents now released at the National Archives, it is now clear that the contribution of spin-off certainly featured in British government discussions, while the Russian threat was very much contextual. Yet this is not the full answer.

Retrospectively Thatcher recalled that for her the 'decisive argument' in favour of SDI was 'that you could not ultimately hold back research into new kinds of offensive weapons. We had to be the first to get it. Science is unstoppable: it will not be stopped for being ignored'.[109] Again in retrospect, Heseltine stated that he saw the military-industrial arms race as unstoppable, and therefore that in his decision to pitch for funding he was simply out to maximise the research money that would come from SDI: if there was going to be 'any kind of involvement by the United Kingdom it had to be at the research level and we must get whatever we possibly could out of it'.[110] Both recollections are misleading. Thatcher had indeed made the point, during her Camp David discussions with Reagan in December 1984, that the research should go on, but the reasons for this seem more tactical than due to the 'unstoppable' nature of science. When Heseltine's bid for $1.5 billion research and development funds failed, it was clearly evident that research could indeed be stopped.

Furthermore, Heseltine's bid for funding was not merely maximising research income to the United Kingdom. I argued that it is at least as plausible – and indeed not entirely contradictory – that Heseltine, like Howe, had severe doubts about SDI. Powell and Thatcher certainly wondered at the time if his approach was a 'wrecking bid'. Science policy here might even be seen as a move in the game of political rivalry. Within months, as the tensions finally surfaced in public, Heseltine would resign on another, apparently far less significant, defence technology issue – the sale of the helicopter firm, Westland. Perhaps that was the final straw, whereas SDI was part of the camel's main burden.

Putting aside such speculations, the SDI affair, during which civil servants and industrial partners conjured up hundreds of millions of pounds of potential research and development work, is a rather unedifying sight. The UK was eager for dollar-funded research; it was being swayed, dependent on the United States and willing to tear up existing important plans for defence research. Recall, as I traced in Chapter 3, that these discussions took place at a time when the consensus among ministers and the chief scientific adviser was that defence R&D spending was far too high and needed to be cut.

Nehring also asks what role did scientific advisers and advisory committees play? The answer here is that they played only a secondary one. Thatcher listened carefully to Richard Norman, the Chief Scientific Adviser at the Ministry of Defence, but her own GCSA, Robin Nicholson, had only a marginal role on this issue. She listened to the US SDI advocates (Abrahamson 'had a regular pass to Number 10 Downing Street for several years, to come and tell us about every latest development,' recalled Powell), read the papers Powell fed her and, apparently, 'devoured' *Aviation and Space Weekly*.[111] She did not listen to academic critics, including hundreds of physicists who signed a petition against SDI research on campus. 'You have to remember that we had a government led by a lady who famously defied the advice of 365 economists in a letter to *The Times* about her Budget,' commented Powell, 'so she was not going to be terribly impressed by 365 physicists.'[112] Thatcher may have been a scientist-turned-politician, but her sense of how science informed politics was her own.

Appendix

Table 6.1 Areas identified for UK SDI research, September 1985. Source: PREM 19/1445. Annex A, Mottram to Powell, 25 September 1985. Sciences identified by the author.

Research area	Description	Sciences involved	Notes
Item 1. Architecture study	An 'examination of the requirements for the European elements of defensive systems'. In other words, adapting SDI to defend Europe	Organisation studies Operational Research Computer science Electrical engineering	Was linked to the Test Bed Facility proposal
Item 2. Directed energy protection programme	An 'understanding of the interaction of new forms of directed energy with weapons materials', with implications for countermeasures.	Material science Physics	
Item 3. Electromagnetic launcher technology	A 'new approach to high velocity missiles'	Physics Electrical engineering	
Item 4. Ion sources	Building on existing work on 'ion beam generators'	Physics	The existing work was at the Culham fusion laboratory

(Continued)

Table 6.1 (Continued)

Research area	Description	Sciences involved	Notes
Item 5. Optical computers	Replacing electronic signals with light beams in computers, promising increased speeds	Materials science Physics Computer science	The UK was regarded by MoD as having 'first class credentials' in this field
Item 6. Advanced thyratrons	Thyratrons are 'high current switching devices, of interest to the US in their directed energy programme'	Electrical engineering	UK, thought MoD, was 'ahead of the US'
Item 7. Non-electronic materials	Generic types of material 'applicable to space mirrors, to large space based structures and to structures that must survive battle conditions'	Materials science	The UK 'has expertise', said MoD
Item 8. Sensor package	Materials and devices for remote sensing	Materials science Electrical engineering	An area of 'exceptional expertise' of MOD R&D establishments
Item 9. Terminal radar	'This is an area in which the UK has novel ideas for future multi-beam radars as well as for significantly reducing the production costs of more conventional phased array radars'	Electrical engineering	The US were the acknowledged leaders in phased array radars
Item 10. Terminal interceptors	The 'application of technologies for the interception of a wide range of threat types – all of them likely to arise in Europe in the next decade'	Electrical engineering	An area of 'very strong UK expertise'
Item 11. Laser radar and vibrometry	'New techniques in data processing, combined with laser radars, show promise in early identification of threats'	Physics Computer science	UK regarded as 'on a par with the US in this work'

(Continued)

Table 6.1 (Continued)

Research area	Description	Sciences involved	Notes
Item 12. Counter-measures	Defensive countermeasures	Various	'The UK has 20 years experience in defensive counter-measures'
Item 13. Software security	This was 'concerned with the maintenance of data processing capabilities which are crucial to the whole system'	Computer science	A 'particularly sensitive area in which to negotiate collaboration'
Item 14. Electronic materials	The 'development and processing of materials for use in the very high rates of computing demanded by future systems'	Materials science	'UK is a world leader'. Also, this work 'has significant civil applications'
Item 15. Phase conjugation	'These techniques apply to the modification of the properties of mirrors in space'	Physics – Optics	US 'probably ahead in many aspects of this work'
Item 16. Battle management/ command and control and communica-tions	The 'timely processing of data relating to the management of defence assets and the problems of command, control and communication'	Computer science Electrical engineering	Required 'the setting up of physical battle models'
Item 17. Signal processing	'Modern approaches to signal processing open up new possibilities for target signature analysis, decoy discrimination and data fusion at extremely high rates'	Electrical engineering Computer science	'Considerable civil potential'
Item 18. Space technology	Problems of materials and technologies	Material science Engineering	

Table 6.2 UK SDI Research validation experiments proposed by the Ministry of Defence, September 1985. Source: PREM 19/1445. Annex B, Mottram to Powell, 25 September 1985.

Research area	Description	Sciences involved	Notes
Item 1. Airborne sensors	'Experiments would be designed to demonstrate the capabilities of sensor materials and devices, for Infra Red and visual detection and for discrimination between threat and non threat objects'	£170 million	
Item 2. Ground based – air based radars	Ground and air based radars, particularly multi-beam phased array radar	£120 million	An area of overwhelming US expertise, but UK companies 'believe they have innovative approaches'
Item 3. Terminal phase weapons	Complex interceptor weapons	£150 million	'There are highly innovative UK approaches'
Item 4. Counter-measures	'These experiments would demonstrate the possibilities for establishing countermeasures to BMD systems'	£80 million	
Item 5. Optical computers	Very fast, non-electronic computing'	£40 million	'UK has leading work in this area … .With very wide civil applications'
Item 6. Electro Magnetic Launcher	A technology for high-velocity interceptors	£120 million	The UK already 'has an extensive investment', including a 'fully instrumented test range … which is of great interest to both UK and US programme'

(Continued)

Table 6.2 (Continued)

Research area	Description	Sciences involved	Notes
Item 7. Culham Ion Source	An ion source	£40 million	'This technology is of great interest to the US and incorporates a very high UK investment of intellect. It represents a high value bargaining counter with the US'
Item 8 C3 Test Bed and Simulator	C3 (Command, control and communication) test bed and simulator	£120 million	'UK expertise is extensive in this field'

Notes

1. Ronald Reagan, Address to the Nation on Defense and National Security, 23 March 1983. Copy of text of speech at: http://www.reagan.utexas.edu/archives/speeches/1983/32383d.htm.
2. Margaret Thatcher, *The Downing Street Years*. London: HarperCollins, 1993, p.247.
3. Thatcher, *op. cit.*, p.463.
4. Charles Powell, in Michael Kandiah and Gillian Staerck, *The British Response to SDI*, Witness Seminar held in the Chancellor's Hall, Senate House, University of London, 9 July 2003. London: Centre for Contemporary British History, 2005, p.32.
5. 'She boasted that her scientific education, contrasted with "Laid back generalists from the Foreign Office", enabled her to grasp the [SDI] concept and run with it. She was certainly not one of those who were appalled. But the very scientific education which she mentioned led her to be "dubious about the practicality",' Robert Armstrong recalled. ' ... I think she instinctively doubted whether it would be as effective as Reagan seemed to think it would be.' Charles Moore, *Margaret Thatcher. The Authorized Biography. Volume Two: Everything She Wants*. London: Allen Lane, 2015, pp.107–8.
6. PREM 19/1188. Untitled document, undated (1979). Frustratingly only two middle pages of the document have been kept and civil servants at the time seemed to be unsure of its provenance, one adding a note, presumably to Michael Heseltine, asking 'Did you put this in? I think it is from the publication "Transnational Security".' I cannot trace a publication of that name from that year. Nevertheless, this document, with its handwritten queries from Thatcher, is the start of her engagement with laser weapons as an issue.
7. Also requested was advice on whether Soviet developments in SAM ground-to-air missiles and AWAC radar would neutralise the Americans' cruise missiles.
8. PREM 19/1188. Norbury to Alexander, 28 January 1980.
9. Although General Keegan did raise the possibility of a 'self-resonating collective generator' in relation to a (probably space-based) accelerator for particle-beam weapons in April 1980.
10. The pattern of advice would often start with a news story in the broadsheets or scientific press. These would be forwarded to the Prime Minister, who might then ask for internal expert advice. Another such episode was from December 1981 in response to Kosta Tsipis's article 'Laser weapons' in *Scientific American*. John Nott, the Junior Defence Minister, asked for the Prime Minister to be shown it. The Ministry of Defence took the opportunity, in the covering note, to re-emphasise its scepticism. Tsipis called the difficulties of laser weapons

'insurmountable'; this view was said to be 'very much in line' with earlier advice and was also 'fully supported by Sir Ronald Mason, our Chief Scientific Officer'.

11. 'US space shuttle to test laser weapons', *The Times*, 7 January 1981. For what the UK knew about the military space shuttle, see: FCO 66/1545. Cuthbertson to Swift, 19 November 1981.
12. 'No future for laser weapons in space', *New Scientist*, 1 January 1981. The MIT study was Michael Callahan and Kosta Tsipis, 'High Energy Laser Weapons: a Technical Assessment', Program in Science and Technology for International Security report no.6. Cambridge, MA: MIT, 1980. See, for a contemporary STS view, Kent D. Lee, 'The role of scientific advisers in the Strategic Defense Initiative', *Technology in Society* 8(4) (1986): 291–8.
13. PREM 19/1188. Dawson to Alexander, 14 January 1981.
14. PREM 19/1188. Heseltine to Thatcher, 15 January 1981, handwritten on Dawson to Alexander, 14 January 1981.
15. Ronald Reagan, Address to the Nation on Defense and National Security, 23 March 1983. Copy of text of speech at: http://www.reagan.utexas.edu/archives/speeches/1983/32383d.htm.
16. PREM 19/1188. Mottram to Coles, 29 March 1983. The private letter is in the Thatcher archive. Thatcher MSS (Churchill Archive Centre): THCR 3/1/29 Part 2 f49 (T37A/83). Reagan to Thatcher, 23 March 1983. Copy here: http://www.margaretthatcher.org/document/131533.
17. PREM 19/1188. Ministry of Defence, 'President Reagan's speech on defensive technology', undated (March 1983). For Britain and various arms restriction treaties see: John R. Walker, *Britain and Disarmament: the UK and Nuclear, Biological and Chemical Weapons Arms Control and Programmes 1956–1975*. Farnham: Ashgate, 2012. Jeremy Stocker, *Britain & Ballistic Missile Defence, 1942–2002*. London: Frank Cass, 2004.
18. PREM 19/1188. Ministry of Defence, 'President Reagan's speech on defensive technology', undated (March 1983). Also: MoD Defence Scientific Staff, 'Defence against strategic nuclear missiles: a technical assessment', undated (March 1983).
19. There were two reports to the President submitted in October 1983, one on strategy and policy (the 'Hoffman Report') and one on defensive technologies (the 'Fletcher Report').
20. PREM 19/1188. Telegram, Washington to FCO, 27 September 1983.
21. PREM 19/1188. Howe to Heseltine, 14 June 1984.
22. PREM 19/1188. Cradock to Powell, undated (June 1984). On the shorter term, ASAT, issue, MoD view was that a ban on ASATS was desirable. Thatcher disagreed with important parts of the analysis.
23. PREM 19/1188. Telegram, 18 July 1984.
24. Not least the need to 'incorporate a hair-trigger response, dependent upon automatic, computer-driven decisions'. The full list of pros and cons is interesting. Pros: 'ethical merit' (MAD perhaps even replaced by 'Mutual Assured Survival'), 'popularity with US public', 'strengthening of the US guarantee to Europe', 'new crisis management options' ('As a deterrent to provocative behaviour in periods of tension, US Presidents could formally notify unfriendly states that the BMD system was being switched over to automatic', 'strengthening of deterrence' (by disarming first strike), 'damage limitation if deterrence failed', 'incentives for deep cuts in offensive nuclear systems', 'nullifying the risk of accidental missile launch', 'avoidance of nuclear threats by small nuclear powers', 'the need for a prudent hedge against Soviet ABM/BMD efforts', 'improved monitoring and verification of Soviet compliance with the 1972 Agreement', 'comparative US advantage in this area of military competition', and 'achievement of a US lead in 21st century weapon systems and the domination of space'. Cons were: 'the enormous technical uncertainty of the project', 'the relative ease and cheapness of countermeasures', 'the risk of saturation by increased numbers of offensive systems', 'increased dangers of automatic response leading to war', 'the danger of strategic destabilisation', the 'particular dangers of transition from deterrence to defence', the 'improbability of accidental nuclear release' (*sic*), 'low likelihood of ballistic missile attack by future nuclear states', 'the worldwide nuclear threat itself would still remain', 'stimulation of a new arms race', 'arms control' issues and 'no hard evidence of Soviet intention to break out of the 1972 ABM Treaty'. PREM 19/1188. 'Ballistic Missile Defence (BMD): implications for UK policy towards the US Strategic Defence Initiative (SDI)'.
25. PREM 19/1188. Howe and Heseltine to Thatcher, October 1984. The substantial 35-page review plus annexes of SDI was 'Ballistic Missile Defence (BMD): implications for UK policy towards the US Strategic Defence Initiative (SDI)'.

26. PREM 19/1444. Powell to Thatcher, 28 March 1985. The full quotation is revealing: 'Particularly in areas such as SDI where you are running our policy directly, it would be really very difficult if I could not keep your most senior advisers informed of points of major importance'.
27. PREM 19/1188. Powell to Thatcher, 11 October 1984.
28. PREM 19/1188. 'Text of Prime Minister's statement on SDI at press conference in Washington on 22 December as agreed with President Reagan'. This was also the occasion of Thatcher's proposal, and Reagan's acceptance of the 'four points', repeated as the basis of Western views on SDI. See: US Information Service, 'The Strategic Defense Initiative', 5 June 1985, copy in PREM 19/1445.
29. McGeorge Bundy, George F. Kennan, Robert S. McNamara and Gerard Smith, 'The President's choice: Star Wars or arms control', *Foreign Affairs* 63 (Winter 1984): 264–78.
30. PREM 19/1188. Powell to Appleyard, 12 December 1984, noted that Thatcher 'does not agree with some of its central propositions, particularly the idea that it is impossible to pursue successfully both ballistic missile defence and agreement on strategic arms control'. On 19 December 1984 Weinberger gave a 'major speech on SDI to the foreign press ... This is the most forceful and fullest public exposition so far given by a senior member of the administration of the rationale behind SDI. Its immediate purpose was probably to refute the arguments, recently put forward in quote Foreign Affairs unquote by McNamara, Bundy, Smith and Kennan'. Telegram, Washington (British Embassy) to FCO, 19 December 1984.
31. PREM 19/1443. Powell to Ricketts, 'Prime Minister's meeting with Mr McFarlane: US/Soviet talks on arms control', 9 January 1985. The previous day Thatcher was briefed on UK knowledge of SDI (based on a 'review of all recent available evidence') by the chief scientist of MoD, the chemist Professor Richard Norman. PREM 19/1444. Powell, 'Strategic Defence Initiative' (note of meeting between Thatcher and Professor Richard Norman, chief scientist, MoD), 8 January 1985.
32. PREM 19/1443. Thatcher to Reagan, 14 January 1985.
33. PREM 19/1443. Budd to Powell, 16 January 1985.
34. On 22 December 1984, concerning SDI among other topics. PREM 19/1443. 'Text of Prime Minister's statement on SDI at press conference in Washington on 22 December as agreed with President Reagan'.
35. PREM 19/1443. Wright to Howe, 29 January 1985.
36. Thatcher, *Downing Street Years, op. cit.*, p.468, quotes the main text. Moore, vol.2, p.245.
37. Stuart Butler has pointed out to me that there are clear parallels between the offer of US SDI collaboration and President Nixon's offer of space collaboration post-Apollo, which also caused tensions in Europe and was ultimately not taken up. See: Stuart A. Butler, 'National prestige and in(ter)dependence: British space researchPolicy 1959–73', PhD thesis, University of Manchester, 2016, p.253.
38. PREM 19/1443. Powell to Thatcher, 'Arms control: briefing meeting, Chequers, 2 February', 31 January 1985.
39. Howe made a speech at the Royal United Services Institute in mid-March 1985 that was interpreted as critical of SDI. In 2003 historian Holger Nehring wondered whether the speech might be a more accurate representation of the UK official line than Thatcher's Camp David support. I think it is now clear that there was a split. Holger Nehring, 'The British response to SDI: introductory paper', in Kandiah and Staerck, *op. cit.*, pp.17–24.
40. PREM 19/1444. Powell to Mottram, 8 January 1985. 'The Prime Minister's inclination will be to respond favourably to Sir Oliver Wright's suggestions'.
41. PREM 19/1444. Thatcher to Weinberger, undated (1985).
42. Donald R. Baucom, *The Origins of SDI, 1944–1983*. Lawrence: University of Kansas Press, 1992. William J. Broad, *Teller's War: the Top-secret story behind the Star Wars Deception*. New York: Simon & Schuster, 1992; Frances FitzGerald, *Way Out There in the Blue: Reagan, Star Wars and the End of the Cold War*. New York: Simon & Schuster, 2001; Gordon R. Mitchell, *Strategic Deception: Rhetoric, Science, and Politics in Missile Defense Advocacy*. East Lansing: Michigan State University Press, 2000.
43. W. Patrick McCray, *The Visioneers: How a Group of Elite Scientists Pursued Space Colonies, Nanotechnologies, and a Limitless Future*. Princeton: Princeton University Press, 2013.
44. Rebecca Slayton, *Arguments that Count: Physics, Computing, and Missile Defense, 1949–2012*. Cambridge, MA: MIT Press, 2013.

45. David Baker, 'The making of Star Wars', *New Scientist*, 9 July 1987, is a well-informed long history of SDI that locates its origins in Teller and his influence.
46. See also 'Comparative US and Soviet BMD attainments and capabilities'.
47. The first, the Hoffman report, was more sceptical of Reagan's long-term goal. See: FitzGerald, *op. cit.*, p.253.
48. A British newspaper quotation repeated in Karen DeYoung 'British seek "Star Wars" answers', *Washington Post*, 2 July 1985: 9.
49. PREM 19/1444. 'Record of discussions at the Pentagon on SDI: 17 July 1984', July 1984.
50. PREM 19/1444. Powell, 'Strategic Defence Initiative' (note of meeting between Thatcher and Professor Richard Norman, chief scientist, MoD), 8 January 1985. Norman took over from Ronald Mason in 1983.
51. Slayton, *op. cit.*, chapter 8.
52. PREM 19/1444. 'US and USSR BMD programmes', undated (October 1984).
53. Ibid. The latter refers to discriminating between decoys and warheads in the upper atmosphere.
54. Ibid, p.2.
55. PREM 19/1444. Powell to Mottram, 8 January 1985.
56. The Fletcher report was 'essentially promotional material, presenting an optimistic picture of the SDI concept without giving details', in the view of Whitehall. PREM 19/1444. Brennan to Powell, 31 January 1985.
57. Especially by European allies: PREM 19/1444. 'Record of discussion at the Pentagon on SDI: 17 July 1984'. 'Herr Ruth: … . Within Europe the Report on SDI by the Union of Concern Scientists had established itself as one of basic sources in the debate. Had any work been done within US Government to analyse and refute it?'
58. Zbigniew Brzezinski, Robert Jastrow, Max Kampelman, 'Search for security: the case for SDI', *International Herald Tribune*, 28 January 1985.
59. Hans Bethe, Richard L. Garwin, Kurt Gottfried and Henry W. Kendall, 'Space-based ballistic-missile defense', *Scientific American* 251(4) (October 1984): 39–49. 'Star Wars', *The Economist* (3 March 1984): 80–1. Kosta Tsipis, 'Laser weapons', *Scientific American* 245(6) (December 1981): 35–41.
60. Slayton, *op. cit.*, pp.186–8.
61. David Caplin, witness seminar testimony in Kandiah and Staerck, *op. cit.*, pp.93–106.
62. Handwritten note, from Powell to Thatcher, on PREM 19/1444. Brennan to Powell, 31 January 1985.
63. For example: Les Allen, 'Star wars: a paradox for our time. Whichever way you look at it, star wars doesn't make sense', *New Scientist*, 1 May 1986, pp.53–5. The Labour Party, led by Neil Kinnock, attacked SDI. See, for example: 'Labour opposition to SDI spelt out by Kinnock', *Financial Times*, 7 March 1985; '"Star Wars" condemned by Kinnock at NATO', *The Times*, 7 March 1985.
64. PREM 19/1444. Heseltine to Thatcher, 15 February 1985.
65. PREM 19/1444. Thomas to Thatcher, 6 February 1985.
66. Ibid. Kissinger, as a vale dictum, also 'wanted to congratulate [Thatcher] on what he took to be the outcome of the miners' strike and thought it a great victory not only for sanity but for the principle of resolution in these matters'.
67. PREM 19/1444. Telegram, Bonn to FCO, 12 February 1985. The conditions related to arms control, alliance unity and two-way exchange.
68. PREM 19/1444. Telegram, 'NATO supports SDI research', to FCO, 26 March 1985.
69. PREM 19/1444. Weinberger to Heseltine, 26 March 1985.
70. PREM 19/1444. Heseltine to Howe, 27 March 1985.
71. PREM 19/1444. Powell to Thatcher, 27 March 1985, Thatcher's emphasis.
72. PREM 19/1444. Powell to Mottram, 28 March 1985.
73. The FCO view was that there were obstacles to participation in research (such as Congress's 'Buy American' preferences, and a technical violation of the 1972 ABM Treaty, namely Article IX which precluded the sharing of technical descriptions or blueprints of ABM systems by the United States or USSR with allies'), but that there was 'little doubt … that the momentum generated by the new US programme will lead to new technologies with far-reaching defence implications'. PREM 19/1444. Budd to Mottram, 4 April 1985.
74. This committee was MISC 117. It is referred to in PREM 19/1445. Armstrong to Thatcher, 5 June 1985 and the minutes and memoranda are contained in CAB 130/1303.

75. John Peterson, 'EUREKA: a historical perspective', in John Krige and Luca Guzzetti, eds, *History of European Scientific and Technological Cooperation*. Luxembourg: Office for Official Publications of the European Communities, 1997, pp.323–345.
76. Peterson, *op. cit.*, p.328.
77. PREM 19/1445. 'Record of a meeting between the Defence Secretary and the US Defense Secretary, Mr Weinberger at 1400 on 22nd May 1985 in NATO HQ, Brussels', 23 May 1985.
78. PREM 19/1444. Powell to Thatcher, undated (March 1985).
79. PREM 19/1445. Heseltine to Thatcher, 28 May 1985. Powell's and Thatcher's handwritten comments dated 29 May 1985.
80. PREM 19/1445. Pattie to Thatcher, 24 June 1985. Pattie, 'SDI and Eureka', 24 June 1985.
81. Solly Zuckerman had offered quick calculations that supported this view. There were also European fears of a brain drain, one influence on the launch of Eureka. PREM 19/1445. Telegram, on Genscher's visit to Paris, 23 May 1985.
82. PREM 19/1445. Nicholson to Thatcher, 10 July 1985.
83. Heseltine's proposals were mauled at the Cabinet OD committee.
84. PREM 19/1445. Mottram, 'Record of a meeting between the Defence Secretary and Mr Caspar Weinberger, the US Defense Secretary, at the Pentagon on 22nd July at 1230', 24 July 1985.
85. PREM 19/1445. Heseltine to Thatcher, 23 July 1985.
86. This is the £1.1 billion noted in Paul Sharp, *Thatcher's Diplomacy: the Revival of British Foreign Policy*. London: Macmillan, 1997, p.130.
87. PREM 19/1445. Powell to Thatcher, undated (July 1985). Powell noted that Heseltine had not mentioned a specific figure when the approach was discussed at OD committee.
88. I have asked Michael Heseltine this question directly, at the Thatcher Network conference 'Thatcherism Now', 5–6 April 2018, but unfortunately his answer, many years of course after the date, addressed generalities.
89. PREM 19/1445. Panton to Armstrong, 30 April 1985.
90. A published discussion of what research was ongoing can be found in the witness seminar testimony of Roy Dommett, in Kandiah and Staerck, *op. cit.*, pp.37–44.
91. PREM 19/1445. Unwin to Mallaby, 26 September 1985.
92. PREM 19/1445. Owen to Thatcher, 25 October 1985.
93. PREM 19/1445. Joseph to Heseltine, 28 October 1985.
94. 'US/UK officials hit snags on SDI co-operation', *Washington Post*, 26 October 1985.
95. PREM 19/1445. 'Discussions on UK participation in SDI research', undated (September 1985).
96. 'US/UK officials hit snags on SDI co-operation', *Washington Post*, 26 October 1985.
97. PREM 19/1445. Nicholson to Thatcher, 25 October 1985. Particular areas of spin-off, thought Nicholson, were optical computing, electronic and non-electronic materials, and software security.
98. PREM 19/1445. Powell to Thatcher, 25 October 1985.
99. PREM 19/1445. Heseltine to Weinberger, 30 October 1985. Heseltine to Thatcher, 31 October 1985.
100. CAB 130/1303. MISC117(85)24, 17 December 1985, contains the agreed Memorandum of Understanding. CAB 128/81/35. Minutes, Cabinet, 5 December 1985 contains the Cabinet approval. CAB 128/81/36. Minutes, Cabinet, 12 December 1985 records Heseltine's signature of the Memorandum on 6 December 1985. The Cabinet also discussed Westland plc.
101. PREM 19/1445. Powell to Thatcher, 29 November 1985.
102. Copy of Reagan's apologetic telegram is here: http://www.thereaganfiles.com/851105.pdf.
103. Thatcher, *Downing Street Years, op. cit.*, p.471.
104. Thatcher, *Downing Street Years, op. cit.*, p.463.
105. See, for example, PREM 19/1444. R. V. Jones, 'Some thoughts on Star Wars', 15 March 1985. PREM 19/1445. Panton to Armstrong, 3 April 1985, p.2.
106. Thatcher, *Downing Street Years, op. cit.*, p.471.
107. Freeman J. Dyson, *Disturbing the Universe*. New York: Harper & Row, 1979.
108. Holger Nehring, 'The British response to SDI: introductory paper', in Kandiah and Staerck, *op. cit.*, pp.17–24.
109. Thatcher, *Downing Street Years, op. cit.*, p.466.

110. Michael Heseltine, witness seminar testimony, in Kandiah and Staerck, *op. cit.*, p.32. Heseltine later recalled the initial reactions to SDI: 'I think that the reaction to the announcement was one of despair – Oh Lord! Here we go again, the next escalation in the arms race – and then realpolitik – Oh help! What's this going to do to Britain's independent nuclear deterrent? Then thirdly – and my golly, US$29 billion worth of research, what sort of competitive advantage is that going to deliver to the United States?'
111. Charles Powell, witness seminar testimony in Kandiah and Staerck, *op. cit.*, p.32, p.51.
112. Charles Powell, witness seminar testimony in Kandiah and Staerck, *op. cit.*, p.72.

7
Environment and science

This chapter builds a picture of Thatcher's response to five major environmental issues, all of which intimately involved scientific knowledge, either in the framing of the issue or in the articulation of the political response. First acid rain, the consequences of pollution of the lower atmosphere by sulphur dioxide and nitrogen dioxide, became an international controversy in the 1980s – not least when the leaders of West Germany and Norway directly appealed to the Prime Minister to take urgent remedial action. Thatcher drew on her experience in chemistry, reviewing, for example, the atmospheric chemical reactions involved. Like the second issue, the discovery of a 'hole' in the ozone layer by British scientists based in Antarctica, the acid rain controversy could not be kept separate from other major events of the Thatcher administration: the miners' strike and the politics of coal-fired power stations in the case of acid rain, and the significance of science in justifying UK presence in the South Atlantic, a factor in the Falklands saga. The third issue, climate change, while slower to gather momentum, also had surprisingly deep roots, but would be emphasised by Thatcher in her 1988 science speech to the Royal Society. The fourth topic is the role of science and scientific evidence in conservation issues. These include the Wildlife and Countryside Act (1981), the disputes over private farming interests and Sites of Special Scientific Interest and the development of the World Conservation Strategy. Finally I examine the growing concern over the release of genetically modified organisms (GMOs) in the environment. The first three issues were regarded by Thatcher's Cabinet as more important than the last two, as measured by the prime ministerial and ministerial attention they received.

Acid rain

The new environmental movement that emerged in the 1960s led to international discussion and limited action in the 1970s. The high-water mark for international agreement on global and regional concerns was the UN conference on the human environment held in Stockholm in 1972. The issues debated included desertification, deforestation and over-exploitation of non-renewable natural resources, as well as worries about the state of the atmosphere – including climate change, damage to the ozone layer and the possible harm caused by transport of acidifying air pollutants. Between 1955 and 1965 Swedish and Norwegian weather stations reported a rise in acidity, with similar research being conducted in eastern North America a few years later. Also by the late 1970s the fact of long-distance atmospheric transportation – of the order of 1,000 kilometres or more – of sulphur, nitrogen oxides, sulphates and nitrates was established. While the increasing acidification of lakes was clear – in Sweden, for example, the acidity of water in 10,000 lakes had dropped to pH6 and in the worst-affected 5,000 to pH5 – other chemical and ecological effects, such as the mobilisation of heavy metals, effects on plankton and the reduced growth of trees were also being reported.

In Geneva in November 1979 ministers of the environment, including the UK's, signed a Convention on Long-Range Transboundary Air Pollution. The convention only dealt with research, monitoring and exchange of information, with the development of policies of mitigation or prevention therefore dependent on agreed progress in these scientific activities. The combined effects of a shift from coal to oil (not least because of North Sea Oil coming on stream) and economic recession meant that, in fact, UK sulphur dioxide emissions had fallen to 1950s levels in the early 1980s. Nevertheless the UK still produced about one-quarter of the European Community's sulphur dioxide, although it deposited proportionately less on other countries.[1]

Scandinavian countries, disadvantaged by the geographic locations of polluting industries and westerly air movements, were particularly vocal in arguing that action must be taken against 'acid rain'. In August 1981 Anders Dahlgren, the Swedish Minister for Agriculture, invited Michael Heseltine, then UK Minister for the Environment, to a conference, to be held in June 1982, on the 'various conceivable solutions to the problems of … long range transboundary air pollution'.[2] 'Although there is no doubt that Scandinavia does suffer from the problems of acidification, and although it is clear that the problem is at least partly due to sulphur emissions in Western Europe, the scope for action to help the Scandinavians is limited,' noted a civil servant in the Ministry for

Environment, adding that achieving a 'reduction in sulphur emissions would be expensive, and would place a substantial burden on the electricity industry'.[3] The Department of Energy, which could be expected to protect electricity producers' interests, took up a posture that was satirised as 'do nothing, and do it in concert with the other major industrial nations'.[4] Therefore a range of departmental and industrial interests conflicted and needed to be consulted further. It is among these different interests that we can see different interpretations about the need for research and the conclusions that could be drawn from research findings.

Indeed, fracture lines among these different interests can be found within research (such as that supported by the NERC and that supported by industry), between government departments (such as Environment versus Energy), between central government and other public bodies (such as the National Coal Board) and between politicians and the civil service. Broadly, however, the UK line was that the country's position should be 'positive, but rigorously logical' and to be 'willing to join any critical analysis which would allow a sensible weighing up of the options and their attendant costs and implications', while avoiding 'any gesture for gesture's sake'. Most importantly, a position was sought that would avoid the UK being singled out or isolated.[5] This balance would be tricky to sustain since, as Martin Holdgate, the Scientific Adviser to the Department of the Environment, noted, 'the United Kingdom [was] probably the largest single foreign source of the sulphuric and nitric acid falling over southern Scandinavia'. From these overarching aims Holdgate concluded that, first, ministerial, not just official, representation was desirable in the 'hot-seat' of the 1982 conference. Second, the logic of the line would depend on the careful deployment of scientific evidence:

> (i) we accept the current scientific evidence that there has been an increase in acidity in certain southern Scandinavian lakes and river systems (as there has been in Canada where similar geological and environmental conditions prevail).
>
> (ii) we show less willingness to accept the arguments on damage to forests and land systems, resting on the various scientific reviews that have recently occurred (we shall of course need for the meeting a really rigorous scientific brief …)
>
> (iii) we reaffirm our commitment to the information exchange and monitoring component under the ECE [UN Economic Commission on Europe] Convention, but stress that for any such action to be effective all European industrialised countries, east and west, must participate vigorously …[6]

Any practical action therefore, would then depend on further research, working back from Scandinavian targets, calculating degrees of abatement required, 'work out in reasonably hard scientific terms the key factors and thus the areas where a campaign to abate pollution would be most likely to succeed' and only pursue these after 'economic evaluations' conducted in such as a way as 'timescales are taken into account' – perhaps leading to 'abatement over a century' and avoiding a 'crash campaign designed to make a measurable effect within 15 years', say.

The interests at play in UK research into acid rain were antagonistic, although not to the extent that had developed in the United States. Under the new Reagan administration a battle over a Clean Air Act had seen US environmental agencies and Canadian politicians coming into sharp conflict with US industrial interests. The latter, under the banner of the 'Coalition for Environmental-Energy Balance', had organised advertisements that stressed the political action should not be taken while 'uncertainty' existed over the causes of acid rain. The Science Attaché at the British Embassy in Washington explained that this strategy was aligned to the US federal government's espousal of 'Sound Science''. By this they seek to exploit the uncertainties and apparent contradictions in what is known about long range transportation'. He did not pull punches in his analysis:

> Compared to the previous Administration's tendency to legislate based on a rumour of possible environmental hazard, it is tempting to regard the new US attitude as more balanced and reasonable. In my view this would be a mistake for the US has swung through the middle ground and is now at the opposite extreme; ... each side unashamedly collects evidence to support its own case and ignores results that run counter to the official line. Committees are staffed with 'scientists' whose preconceived views are already well known, and thus sound science has become a euphemism for numbers selected to support a political position.[7]

In the UK the industrial interests were represented by the Central Electricity Generating Board (responsible for coal-powered power stations) and the National Coal Board (responsible for producing, and promoting the use of, coal). An internal strategy document at the National Coal Board concluded that the 'Board's first line of defence (that the scientific case has not been made) is looking vulnerable to political pressures'; it argued that work needed 'to begin in earnest on a second line', specifically cheaper technological options than scrubbing ('scrubbers' are devices added to power plants to remove pollutants). A CEGB-sponsored meeting of experts on 'Forests and acidification' stressed that a decade's research into the phenomenon of acid rain had found that the 'relationship between air quality and rain on the one hand, and the quality of biotic response in surface waters' was not a 'simple one'.[8] This level of uncertainty was not unusual, however, although the extent to which it was emphasised or seen as a cause for 'precautionary' action mattered immensely.

The foremost UK sceptical expert was Peter Chester, director of the Central Electricity Research Laboratories.[9] Chester's research papers and public talks emphasised the uncertainties over acid rain. For example, a 1983 speech at the Royal Society of Arts argued that 'the politics of acid rain have run ahead of the science from the outset'.[10] Chester wrote to J. M. Doderlein of the Royal Norwegian Council for Scientific and Industrial Research (following their meeting at a symposium organised by Exxon). Clearly continuing an earlier argument, Chester listed 'three different research findings which separately or together do not support the simple expectation that a reduction in SO_2 emissions in Western Europe would bring about a corresponding amelioration of fishery problems in Southern Scandinavia'.[11] Chester argued that it would be 'heartbreaking for Europe to invest £ billions in measures which then turned out to be ineffective'. Norway's 'foremost scientific specialist on the acid rain issue', Hans Martin Seip, hit back with a detailed critique of Chester's data, assumptions and argument. He concluded that:

> You may find it heartbreaking for Europe to invest an enormous amount in measures which turned out to be ineffective. Others will find it heartbreaking if nothing is done, and the consequences turn out to be as serious as feared not only for aquatic systems but also for vegetation and perhaps health … In summary we agree that many of the questions you raise are important, but if all details have to be cleared up before action is taken to reduce emissions, the damage done may become very large and difficult to repair.[12]

We know about this exchange because the Norwegians alerted and shared the correspondence with the Department of the Environment.[13] It was a confirmation of something they already knew: that there was a divergence of interests, and conclusions from research, on the acid rain issue. The Department of the Environment assumed that CEGB, on the basis of another of Chester's papers, would dispute the Department's statement that 'evidence, albeit circumstantial [of a link between long-range transport and acid rain effects] is beginning to accumulate'.[14]

The UK response to this division among researchers was to seek a balance of bodies. The experts chosen to attend the Stockholm conference were carefully matched not only to subjects, but also to obtain a 'reasonable balance' between government and research council nominees on the one hand and CEGB scientists on the other. In so doing they sought to ensure that the 'UK expert delegation does not become CEGB dominated while recognising at the same time the important role CEGB scientists are playing in the acid rain issue'.[15]

However, there was a problem. When Holdgate had to assess the arguments of Chester he found that there was 'nobody in Whitehall who is competent to give us an authoritative referees report'.[16] The Department of the Environment had funded research on new abatement technologies (to gauge the rate at which they were likely to improve and therefore be an option in the future), research on acidity of rainfall,[17] research on dispersion and research on the relationship between SO_2 and NO_x emissions and damage to vegetation.[18] Possible damage to animals and human health had not been investigated as it was seen to be less likely. There were other gaps too. The Department of the Environment, under pressure to cut public expenditure, had reduced funding at government laboratories and universities; the Natural Environment Research Council, one of the victims, had warned that the cuts were 'doing serious damage to the credibility of the DOE and the "independent scientists" in Europe and N America'.[19] Some within the department were comfortable with leaving the research to the CEGB. Holdgate warned against this approach:

> I do not believe that Government will be comfortably placed if the only source of expertise in the country is seen to be the organisation with the greatest financial interest in avoiding emission control. … I certainly think that we shall be ill-placed to negotiate with confidence in the international world if it is known that all our data comes from our public utility (or indeed from any other industrial corporation).[20]

Likewise the NERC agreed: 'CERL is distrusted here and abroad, however much it protests neutrality in science. Lack of DOE support is giving CERL a much higher profile at international level which can only be damaging to the UK especially in Europe'.[21]

The UK, then, went into the Stockholm conference on acid rain in June 1982 with the line that, while the link between acid rain and environmental damage in Scandinavia had become slightly more certain, the continuing uncertainty (and the high cost of technological fixes, such as flue gas desulphurisation)[22] meant that the policy should still be more monitoring and research, even while knowing that the interests behind the research were conflicted.[23] Heseltine wrote to Lawson confirming that the line to be taken by the UK ministerial representative, Giles Shaw, was to use science to restrict UK liabilities:

> it is important that we should not be carried on a wave of assertion about the effects of acid precipitation beyond what the scientific evidence establishes. It would be best if we were able to avoid commenting on these matters at all, but this may prove to be impossible. If our delegation is pressed, they might accept the causal link between the emission of various gases and the acidity of rain, and the capacity of that acidity to cause damage to certain types of fresh water system and to alter the chemistry of certain soils. But the cause and effect chain is complex … This acceptance might be placed in the context that energy, economic and technical issues as well as environmental ones must be involved in any attempt to find a solution to the acid rain problem.[24]

Shaw returned from Stockholm to report that the 'UK achieved its broad aims'; a more informal feedback was that 'the pace is quickening on this whole subject' and what was needed was not only 'more, and better balanced, research', but also policies adopted that actually controlled sulphur emissions.[25]

At this point a new configuration of organisations had emerged that shaped acid rain research in the UK. In January 1983 Nigel Lawson, then Secretary of State for Energy, and Tom King, who had stepped up to become Secretary of State for the Environment on Heseltine's move to Defence, invited views on a new review of acid rain policy. Citing the Stockholm conference as evidence of growing international concern, the joint review aimed to deliver 'a clearer idea of the controls which the UK might accept over the next fifteen years or so, leaving us better placed to exercise a constructive influence on international discussions and decisions'.[26] The consultation lasted three months.

In this consultation the National Coal Board repeated the view, citing Shaw at Stockholm, that '"acid rain" and its effect are not properly understood' and that 'the cost which on present knowledge would be involved in securing even a measure of reduction in emissions is so great that a much more precise evaluation of the effect of such reductions is essential and they would help in such a study'.[27] Walter Marshall, FRS, the new chair of the CEGB, turned to the Royal Society to help. In

September 1983 the Royal Society (alongside its sister scientific academies of Norway and Sweden) announced a collaborative programme of research into the causes of acidification of surface waters in Norway and Sweden and the implications for fisheries. John Mason, who had just retired from the Meteorological Office, was to direct. The funders of this programme – over £5m – were the National Coal Board and the CEGB.

This unusual combination deserves explanation. In addition to diplomatic pressure from Scandinavia, the CEGB and the National Coal Board were certainly under public pressure as acid rain became a matter of public controversy. *Time* magazine, for example, had declared acid rain 'the silent plague', 'the scourge' and the 'ecological issue of the 1980s'.[28] The British press offered similar apocalyptic imagery (one example, a *Mirror* editorial of 1983 was titled 'Death from the skies').[29] Acid rain demonstrators protested outside the headquarters of CEGB and the Department of the Environment in the same year,[30] and dumped a coffin of dead fish outside the Royal Society.[31] The appeal of asking the Royal Society to lead the research was that it had the reputation of being, in the words of its historian, an 'honest broker'.[32] Indeed the press releases stressed that the 'content and direction of the [research] programme will be entirely in the hands of the Royal Society and the academies', while the 'results will be published without restrictions'.[33] The trade press *Energy Daily* called bringing in the Royal Society 'a master stroke of diplomacy'.[34]

Not everyone was convinced by the Royal Society as honest broker, however. Des Wilson, chair of Friends of the Earth, in a letter published in the *Guardian*, cast doubt on the objectivity of the research, remarking that it was 'extraordinary how often industrially-financed research happens to support the view of industry'.[35] The Royal Society rejected the accusation, citing past examples of industrially-funded research that showed 'no signs of industrial bias'.[36] Another factor might be nuclear politics: journalists noted that Marshall, the long-standing advocate of nuclear power, had emphasised that desulphurisation raised the cost of coal power.[37]

Other acid rain research was expanded in this period, despite the cuts. The Department of Energy sponsored research at its Energy Technology Research Unit at Harwell.[38] The Department of the Environment and the Nature Conservancy Council collaborated to ensure that a network of monitoring stations to assess the composition of precipitation gave more complete coverage of the UK.[39] The Nature Conservancy Council also commissioned Imperial College to review research and arbitrate between two conflicting studies of damage to native tree species by air pollution and acid precipitation. The first, conducted by the Forestry Commission, had found little damage. The second, an example of citizen science *avant la lettre*, had been organised by Friends of Earth. Volunteers had submitted observations of yew and beech trees from around the country. The Imperial College report made methodological criticisms of both projects, but concluded that while further research was needed there were 'grounds for concern'.[40] Like the Royal Society initiative, it was an interesting intervention to attempt to resolve accusations of interest-driven research.

As the acid rain controversy rose up the political agenda, so we see more discussion at the centre. Furthermore, this discussion played environmental issues against other, including party political, concerns. The respect within Number 10 accorded to the Department of the Environment was not high. Ferdinand Mount, head of the Number 10 Policy Unit, told Thatcher that it was 'in reality only a glorified Ministry of Housing and Local Government' and shared William Waldegrave's rating that its environment staff was 'small and of poor quality'.[41] Yet Mount also insisted that environmental matters were 'going to provide some of sharpest political challenges in this Parliament', while worrying that 'the Conservatives tend to be branded as uncaring Philistines [on environmental issues] – thus creating a breeding ground for the SDP and the Liberals'.

Thatcher was also 'disturbed about inadequate public understanding of the problem of acid rain'.[42] On a Sunday in late May 1984 she hosted at Chequers a series of presentations on acid rain for the benefit of her ministers. The aim was 'to present the scientific evidence' and 'to describe the state of the art … in abatement technology'.[43] Her Chief Scientific Adviser, Robin Nicholson, spoke on sources of emissions, John Mason on chemical changes in the atmosphere, Hermann Bondi on lakes and streams, Martin Holdgate on forest damage and Peter Chester on abatement technologies for power stations (emphasising the high costs and uncertainty of results). Finally Nicholson returned with an account of abatement technologies for vehicles (such as catalytic converters, already introduced in the US, and 'lean burn' engines, favoured by Europe-based manufacturers). Thatcher, the former chemist, asked to see the chemical equations, which Chester subsequently provided – a list of 93, a 'glance' at which, he suggested to her, 'will give you some idea of the complexity and the role played by hydrocarbons'.[44] Even a list of chemicals could serve a political purpose between chemists.

Following the Chequers meeting there was extensive ministerial and official discussion of the acid rain issue. Patrick Walker, the Energy Minister, wrote that while the CEGB had assumed 10GW – nine Sizewells – of new nuclear power would be in operation by 2000, he now considered four to five new stations more likely.[45] This would mean that 'we should be well short of achieving a 30% reduction in emissions by 2000'. Yet he also told Thatcher that retrofitting large plants with desulphurisation technology was too expensive. In the light of 'current scientific uncertainty', Walker urged a continuation of present policy.

William Waldegrave, the Undersecretary of State for the Environment, writing to Thatcher, gave a more substantial set of options for a problem he noted had 'both scientific and political components'.[46] He urged rejecting three options: the existing policy ('pursue a vigorous and well published research programme ... but take no other special action'), the Large Plant Directive stemming from the European Commission (in other words gas flue desulphurisation) and the one urged by West Germany and the Scandinavian countries (join a '30% club' of firm commitments to cut sulphur dioxide emissions by this figure). Instead Waldegrave argued that a mix of new nuclear (but at least four to five new Sizewells), NO_x burners, action on vehicle emissions and efforts on sulphur short of the expensive refitting would be an appropriate response. Nicholson supported the plan, citing 'scientific evidence'; so did David Pascall, a member of the Number 10 Policy Unit seconded from British Petroleum.[47] The plan was agreed, with minor changes, on 19 June 1984. It was the position taken by the UK delegation, led by Waldegrave, to the Munich Air Pollution conference held later in the month.[48]

In September 1984 the Select Committee on the Environment issued a substantial report on acid rain.[49] It was severely critical of the government and advocated retrofitting power stations and joining the '30% club'. Nicholson did not think its arguments or evidence would have changed the decision taken on 19 June, but also noted that the 'fact that a Committee of MPs has come out so strongly for severe abatement measures will increase international pressure on the United Kingdom to take more action than is envisaged under current Government policy'.[50] The government's response largely confirmed the existing policy – in particular that the fitting of desulphurisation technologies on coal power stations was too expensive.[51] Specifically this was justified by an appeal to uncertainty and science:

> [The Government] does not believe that the very substantial expenditure (running into hundreds of millions of pounds) which would be required to install flue-gas desulphurisation plant at existing power stations can be justified while scientific knowledge is developing and the environmental benefit remains uncertain ...

Pollution is dealt with by political action, but it is explained by science. Science is dynamic, and the policies of this and other Governments must evolve to meet new evidence.[52]

In the more detailed responses to the select committee's recommendations, the government again promised more research.[53] Likewise the Royal Commission on Environmental Pollution's 10th Report was met with a restatement of 'pollution control achievements and philosophy'.[54]

The government was also increasingly feeling the pressure from Europe. Jenkin confessed to Howe, the Foreign Secretary, that the UK was faced with 'a difficult and uncomfortable balancing act; we need to avoid killing negotiations while making our reservations about a commitment clear'.[55] 'We will find it difficult to tie ourselves to inflexible reductions and time scales,' Jenkin wrote. 'However, we should indicate that we do not rule out ultimate consensus.' John Redwood, then head of the Number 10 Policy Unit, was outraged: 'the "Yes Minister" script of Patrick Jenkin to Geoffrey Howe will not do,' he told Thatcher. What was needed instead, he argued, was for the agreed acid rain policy of 19 June to be stuck to and sold:

> If we do not come out soon with a clear and forthright statement of our intent here in the UK, we will find that the pressures represented by the Environment Committee will build up further and may force us into a more expensive manoeuvre on Patrick's high wire …
> The Government has to be seen doing more than just singing in the acid rain, and if it delays any longer, it will find it too expensive to buy an umbrella.[56]

The international pressure continued when the Prime Minister of Norway, Kåre Willoch, wrote directly to Thatcher, having recently agreed a joint declaration with Helmut Kohl of West Germany.[57] Willoch, perhaps pointedly, addressed Thatcher as 'FRS, MP'. Thatcher replied, restating the UK position and adding 'We shall naturally stand ready to take further action in the light of changing scientific evidence'.[58] A few months later, in conversation in Thatcher's House of Commons office, Willoch told her that 'the United Kingdom argument that there were scientific uncertainties about the effects of acid rain did not carry great conviction'.[59]

The pressure began to tell. German insistence on vehicle emissions led to new standards being conceded by Waldegrave in Brussels in June 1985 ('a good week for Germans and Japanese,' wrote an incandescent Norman Tebbit).[60] Patrick Jenkin raised the possibility of the UK making the public act of joining the '30% club', since the likely drop in sulphur dioxide was looking more approachable.[61] Nicholson advised that the 'scientific understanding of the processes and effects of acid deposition' had shown 'no significant change' in the 19 June 1984 ministerial agreement; any commitment to 30 per cent must therefore imply a willingness to adopt, if missed, either retrofitting (still regarded as too costly) or an accelerated nuclear build (which Nicholson preferred).[62] Officials cast doubt on whether the decline in sulphur dioxide emissions could be maintained – the recent drop was mainly due to the closure of steel works.[63]

We have now reached the moment of major policy change, one which has been described as 'the first major policy decision on acid rain' and one which 'represented a complete change in direction in UK air pollution policy'.[64] What is fascinating to me is that the decision was brokered by an appeal to shared scientific understanding between two politicians, yet the role of scientific evidence was ambivalent.

In late June and early July 1986 Walter Marshall, recently ennobled as Lord Marshall of Goring, allegedly in appreciation of his work to 'keep the lights on' during the miners' strike,[65] visited Norway and Sweden. Here he was the guest of the two academies with which the Royal Society was conducting the acid rain research sponsored by the CEGB and the National Coal Board. In Norway he met Gro Harlem Brundtland, at her request. Brundtland was starting her second term as Norway's Prime Minister, having replaced Willoch in May. Also in attendance was Professor Lars Walloe, a close friend, indeed ex-supervisor, of Brundtland, and a previous Minister for the Environment under her.

Knowing about the academies' project, Brundtland asked Marshall about acid rain.[66] Marshall replied that he saw two 'historical reasons' why it was difficult to assess.[67] With these resolved sympathetically, Brundtland argued that she now thought the 'Norwegian scientific argument was now much stronger and better established scientifically than it was some years ago, and she would rely on the British to acknowledge that scientific evidence and take appropriate actions in the near future'.[68] Marshall replied that he was 'unable to detect any serious difference in scientific approach' between them, and that the 'present Joint Research programme was a good beginning to getting a fresh understanding between our two countries on this important subject'. Brundtland, reported Marshall to Thatcher, 'was very much looking forward' to Thatcher's forthcoming visit to Norway, adding:

> She was proud of the fact that you both had a scientific training before entering politics and said that 'we scientists must stick together and set an example to other people' [the underlining is in Thatcher's hand].[69]

On returning to Britain, Marshall wrote to Peter Walker, Secretary of State for Energy. His letter indicated that he was now convinced that there must be a 'fundamental shift in the CEGB's attitude towards retrofitting of emission control equipment in existing power stations', citing the joint academy research as 'sufficiently convincing scientific evidence':[70]

> We cannot sustain the position (that scientific evidence was incomplete and inconclusive) once the Royal Society has reported to us in a year's time and we see great merit to anticipate that position by taking our first steps now.[71]

Several analytical comments can be made. First, this change of heart by Marshall was the new view that triggered the new policy. Second, no other agent was responsible for overturning the existing policy (Cabinet committee documents that are exactly contemporary show that the intention was to defend the existing line.)[72] Third, while it was made supposedly in the light of the joint academy research, the decisive moment was in advance of this evidence being presented and discussed (although interim results may have been available). Fourth, since, as Thatcher's private secretary advised, 'Lord Marshall had not provided the scientific evidence to support his change of views and indeed he has not discussed his position even with other people in the CEGB; Lord Marshall's views can and do change and could change again', Marshall's word alone was not enough to change policy.[73] The extra element, plausibly, was Brundtland's canny tactic of framing this decision as one to be taken by a sisterhood of ex-scientists.

The new policy was announced in September 1986, Thatcher having accepted Number 10 Policy Unit advice that politically 'it would be dangerous for the Government to do less than endorse the CEGB's proposal'.[74] All new coal-fired stations – and even three 2000 MW of existing plant, starting with Drax – were to be fitted with flue gas desulphurisation, the technology previously deemed too costly to install and unjustified by scientific evidence. A summary of the evidence was prepared by the chief scientists at the departments of Energy and Environment, and at the CEGB.[75] Meanwhile John Fairclough, the new Chief Scientific Adviser to the government, gave his Prime Minister his assessment:

> Although there has been no dramatic breakthrough, I am satisfied that the weight of evidence is now sufficient for action. … Absolute certainty is not the currency in scientific issues as complex as this. Lack of such watertight certainty should thus not deter us from taking action .[76]

Of course, lack of watertight certainty had indeed been the reason given for not taking action before. The announcement took place to coincide with Thatcher's visit to Brundtland in Norway. Her press officer, Bernard Ingham, told the Prime Minister that 'I have been trying to get over the idea that … the decision will be based on scientific evidence and not because you are visiting Norway or Germany next week'.[77]

To sum up: acid rain became steadily more prominent as a political issue through the 1980s. The government was slowly dragged into action, although when the change of policy happened in 1986 it came very suddenly. Scientific evidence was cited as a major part of the justifying arguments throughout, both as a reason for resisting major change and then, abruptly, for making it. The research led by the Royal Society was essential to Walter Marshall's change of heart in 1986. A shared scientific background was also used by the Norwegian Prime Minister, Gro Harlem Brundtland, in persuading Thatcher, at one remove via Walter Marshall, of the new acid rain policy.

While the policy changes followed the miners' strike, there was no direct connection. *The Times*, in an editorial on the event of the publication of the Select Committee report on acid rain in September, written as the strike was intensifying, noted that 'Curiously enough, the MPs show no eagerness to see our highly sulphurous home-mined coal replaced by imports or by more nuclear power'.[78] Yet there is no evidence that the government used acid rain policy as a tactical option in the miners' strike – not least because there were other reasons for reluctance to promote in public a new nuclear build. Nor is there documentary evidence to suggest that Chernobyl, by making the policy choice of more nuclear power stations less likely, contributed to the acceptance of flue gas desulphurisation as the alternative route to lowering sulphur dioxide emissions, although one can speculate about Walter Marshall's reasoning. Acid rain remained a European issue after 1986, with further pressure to reduce SO_2 (perhaps by 70 per cent by the 2000s) and NO_x emissions and to adopt the Large Plant Directive. But it took second place to other environmental issues, even as green politics briefly flourished in the late 1980s.

Antarctic research and the ozone hole

The 1959 Antarctic Treaty, which froze territorial claims, formalised the situation that political influence on the southern continent was contingent on the active conduct of scientific research.[79] However, by the early 1980s Antarctic research was unloved and, given its expense, an understandable target for cuts in research budgets. Options circulating in 1980 included at one extreme the closure of research stations and the withdrawal of the research ship *John Briscoe*.[80] In September 1981 the Natural Environment Research Council, which had assumed responsibility for the British Antarctic Survey (BAS) in 1967 from the Foreign Office, proposed closing the Grytviken base on South Georgia.[81] After the Foreign & Commonwealth Office expressed alarm, a compromise was reached whereby the Falkland Islands Government agreed to pay to meet the cost of maintaining 'a scientific presence' at Grytviken.[82] Nevertheless in December the Advisory Board for the Research Councils, being 'responsible for recommendations on the allocation of the Science Budget' and of the opinion that there should be a 'curtailment of the most expensive areas of science' at a time when there were 'numerous claims on the science budget' (see Chapter 3), suggested that NERC might hand back responsibility to the FCO.[83]

On 3 April 1982 South Georgia was seized by Argentine naval forces, and 'the 13 BAS staff present at the station were forcibly removed to the Argentine ship *Bahia Paraiso*' (a further nine BAS scientists, as well as two visiting photographers from Anglia TV, remained at large, hiding out in field huts on the island).[84] Thus the Falklands Islands conflict began. The reduction in scientific staff had been interpreted in Buenos Aires as a signal of diminishing political will to keep Las Malvinas.

Symbolically, then, British Antarctic Survey research had to be expanded again. Within weeks of the end of the Falklands War in June 1982 the Foreign Secretary, Francis Pym, proposed a virtual doubling of the BAS grant; a Cabinet committee agreed.[85] The decision was described as 'purely political'.[86] The question was where the money should be found: should this be by rearranging priorities, finding additional money (say from the contingency fund) or even by effectively abolishing the Social Science Research Council?[87] Thatcher demanded that she talk to 'those who decide the allocation of the research money', specifically the chair of NERC, Hermann Bondi. In the event Thatcher met with Bondi, Alex Merrison (chair of the ABRC) and Keith Joseph, the responsible minister, where she opened with the strategic reasons to expand BAS:

> the value of the activities of the British Antarctic Survey had not been fully appreciated until the Falkland Islands crisis. She had discussed the work of the Survey with some of its principal scientists and she found it very impressive. We needed to ensure that we were in the strongest position in the Antarctic region when, in 1991, the possibility of modifying the Antarctic Treaty would first arise. The

> area was one of great strategic importance with extensive natural resources. In the past we had followed the policy of backing good scientists, for example in the field of nuclear physics. She believed that another £5 million should be found for BAS, from either the existing NERC budget or from within the total Science Budget. She did not wish to become personally involved in discussion of priorities which was for the bodies concerned.[88]

But with a NERC total budget of only £57.5m, and a total research council budget of £234m, neither Bondi nor Merrison wanted to accept BAS activities in place of research that they considered had higher scientific justification. Merrison took the opportunity to complain about his research council budget declining in 'real terms'. 'On scientific grounds,' he said, 'we should, before increasing expenditure on BAS, support new projects in areas of greater scientific priority and provide more adequate support for existing projects.' Thatcher in turn insisted 'we now needed an extra £5 million for … work which was of great importance to the country as a whole'. When Merrison said that he would accept being told to take 'strategic money' out of the science budget, Thatcher repeated that she did not want to do this, but there 'had to be some means of adjusting priorities to take account of changing circumstances'. The meeting ended with Bondi suggesting that the science budget be increased by £5 million, supported by Merrison. The two went away thinking this had been agreed. The misunderstanding was not revealed until August 1982. Then, even though Joseph warned that to 'earmark the money within the Science Budget would … run a serious risk of souring relations' between Government and the research councils', Thatcher did just that: 'The Government will earmark £5 million annually for BAS by setting aside that sum from within whatever provision is made for science in our cash plans'.[89]

The affair is interesting for several reasons. First, it shows that Thatcher was willing, despite her opening remarks, to intervene as a politician on research council priorities. The context here was both immediate (the Falklands War) and strategic (looking ahead to UK interests if the Antarctic Treaty was revised). It was a breach of convention – the so-called Haldane principle, which stated that politicians should not direct research councils' decisions on research. Nevertheless, it should also be noted that this intervention was reluctant and rare. The existence of the convention was shown by the breach (witness Joseph's response).

Second, notice that the scale was modest. *Despite* the shock and international significance of the Falklands conflict, and its reputational implications for Thatcher, and *despite* the fact that scientific research was essential to being a player in Antarctic politics, the increase in BAS activity was relatively tokenistic. More important than either consideration was the commitment of Thatcher's administration to reducing public spending. For this reason even a modest increase in overall research council funding was rejected. The political considerations governing science policy and funding discussed in Chapter 3 were, when the whole is taken into view, more important than the research funding decisions raised in this chapter.

In 1981, the areas of Antarctic research that were perceived to be of growing importance were first, geology ('as the search for hydrocarbons, metals and other resources will intensify around the world'); second, life sciences (the Antarctic, like the Arctic, offered unique and relatively closed ecosystems for study); and third, climate, about which 'concern … was increasing all the time'.[90] The British Antarctic Survey's three geophysical observatories, at Halley Bay, Faraday and Grytviken, were 'strategically placed in a zone which forms a unique natural laboratory for the study of many atmospheric phenomena', not least because of freedom from local pollution and access to polar features.[91] Under the conditions of expanded research, various topics were listed in atmospheric geophysics, such as the interaction of the solar wind with the magnetosphere, 'plasma waves, plasma convection and the dissipation of auroral substorm energy'.[92]

Not listed was the research that led to perhaps the most dramatic discovery of the period – one which, although not enabled by the post-Falklands expansion, would not have been possible if BAS had been shut in 1981 and one which would be folded into Thatcher's post-Falklands late-1980s embrace of environmental science. Joe Farman was in charge of BAS's Antarctic Dobson meter, which measured ozone levels above the Halley Bay base. In 1982, and again in 1983, Farman and his junior colleague Jonathan Shanklin noticed abrupt fluctuations in ozone readings. On both occasions he checked with NASA, comparing BAS findings with their NASA satellite observations. NASA had seen nothing, although when they later checked its scientists realised that their computers had been 'programmed to throw out any wildly abnormal readings'.[93] Farman and BAS colleagues (including Shanklin) published the results – the discovery of the 'ozone hole' – in *Nature* in May 1985.[94]

By 1985 some of the Falklands tensions were lessening, to the extent that the Royal Society tentatively began to reopen formal relations with Argentina. The process started with a visit by Argentina's Foreign Secretary, while the pharmacologist Arnold Burgen met with his academy counterparts to discuss resuscitating a bilateral agreement on scientific exchanges signed in 1977. The Foreign Office was nervous, not least because 'the PM is also FRS'.[95] While several individual scientists had visited Argentina, and César Milstein, the Argentine co-discoverer of monoclonal antibodies based at Cambridge University, had felt it necessary to withdraw 'because of the bad publicity his presence would create in Argentina', a Royal Society delegation successfully visited in March 1985.[96] However, with these steps towards normalisation of relations, it was also possible for NERC to misjudge the political significance for continued BAS support.

In 1986 Lord Shackleton (geographer, Labour minister under Wilson and the son of Ernest Shackleton) raised with Thatcher resurgent concerns about the funding of Antarctic science.[97] He reminded her of his support for the Falklands campaign and the quality of BAS research:

> How important this work is has recently been shown by the discovery of the 'ozone hole' which forms over the Antarctic in early Spring. Analysis of ground-based measurements showed ozone amounts over the UK base at Halley to have diminished by 40% over a decade. The BAS findings stimulated a search through NASA satellite records by US workers, who confirmed the seasonal drop. This may be the first real evidence that atmospheric pollution is damaging the ozone layer.

After delivering what was the earliest document that flagged the ozone hole discovery to the Prime Minister I have found (although surely she was aware of it from the 1985 news coverage), Shackleton expressed his worry about 'a change in funding which could mean we lose our eminently visible position ... We cannot risk being seen to be weak in one [South Atlantic territory] without possible effects on the others' (in other words, the Falklands). The cause was NERC deciding that it had to constrain its Antarctic funding to a level £12m per year, and therefore funding would drop in real terms.[98] Possible consequences included the closure of two or even three bases and the failure to replace the ageing research ship *John Briscoe*. Thatcher demanded to see all recent correspondence. She considered 'it important to give no signal of a reduced commitment to the British Antarctic Territory and to the Falkland Islands and their Dependencies' and let it be known that she was 'strongly of the view that the higher level of activity by BAS, on which Ministers agreed in 1982, should be maintained'.[99]

Charles Powell conveyed Thatcher's view to the Department for Education and Science in no uncertain terms: she considered it 'inexplicable and regrettable that a collective decision by a Cabinet committee had not been implemented. In her view, the episode raises questions about the management of the Science Budget'.[100] Kenneth Baker, who had replaced Joseph in May 1986 as Secretary of State at the Department of Education and Science, wrote back, chastened, saying that he would give NERC the 'direction' to allocate the money.[101] The significance of this moment – a politician telling the research council directly where research money should go – was not lost on Baker:

> This will be, I believe, the first time that the holder of my office has given a formal direction under Section 2(1) of the Science and Technology Act 1965. Even so, I am convinced that it is the right thing to do. The Government has decided – for territorial and strategic reasons – to overrule the judgement of the Council on the relative scientific merits of its expenditure programmes. We cannot expect the Council to take on their own shoulders responsibility for a decision that will be unpopular within the scientific community and rightly belongs to Ministers. We need to ensure that the Research Councils continue to be very tough-minded about their scientific priorities. It would damage the credibility of our selectivity policies if we tried to twist their arms behind the scenes.

While not expecting to keep such a move 'out of the public domain', and at a time when Baker assessed the research councils to be 'in serious difficulties' and their relation with central government 'in a very delicate' state, he did promise to 'endeavour to minimise the adverse publicity'.

Political action, however, was taken over the ozone hole surprisingly quickly. The link between chlorofluorocarbons (CFCs) and ozone depletion had been raised and investigated in the late 1970s. Indeed the US Environmental Protection Agency had pushed in 1977 to ban CFCs for non-essential uses, such as aerosol cans (but not refrigerants). Callaghan's Labour government, worrying that a ban would hit industrial interests, such as those of ICI and Rio Tinto Zinc as well as other manufacturers, resisted such action.[102] As with acid rain the reason given was a lack of scientific certainty. 'The case against them [CFCs] falls a long way short of proof.'[103]

The revelation of the ozone hole changed this calculation. Although surprisingly absent from the environmental policy files placed in Thatcher's overnight box,[104] international political action proceeded at speed. The Vienna Convention for the Protection of the Ozone Layer was agreed in 1985,[105] while the Montreal Protocol, which phased out named substances including CFCs, was agreed in 1987.[106] Thatcher's speech to the United Nations General Assembly in 1989 was a milestone in her international reputation as a science-trained world leader, and it (discussed below) addressed both the ozone hole and climate change through the prism of Antarctic science.[107] Thatcher's intervention with President George H. W. Bush was important in keeping the United States committed to the Protocol,[108] while her 1990 speech to the United Nations, with its promise of industrialised countries' assistance to industrialising nations, helped to make a truly international agreement stick.[109]

Climate change

Thatcher's 1988 Royal Society speech – which, as discussed in Chapter 3, consisted for the first half of a statement of the new science policy of curiosity-driven basic research and cuts in government-funded near-market research – was famously, in the second half, devoted to the environment. She introduced the subject with the image of the Earth as experimental subject:

> For generations, we have assumed that the efforts of mankind would leave the fundamental equilibrium of the world's systems and atmosphere stable. But it is possible that with all these enormous changes (population, agricultural, use of fossil fuels) concentrated into such a short period of time, we have unwittingly begun a massive experiment with the system of this planet itself.[110]

She then turned to what she saw as the three major environmental challenges, anthropogenic climate change, the ozone layer (in which BAS research was highlighted) and acid rain:

> Recently three changes in atmospheric chemistry have become familiar subjects of concern. The first is the increase in the greenhouse gase – carbon dioxide, methane, and chlorofluorocarbons – which has led some to fear that we are creating a global heat trap which could lead to climatic instability. We are told that a warming effect of 1°C per decade would greatly exceed the capacity of our natural habitat to cope. Such warming could cause accelerated melting of glacial ice and a consequent increase in the sea level of several feet over the next century. … It is noteworthy that the five

warmest years in a century of records have all been in the 1980s – though we may not have seen much evidence in Britain!

The second matter under discussion is the discovery by the British Antarctic Survey of a large hole in the ozone layer which protects life from ultra-violet radiation. We don't know the full implications of the ozone hole nor how it may interact with the greenhouse effect. Nevertheless it was common sense to support a worldwide agreement in Montreal last year to halve world consumption of chlorofluorocarbons by the end of the century. As the sole measure to limit ozone depletion, this may be insufficient but it is a start in reducing the pace of change while we continue the detailed study of the problem on which our (the British) Stratospheric Ozone Review Group is about to report.

The third matter is acid deposition which has affected soils, lakes and trees downwind from industrial centres. Extensive action is being taken to cut down emission of sulphur and nitrogen oxides from power stations at great but necessary expense.

I have traced elsewhere in detail the surprisingly early UK governmental response to climate change.[111] In short, correspondence between the Department of the Environment and the acting chief scientific officer on the subject dates to 1974, while the Heath administration's opening up of long-range horizon scanning and forecasting, typified by the Official Committee on Future World Trends, beginning in 1972 partly in response to the Stockholm environment conference, created a space within which long-term climate change could be perceived and discussed. Despite sceptical views on the possibility of climate change (not least from the Met Office and its head John Mason), a report on global warming and its possible impact on the UK was ready by early 1979.

While it offered a modest but significant minister-level recognition that climatic change was possible, publication of the report was delayed until 1980 by the incoming Conservative government. One reason for the delay was anticipated public response. Angus Maude, a senior figure in the party, thought it had 'no presentational advantage' and might even provoke 'hilarity' in the press.[112] Another reason was the Prime Minister's attitude. At the first meeting with her Chief Scientific Adviser, CPRS, John Ashworth, she had said 'incredulously, "Are you telling me I should worry about the weather?"', when he raised the issue.[113] A civil servant at the time noted, somewhat cryptically, 'Ministerial (and especially Prime Ministerial) coolness towards "Climatic Change"'.[114]

The diplomat Crispin Tickell has claimed responsibility for the presence of climate change in Thatcher's Royal Society speech:

> I went to see her when I was on holiday. … I then suggested three ideas to her which she might try. I didn't know which, if any, of them she was going to follow. Then I heard about three weeks later that she was interested in the one about climate change, and we started toing and froing about what she might say and how she might say it. She's always been very interested in science and felt she had that particular contribution to make.[115]

Tickell's involvement in shaping the Royal Speech is confirmed by other documents.[116] But neither the presence of Tickell's advice, nor Thatcher's scientific interest, explain the timing of why she embraced the issue of climate change in 1988. Tickell himself had been warning about climate change since the late 1970s, and had been advising Thatcher informally since 1984.[117] The documents are silent on the question.[118]

However, Thatcher's highlighting of climate change at the Royal Society had two major effects. The first was historiographical, in that most subsequent historical analysis of the UK government's response to climate change begins with 1988.[119] Second, it did give a considerable impetus and urgency to expanding climate research, especially modelling. Tickell advised that the 'first requirement seems to me to isolate the significant areas of uncertainty, and then put real impetus behind research into them. Realistic policy cannot be made until more is known'. Tickell's second requirement was international action, returned to below.

In terms of uncertainty and research, the Cabinet Office quickly pulled together departmental and government expert views on climatic change for ministerial discussion.[120] This recorded:

> There is as yet no firm evidence of climatic change resulting from the greenhouse effect. But there is no serious disagreement within the scientific community that man's activities will lead to global warming. Prediction of the magnitude of the change is subject to considerable uncertainty… But current estimates suggest that the following changes would in time be inevitable: i) global warming by an average of 1.5–4.5°C … ii) a rise in sea levels by at least 20–140 cm … iii) regional climate change …

In preparation for a meeting of relevant ministers, Richard Wilson, the Cabinet Office head of the economics secretariat, summarised for

Thatcher the policy options.[121] Likewise George Guise sent a set of comments.[122] He followed this commentary up with two pages selected from a report that had been sent to him by George Porter, the President of the Royal Society; one page suggested that sea levels might rise and the other that they might fall.[123] Emphasising the doubt, Guise told Thatcher that it 'reinforces my belief that we cannot formulate a robust policy until scientific advice is more consistent. The priority therefore continues to be more research, analysis and computer modelling'.[124] Guise later also questioned the assumption made in a paper for ACOST on adaptive biology that climate change might lead to a warming of 5 degrees Celsius (he drew Thatcher's attention to the National Academy of Sciences consensus of 2 degrees Celsius). However, Guise also went further, hinting that perhaps causation was not shown at all:

> a recent paper from the AT&T Bell Laboratories, gives a thorough statistical analysis of the evidence for correlation between CO2 increase and global warming. ... CO2 and temperature demonstrate correlation between 1958 and 1988 but the paper warns against the conclusion that there is a causal link. This work is useful in combatting the more hysterical arguments that the present climate problems are all part of greenhouse warming. John Mason would approve![125]

A second, clearer case of the 'merchants of doubt' strategy at work is Charles Powell's forwarding of a pamphlet from the George C. Marshall Institute.[126] Powell, as adviser and gatekeeper to Thatcher, is a particularly important carrier. Adding that its 'authors are eminently respectable', his summary of the Institute's argument again brings uncertainty to the fore:

> Against the current fashion, it predicts that, far from warming the earth in the next century, the greenhouse effect will have the benign effect of halting a new mini-Ice Age. It argues that there are far more powerful forces acting on the earth's atmosphere than man-made pollution, principally the periodic brightening and dimming of the sun. If past patterns are followed, the sun is likely to be less active in the 21st century than in the current one, and this would lead naturally to a cooler earth. 'It is possible ... that the combination of natural and solar variability is the cause of the entire temperature increase of 0.9 degrees F observed since 1880 with the greenhouse effect relegated to a negligible role.' The report also points out how

> difficult it is to make accurate predictions of the greenhouse effect because of the highly variable effect of ocean currents and cloud cover.[127]

Nevertheless, despite the admittance to Number 10 of climate change denial arguments, in general they were overwhelmed by a much more substantial, evidence-based approach to the global warming issue.

The Prime Minister chaired a gathering of experts on 26 April 1989.[128] She took notes, in preparation for summing up at the end of the day, which are fascinating but somewhat cryptic: '24 PWR',[129] 'Targets and standards', 'Radioactivity', 'Loving not cutting their forests', 'Global problem', 'Cold Fusion', 'Brazil', 'Avoid Xssive ambition', 'Loose framework convention' and 'Solutions – Silicon Valley of Energy'.[130] Meanwhile John Fairclough, as Chief Scientific Adviser, suggested the 'establishment here of an international Centre for Climate Modelling' – an idea approved by a meeting of ministers on 19 April.[131] On the broader research programme the Royal Society's British National Committee for the World Climate Research Programme, which convened to 'discuss the scientific community's response to the speech' and was chaired by John Mason, took a coordinating role.[132]

The new Director General of the Met Office, John Houghton, pitched to the Department of the Environment an ambitious national plan for climate change, built around a new centre to be formed at Exeter. The Centre would house the Met Office's existing 'core work in climate', 'additions ... specifically aimed at improving our knowledge of climate change as a result of man's activities', 'work on ocean modelling' and 'work by university personnel on climate modelling' (both the latter mostly funded by NERC).[133] A new supercomputer (in first instance, an ETA-10 or a CRAY Y-MP, either of which would be eight times faster than its Cyber 205) would run coupled models (of atmosphere and ocean), with each run leading to improved, fine-grained predictions of future climate change. The Centre would liaise with other groups in universities (such as Reading, Southampton and the University of East Anglia) and institutes (such as the Scott Polar Research Institute and the Institute of Hydrology).[134] By the time of Margaret Thatcher's address to the United Nations General Assembly in November 1989, it had been named the Hadley Centre. She gave another speech at the Hadley's opening in May 1990.[135]

Recall that Tickell's second requirement was the need for international action. 'I think we need to take some international initiative quickly,' Thatcher had jotted on Tickell's letter, 'as the French will try to take the lead.'[136] Indeed there was a flurry of international moves, including the announcement of a March 1989 conference on Saving the Ozone Layer.[137] Gro Harlem Brundtland wrote to congratulate Thatcher, while green NGOs stepped up the pressure.[138] The wrong-footed French Prime Minister, Michel Rocard, was indeed reported to be 'incensed'.[139] Thatcher also detested Rocard's moves to secure a declaration, issued from The Hague, calling for a new international agency ('GLOBE') to take action on major environmental issues, seeing it as mere words:

> The declaration is pathetic. If you don't know what to do – make a Declaration!
>
> We are doing things now. They aren't – the declaration is pathetic.[140]

The head of the United Nations Environment Programme (UNEP), Mustafa Kamal Tolba, was also upset by the French initiative – not least because it cut across the work of the Intergovernmental Panel on Climate Change (IPCC) that had been set up in 1985 by UNEP and the World Meteorological Office and the International Council of Scientific Unions.[141] The IPCC would make its First Assessment Report in 1990. Behind the scenes Tickell was critical of UNEP's capacity, largely because it was underfunded, isolated (in Nairobi) from other UN institutions and 'worst of all … not taken very seriously'; he pushed for a 'new institutional authority', a possible 'future International Convention on Climate'.[142] Again the point being made was for the UK to take the initiative, otherwise there was the 'risk that the Americans, the Russians, the Signatories of The Hague Declaration [led by the French] and others, will come forward with ideas we may find less palatable'. While Thatcher was initially lukewarm about the idea of a Convention on Climate Change – not least because a comment by John Mason stuck in her mind (that there were 'so many meetings now [that] scientists haven't enough time to "do" the science'),[143] she was amenable to the idea that a Convention would 'pre-empt the interventionist ideas discussed at the Hague conference'.[144]

With what the government saw as a good track record, in 1989 the UK pushed forward on both international and national fronts. On the advice of Chris Patten, Secretary of State for the Environment, urged on by Tickell ('Let boldness be our friend'), Thatcher agreed to give a speech to the UN on the environment.[145] Guise lobbied for the speech to include James Goldsmith's plan to link protection of rainforests to developing countries' receipt of aid, but was rebuffed.[146]

More successful was a report faxed from the research ship *Polarstern* by Peter Wadhams, director of the Scott Polar Institute and an eyewitness to 'what may be early signs of man-induced climatic change'. In the report he suggested that a 'valuable role which we could play, in collaboration with the other great scientific nations of the developed world, would be to undertake the monitoring of climate-related processes and changes which are occurring in the polar regions, in order to take advantage of the opportunity which this early warning offers'.[147] In her speech, on 8 November 1989, Thatcher quoted the unnamed Wadhams, drawing parallels with Darwin's voyage on the *Beagle* and urging that on 'the basis then of sound science and sound economics, we need to build a strong framework for international action'.[148] The speech used Wadhams' observations to testify to the need for international action both on the ozone layer and climate change – and there is a strong sense that what she had learned from the former applied to the latter. Antarctic research, and its importance for Thatcher post-Falklands pointed out a direction for travel and a mode of action for an science-trained world leader on the international stage.

At home Patten won agreement to publish a White Paper, on the proviso that it would be 'eminently readable' and 'have a strong scientific base'.[149] Patten's thinking here was influenced by his special adviser, Professor David Pearce, an economist at University College London. In both published works such as *Blueprint for a Green Economy* and in private advice to the government, Pearce operationalised the idea of pricing environmental goods and harms.[150] It might be 'in the tradition of learned economic tracts – fairly turgid and repetitive', Thatcher's private secretary summarised for her, but its 'philosophical basis was right: price must be a better mechanism than armies of regulators to secure a sound environment'.[151] Nigel Lawson, Chancellor of the Exchequer, was more sceptical:

> The Report … conveys the impression that sustainable development is an operational concept. Regrettably, this does not stand up. … there are severe difficulties in the valuation of environmental resources and impacts; often they cannot even be quantified. …
>
> The references to taxation … raise very difficult issues, both practical and political. It is very important not to encourage any assumption that future policy is directed towards the introduction of pollution taxes … to introduce any pollution tax unilaterally would merely disadvantage UK industry vis a vis overseas competitors, without making any significant difference to the greenhouse effect.[152]

Nevertheless, as the first IPCC report neared publication, it was becoming clear that climate change was a peculiarly difficult problem – not only because of countries' self-interest, but also because of the range of national actors affected. As the Energy Minister summarised:

> First, [the analysis conducted to prepare for IPCC] demonstrates that, set against the rising trend in UK CO2 emissions modelled in the study, CO2 emission control will require action right across the spectrum of energy suppliers and users, including transport, with Government, industry and individual consumers all playing a part. Second, no single technology will prove dominant in our search for solutions, although enhanced energy efficiency in all sectors and the increased use of gas for power generation are among the most promising for the short and medium term. Nuclear power has a potentially important role to play but, as recent events have shown, is subject to special difficulties and needs to improve its economic performance if it is to achieve its full potential. Third, none of the options, apart from energy efficiency measures, comes cheaply, and costs rise markedly as the technologies are made to penetrate less and less favourable niches.[153]

The problem was rightly described as the 'most important and difficult issue' that faced a newly convened Cabinet committee (MISC 141) charged with shaping environment policy for the 1990s.[154] Nevertheless, a commitment to stabilising carbon dioxide emissions by 2005 was announced in May 1990. In September 1990 Patten's environment paper, *This Common Inheritance,* was published.[155]

Conservation

Like acid rain and global warming, conservation policies had national and international aspects. In 1981 Parliament passed a new Wildlife and Countryside Act. It continued the postwar 'voluntary' approach to landowners and their roles in protecting sites of wildlife or landscape value, but also encouraged claims for compensation. In particular, a perverse effect of the Act required the Nature Conservancy Council to offer direct notification to landowners and occupiers of Sites for Special Scientific Interest status, thereby alerting them to the possibility of recompense. (Sites of Special Scientific Interest, or SSSIs, had been introduced by postwar legislation and had grown in number in the following decades.

SSSIs were far more likely to be designations of the wildlife interest of private land than the overlapping system of National Nature Reserves. It was therefore over SSSIs that the conflict between development and nature was fought.)

By 1983 it was clear that the legislation was not working well. In particular there was public revulsion at an emerging pattern of large landowners and farmers receiving considerable compensation for not destroying wildlife sites. For example, a tenant farmer in the Swale, on the Isle of Sheppey, was in line for £340,000 per annum, plus a back payment of £500,000, for not draining 1,800 acres of wetland. At Kings Sedgemoor in Somerset 88 acres were purchased for £183,000 after the farmer threatened not to follow an agreement. Aristocratic actions generated special anger: Lord Cranbourne was to be paid £20,000 a year not to replace deciduous trees with conifers, while Lord Thurso in Scotland was paid £250,000 for not draining a peat bog. Halvergate Marshes, an extensive Site of Special Scientific Interest in Norfolk, was another prominent case. The rising costs and public disquiet made it a political issue. Thatcher supported proposals to block the loophole (Section 28 of the 1981 Act) that allowed three months to elapse, during which destruction of an SSSI might proceed.[156]

A working group of interested government departments produced a substantial document with a range of options from overhauling planning controls and new legislation on one extreme to continuing the status quo on the other.[157] Most ministers supported only very minor change (such as closing the loophole) while bemoaning that the Conservatives received little credit for the actions it had taken (accusing the Labour Party in the 1970s, for example, of ignoring the trend towards prairie farming). The Minister for Agriculture, Michael Jopling, representing the pro-landowner interests of MAFF, went further; he wondered if 'there should be some limit on the number of SSSIs'.[158] The Nature Conservancy Council on the other hand had the aim of designating 10 per cent of the total land area of the United Kingdom as being necessary for conservation – an increase of 4 per cent. Yet Jopling too fell back to the status quo (although he expressed the wish that 'perhaps informally the Nature Conservancy Council should be asked to be more self-restraining'). In January 1985 a ministerial meeting, chaired by Thatcher, concluded that the voluntary principle should continue.[159]

Thatcher did not take an enthusiastic interest in international conservation initiatives. In 1980 Heseltine, then Environment Minister, wrote to her about the launch by the United Nations Environment Programme (UNEP) and the International Union for the Conservation of Nature

(IUCN) of the World Conservation Strategy. The launch had taken place on 5 March 1980 in 31 countries simultaneously; in Britain the unveiling was at BAFTA in Piccadilly, London, with a panel that included Heseltine and David Attenborough.[160] 'I was expecting a document couched in the usual emotive terms,' Heseltine observed. He was to be surprised, noting that 'on the contrary, the Strategy effectively equates conservation with sustainable development', and he urged a review of the 'whole range' of connected government policies 'to see whether any short or long-term shift in emphasis is appropriate'.[161] Thatcher's private secretary thought it to be 'just the kind of interminable internal Government study against which you and he have set your faces'. Thatcher agreed, stating 'we have other things to do'.[162]

Nevertheless, she instructed the considerations to be taken into account in decision-making. There were indeed some direct consequences, such as the Department of the Environment being able to push a reluctant Foreign and Commonwealth Office into taking some leads on environmental issues.[163] Furthermore, the undertaking to complete a National Conservation Strategy was completed.[164] However, it took six years for the UK government to publish its official response to the World Conservation Strategy, a glossy brochure with a foreword from Thatcher.[165] In general, conservation matters, though partly science-based, were not a subject of great interest for the Prime Minister.

GMOs in the environment

I have discussed genetic modification in Chapter 3 as part of the commercialisation of the life sciences and the UK government's response to the growth of the new biotechnology. We also saw in the Coda to Chapter 3 that Thatcher intervened personally to prohibit genetic engineering of human embryos. In 1982 an ACARD report on the food industry and technology considered the 'impact of social and technological change on the production, processing and distribution of food'. Robin Nicholson summarised its views on biotechnology for the Prime Minister:

> Biotechnology promises to provide a major increase in land productivity for food production (by a factor of 10 according to some estimates) which, at least in industrialised countries, will far outstrip increase in demand for food. Additionally, biotechnology will allow the use of other feedstocks, eg natural gas, for food production.

> Land which is surplus for food production may either be used to produce agricultural products for use in other industries, eg energy, chemicals, or be converted to non-agricultural use.[166]

Genetic engineering was, for ACARD in 1982, among the 'more speculative but very significant longer-term possibilities … and could have a major long-term impact on the food industry'.[167] Yet, as Nicholson's summary shows, it opened a vision of the transformation of land use.

The regulation of genetic engineering had been established, after rising concern and controversy, by the late 1970s. It involved a Genetic Manipulation Advisory Group (GMAG, established 1976) to advise and assess risks and, from 1978, the Health and Safety Executive, which enforced a notification system. Thatcher doubted the continued need for GMAG in 1981, but she was persuaded to keep it going – 'for 2 years only', she instructed.[168] In 1984 GMAG was replaced with an Advisory Committee on Genetic Manipulation (ACGM). GMAG had done a good job, Nicholson considered, confident that 'we have neither over-reacted nor under-reacted'.[169] Applications of genetic engineering had appeared in medicine, were being developed in agriculture and were expected in diverse areas from mining to food processing. While the engineered organisms were largely confined to contained spaces – essentially laboratories – in the late 1970s and early 1980s, by the mid-1980s, as companies and institutes pushed agricultural applications, the central question became what to do about the release of genetically modified organisms (GMOs) in the environment.

In the UK the existing regulatory structure – the Health and Safety Executive, supported by its Advisory Committee on Genetic Manipulation – produced guidelines, issued in 1986. The guidelines, which included notification, local consultation and risk assessment, were voluntary. A Planned Release (later Intentional Introduction) Sub-Committee of the ACGM monitored this work. This was the regulatory structure under which the first GMOs were released into the UK environment. In 1986 scientists released a genetically marked baculovirus, AcNPV, which infects and kills the caterpillars of the small mottled willow moth, at the Oxford University Field Station at Wytham, Oxfordshire.[170] At the end of the experiment the site was decontaminated with formalin. In 1987 researchers from the Rothamsted Experimental Station inoculated plants with an engineered *Rhizobium* bacterium in a field trial site in Hertfordshire, while the first genetically engineered plants, genetically marked potatoes, were trialled in Britain by the Institute of Plant Science Research, Cambridge; they were subsequently 'manually deflowered', dug up and disposed of. By March 1989 12 proposals for release had been considered, and six releases had taken place. These were contemporary with releases in the United States.

In 1989 the HSE proposed compulsory regulation. The Royal Commission on Environmental Protection, in a substantial report, went further, calling for a statutory body, licences rather than mere compulsory notification, close case-by-case scrutiny and public access to GMO release information. They were responding to well-articulated concerns: genetically modified viruses might jump host, insects might become resistant to insecticide if the genes for generating the toxin spread to other plants, there might be issues of herbicide resistance and various unforeseen consequences. For the RCEP non-GMO 'aliens' – such as invasive species and diseases – provided case studies of analogous situations that could influence thinking about the likelihood of GMO impact. In Britain, recent instances were the devastating spread of Dutch elm disease and Rhododendrons outcompeting native plants; elsewhere in the world, an important example was the introduction of Nile perch into the African great lakes, which had resulted in a range of unforeseen consequences. Of 1,058 documented aliens, about 1 in 10 had become established in the British Isles.[171] The RCEP rejected the kind of moratorium, or even outright ban, on the release of GMOs being proposed in Europe, notably West Germany. Unsurprisingly, given the RCEP's traditional expertise in industrial pollution, it took an existing procedure for scrutinising chemicals ('HAZOP') and modified it for GMOs ('GENHAZ').[172]

In 1989 the Secretary of State for the Environment, Chris Patten, after consultation with ministers, reached agreement to include a general duty to protect the environment against GMOs in the Environment Protection Bill.[173] Also included were systems for release consent, established by regulations, which also spelled out the circumstances in which disclosure of information might take place. The Act, which received royal assent in November 1990, after Thatcher's downfall, required that notification be given and risk assessments completed before GMOs could enter the environment. The Secretary of State then had the power to prohibit the import, acquisition, releasing, marketing or even 'keeping' of GMOs, if 'he is of the opinion that doing any such act in relation to those organisms or continuing to keep them, as the case may be, would involve a risk of causing damage to the environment'.[174] In general Thatcher did not consider the release of GMOs to the environment to be a major issue. She was content to leave policy on changing regulations to others, with her main concern being to reduce the central bureaucratic machinery.[175] On this issue, the Royal Commission on Environmental Protection was of greater influence than the Prime Minister.

Conclusion

When political scientists reviewed Thatcher's environmental policy at the end of the 1980s, their focus was largely on ideological influence on land development issues. Andrew Blowers, for example, concluded that under 'Thatcher's government, environmental policy has exhibited a pronounced ideological change … favouring private development in the creation of our surroundings' and through which 'large-scale developers and, until the mid-1980s, big farming interests have been major beneficiaries'.[176] Although the contested designation of some land as Sites of Special Scientific Interest is one exception, in this chapter I have focused on the environmental politics where science has been a major factor.

It is also the case that the subjects of my case studies – acid rain, Antarctic science and the ozone hole, anthropogenic climate change, conservation and the outdoor release of GMOs – more than match land development as subjects for attention in the Number 10 files.

Another growing influence in 1980s environment policy matters, including on land development, was Europe. The then European Economic Community issued directives on waste, water quality, noise and chemicals, as well as a Directive on Environmental Impact Assessment. This last was issued in 1985 and required 'the interrelationships between major developments and environmental consequences to be identified' and assessments made compulsory for 'developers for major projects – oil refineries, power stations, asbestos manufacturing plants, integrated chemical installations, major transportation projects, ports and toxic waste disposal facilities'.[177] The European political arena was also, as I have shown, the main stage for discussions over acid rain.

Thatcher's 'out-of-the-blue Green' speech to the Royal Society was part of a wider upswing of environmental concern in the late 1980s.[178] For many people 1988 was a 'year of drought, floods, hurricanes and other disasters' as well as a 'spate of speculation about global warming', noted Tickell, listing the reasons for 'rising public concern'.[179] By making it personally part of her and the Conservatives' political image, Thatcher necessarily opened herself up to campaigns from NGOs claiming that the government was either not going far enough or had not delivered promises. Friends of the Earth, for example, issued an 80-page report critiquing the government's record over 10 years in February 1989. Thatcher was advised to make a public riposte, attacking it on a 'broad front'.[180] Greenpeace ran a campaign based on the unhealthy state of trees suffering from atmospheric pollution near where Thatcher had lived, a sort of localised eco-shaming.[181] The response from within Number 10 to the NGOs was largely to try and keep them at arm's length. Tom Burke, the head of Green Alliance, despite being seen as 'more reasonable than Jonathon Porritt and others in the Friend [*sic*] of the Earth', was bumped from the meeting of experts on global climate because he did not possess 'the right sort of liveliness'.[182] Thatcher did finally agree to meet Porritt.[183]

However, activists' voices were rather distantly heard at the centre of government in the case studies I have traced here. Foregrounded instead was scientific evidence, although research findings were as much cited as evidence of uncertainty and reason for delaying action as they were the cause of policy change. At several points industrial interests can be seen shaping the interpretation of evidence as it was passed into the centre of government: the electricity producers in the case of acid rain and the emphasis on uncertainty – what Oreskes and Conway have called the 'manufacture of doubt' – by the gathering forces of climate change denial. Nevertheless, departmental scientific advisers made important contributions, while the Chief Scientific Adviser's role in many of the environmental issues was largely a secondary one. Nicholson contributed in a minor way to the acid rain discussions; Fairclough offered guidance on the scope and substance of environmental research when climate change became a leading issue from the late 1980s.[184] Furthermore, in the cases of acid rain, Antarctic research priorities and ozone hole and climate change initiatives, the Prime Minister's decisions were most consequential.

Notes

1. European Community statistics for 1983 show the UK generating 5.122m tonnes of SO_2, 24 per cent of the EC total. The next biggest emitters were Italy (20 per cent), Germany and France (17 per cent each). The UK deposited 28 per cent of emissions in other countries, compared to 57 per cent for Germany.
2. AT 82/203. Dahlgren to Heseltine, 6 August 1981. The invitation, for reasons that are unclear, was forwarded to Heseltine by the Swedish ambassador to the UK in October 1981.
3. AT 82/203. Rowcliffe to King, 12 November 1981.
4. AT 82/203. Wedd to Reed and Rutterford, 23 March 1982. Adds: 'I cannot say I like the emerging prospect of appearing in Stockholm as the spokesman for a rather dirty little country upwind of Northern Europe, which burns coal because (a) coal is what it has got and (b) it did not have the sense to go nuclear to the extent that the French have done, and really does not care what it does to its neighbours'.
5. AT 82/203. Holdgate to Rowcliffe, 7 January 1982. The 'rigorously logical' phrase was attributed to Heseltine, see Letendrie to Rowcliffe, 12 January 1982.
6. AT 82/203. Holdgate to Rowcliffe, 7 January 1982. A particular tactic was suggested: to avoid the UK being isolated 'which will be crucial at this meeting ... while it may be unkind to point too straight a finger at Poland in their present circumstances, there can be little doubt that they are one of the worst illustrations of uncontrolled air pollution in Europe at the present time'.

7. AT 82/203. Rolt to Reed, 15 January 1982. Reed was the UK's Chief Alkali Inspector. Appropriately enough 'acid rain' had been identified by the very first Alkali Inspector, Robert Angus Smith, in the early 1870s. R. A. Smith, *Air and Rain: the Beginnings of a Chemical Climatology*. London: Longmans, Green and Co., 1872.
8. AT 82/203. G. D. Howells, 'Discussion group on acid waters and soils. Forests and acidification. Report of a meeting held on 13/14 October 1981 at Freshwater Biological Association Laboratory, Ferry House and the Burnside Hotel, Bowness'.
9. See British Library, Peter Chester interviewed by Thomas Lean, *An Oral History of the Electricity Supply Industry in the UK*, C1495/15. Chester calls the attitude to acid rain 'hysteria'. The CERL conducted 'flying chemistry' experiments, co-funded by the US electricity industry, in which instruments in Met Office aircraft tracked the atmospheric chemistry of smokestack plumes. See: Peter Fishlock, 'Acid rain', *Financial Times* (6 September 1983).
10. COAL 74/1716. P. F. Chester, 'Perspectives on acid rain. The Sir William Jackson Pope Memorial Lecture, Royal Society of Arts, 13 April 1983'. *Journal of the Royal Society of Arts* 131 (1983): 587–603. Chester attacks the Swedish science in detail in this speech, and in passing quotes Enoch Powell in support of a claim about how political language twists facts. Area directors of the National Coal Board were handed copies of this 'important speech'. NCB, *Environmental Newsletter* (5 July 1983) p.2.
11. AT 82/203. Chester to Doderlein, 28 October 1981.
12. AT 82/203. Hans Martin Seip, 'Comments to letter and paper by P. F. Chester', undated.
13. AT 82/203. Hans Christensen to Burgess, 26 January 1982.
14. AT 82/203. Rutterford to Rowcliffe, 21 January 1982. Chester's paper in this case was 'Acid rain, SO_2 emissions and fisheries'.
15. AT 82/203. Rutterford to McConnell, 12 January 1982. Researchers from the Institute for Terrestrial Ecology (ITE), Warren Spring Laboratory and the Macaulay Institute were offered to balance those of CERL. A Met Office scientist was also invited later.
16. AT 82/203. Holdgate to Everest, 7 May 1982. Holdgate judged Chester's paper to be 'self-consistent', but was a 'little worried about whether its conclusions and self-consistency are not the result of an inadvertent selectivity. I am not a chemist, and I cannot therefore probe the information further'.
17. Warren Spring Laboratory, UK Review Group on Acid Rain, 'Acidity of rainfall in the United Kingdom – a preliminary report', June 1982.
18. The full network of research directly or indirectly relevant to acid rain is captured in papers of the UK Acid Rain Research Seminar held on 11 June 1982, just prior to Stockholm. Work was being done at Warren Spring Laboratory, AERE Harwell, NERC establishments, DAFS (Department of Agriculture and Fisheries Scotland) establishments, the Forestry Commission, the Nature Conservancy Council, 15 universities (especially Nottingham, Lancaster and Imperial College), 5 water authorities and CEGB.
19. AT 82/203. Saunders to Holdgate, 28 May 1982. See also Marek Mayer, 'Britain slashes research on air pollution', *New Scientist* (29 April 1982), p.271, which noted cuts to Harwell's Atmospheric Pollution Group, the Institute for Terrestrial Ecology and Nottingham University. One of these grants, £120,000 to the Institute for Terrestrial Ecology, was reinstated after 'lobbying from acid rain investigators', according to Fred Pearce, 'Warning cones hoisted as acid rainclouds gather', *New Scientist* (24 June 1982), p.70. The Department of Environment papers seem to dispute this reinstatement.
20. AT 82/203. Holdgate to Burgess, 16 February 1982. Holdgate did state that 'In writing that I have no wish to imply anything other than the highest standards of scientific integrity among our colleagues at CEGB, for whom I have the highest regard. But I believe that the public will expect Government to be independent in its analysis of these matters'.
21. AT 82/203. Saunders to Holdgate, 28 May 1982.
22. Flue gas desulphurisation was in fact an old technology, invented in Manchester in the 1880s, and fitted to some small urban power stations following the crisis over smog in the 1950s. Fred Pearce, 'Acid rain: new fears prompted cleanup', *New Scientist* (18 September 1986), pp.22–3.
23. AT 82/203. Wedd to PSs of Shaw, King and Heseltine, 14 June 1982.
24. AT 82/203. Heseltine to Lawson, 23 June 1982. Shaw was Parliamentary Undersecretary of State, Department of the Environment.
25. AT 82/203. 'Acidification conference, Stockholm 28–30 June 1982. Delegation report', July 1982. The more informal report is Wedd to Harrop, 23 July 1982.
26. COAL 74/1716. Lawson to Siddall, 28 January 1983.

27. COAL 74/1716. National Coal Board, 'Government policy review on acid rain', 11 April 1983.
28. *Time* (8 November 1982).
29. Editorial, 'Death from the skies', *Mirror*, 9 September 1983. Geoffrey Lean, the *Observer*'s environment correspondent, also wrote a series of critical articles.
30. COAL 74/1716. 'Statement by CEGB prepared for anti-acid rain demonstration due on Thursday December 15 at CEGB headquarters and DOE'. December 1983.
31. Andrew Moncur, '£5m research on ways to curb pollution and acid rain', *Guardian* (6 September 1983).
32. Peter Collins, *The Royal Society and the Promotion of Science since 1960*, Cambridge: Cambridge University Press, 2016, p.140.
33. COAL 74/1716. Royal Society press release, 'Acidification of surface waters in Norway and Sweden', 5 September 1983. Walter Marshall's statement on the same occasion contains a line that is clearly reminiscent of Chester's 'heartbreaking': 'Since electricity is essential for everyone, this [expense of desulphurisation] effectively lowers everyone's standard of living and it would be tragic to do so without understanding exactly what we are accomplishing'. Marshall, 'Launch of the acid rain research project', 5 September 1983.
34. David Fishlock, 'Acid rain: UK's electric and coal industries find an impartial judge', *Energy Daily* (8 September 1983), p.3. Copy in COAL 96/91.
35. Des Wilson, letter to the *Guardian* (7 September 1983). Wilson thought the ulterior motive was simply to delay action by five years.
36. COAL 74/1716. Sugden (Vice-President, Royal Society) to Editor of the *Guardian*, 7 September 1983.
37. David Fishlock began his long article covering the launch of the programme by recollecting a bumper sticker at the Atomic Industrial Forum that read 'Acid rain causes nuclear power'. David Fishlock, 'Acid rain', *Financial Times* (6 September 1983). Moncur's *Guardian* piece was subtitled 'Study could give ammunition to nuclear power lobby'. See also R. D. Pryke (Anti Nuclear Campaign), letter to the *Guardian* (12 September 1983).
38. PREM 19/1217. Reidy to Barclay, 15 June 1984. The research examined 'the origins, transport, chemical transformation and deposition of acidity and its movement through the soil and into waterways', as well as the 'evidence linking observed environmental damage with acid deposition'.
39. FT 42/29. Irwin (Warren Spring) to Cooke (NCC), 19 October 1984.
40. FT 21/659. K. A. Ling and M. R. Ashmore, 'Acid rain and trees. An appraisal of the evidence for damage to native tree species by air pollution and acid precipitation in the United Kingdom', report commissioned by NCC, April 1986, p.113. FT 42/29 contains Friends of the Earth's press releases on the 'dieback' project and also evidence that the NCC helped promote FOE citizen science (for example, by distributing pamphlets to regional offices).
41. PREM 19/1216. Mount to Thatcher, 9 December 1983. Mount obtained Thatcher's agreement to strengthen the environmental parts of the select committee machinery.
42. PREM 19/1369. Turnbull to Hodgkinson, 4 May 1984. The comment came at the end of a meeting between Thatcher, her ministers responsible for the science budget, her GCSA (Nicholson), the chairs of the ABRC (Phillips) and SERC (Kingman), discussed in Chapter 3. Nicholson responded that the 'priority was better understanding of the scientific processes involved' and offered to send the Prime Minister a note on the subject.
43. PREM 19/1217. 'Record of a presentation on acid deposition given at Chequers on Sunday 27 May 1984', 8 June 1984.
44. PREM 19/1217. Chester to Thatcher, 5 June 1984. One wonders whether Chester's highlighting of hydrocarbons was a way of downplaying the chemicals emitted by CEGB's power stations.
45. PREM 19/1217. Walker to Thatcher, 15 June 1984.
46. PREM 19/1217. Waldegrave ('for Patrick Jenkin') to Thatcher, 15 June 1984. Jenkin was the senior Environment minister, but Waldegrave did the work on this issue.
47. PREM 19/1217. Nicholson to Thatcher, 18 June 1984. Pascall to Barclay, 18 June 1984.
48. PREM 19/1217. Barclay to Ballard, 20 June 1984. Nicholas Ridley, as Minister for Transport, attempted to push back on car emissions. Nicholson noted that he was following an ingrained departmental view: 'the line that no concession to environmental improvement is worth the cost. They fought and lost a rearguard action on lead in petrol but now want to start one on "lean-burn"'. Nicholson to Barclay, 3 August 1984.
49. House of Commons Select Committee on Environment, *Report on Acid Rain*, 2 vols. HMSO: London, 1984.

50. PREM 19/1217. Nicholson to Thatcher, 7 September 1984.
51. PREM 19/1217. 'Environment Committee report on acid rain. Statement by CEGB', 6 September 1984, shows that the CEGB rejected many criticisms made of it by the select committee.
52. Department of the Environment. *Acid Rain. The Government's Reply to the Fourth Report from the Environment Committee*. HMSO: London, 1984.
53. The first concern was for corrosion of stone and concrete, especially with regard to historic buildings and monuments (the committee had witnessed the effects of corrosion during a visit to Cologne Cathedral). It was recommended that the Building Research Station conduct urgent research and surveys.
54. PREM 19/1477. Jenkin to Thatcher, 2 November 1984. The report was published in February 1984. Royal Commission on Environmental Pollution, 10th report, *Tackling Pollution – Experience and Prospects*, Cmnd 9149. London: HMSO, 1984.The RCEP supported a 'modest increase' in nuclear power 'as part of a strategy for reducing dependence on fossil fuels ... and for reducing the polluting effects of their combustion'.
55. PREM 19/1217. Jenkin to Howe, 24 September 1984.
56. PREM 19/1217. Redwood to Thatcher, 28 September 1984. Redwood cited Nicholson's support that the 19 June agreement was being 'lost in bureaucratic prose and half-heartedness'.
57. PREM 19/1217. Willoch to Thatcher, 2 October 1984, enclosing a 'Joint declaration on the preservation of air'. A similar plea arrived a few months later, this time from a meeting of Nordic country prime ministers. The message this time was sent by the meeting's host, the Prime Minister of Iceland. PREM 19/1477. Hermannsson to Thatcher, 17 December 1984. Thatcher to Hermannsson, 11 January 1985. In December 1985 Willoch wrote to Thatcher again after they had had a conversation during which she had asked 'what the problem is' about acid rain. PREM 19/1742. Willoch to Thatcher, 5 December 1985.
58. PREM 19/1217. Thatcher to Willoch, 24 October 1984.
59. PREM 19/1742. Powell to Budd, 27 November 1985.
60. PREM 19/1742. Telegram, Tebbit to Thatcher, undated (June 1985). One implication was that it was a blow to the British Leyland corporate plan.
61. PREM 19/1742. Jenkin to Thatcher, 24 June 1985.
62. PREM 19/1742. Nicholson to Thatcher, 28 June 1985.
63. PREM 19/1742. Unwin to Thatcher, 28 June 1985.
64. K. A. Ling, M. R. Ashmore, R. B. Macrory, 'The use of word-based models to describe the development of UK acid rain policy in the 1980s', *Environmental Science & Policy* 3(5) (2000): 249–62.
65. https://www.independent.co.uk/news/obituaries/obituary-lord-marshall-of-goring-5626891.html
66. PREM 19/1742. Marshall to Thatcher, 9 July 1986.
67. The first was the quality of early (1950s and 1960s) Scandinavian research which was of 'doubtful scientific validity' and the second was that the Norwegian government had abandoned research for a period.
68. Brundtland also spoke to Marshall's core values by stressing her pro-nuclear beliefs. She disparaged Norwegian suggestions that Dounreay was a menace, said she was 'appalled at the emotional outcries and the public's fear of radiation' after Chernobyl (which occurred barely months before) and 'indicated she was a strong supporter of nuclear power'. There was no better way of appealing to Marshall.
69. Gro Harlem Brundtland had trained in medicine at Oslo University and public health at Harvard. Brundtland, *Madam Prime Minister. A Life in Power and Politics*. New York: Farrar, Straus and Giroux, 2002. See also p.253 for her meeting with Thatcher.
70. PREM 19/1742. Wybrew and Booth to Thatcher, 23 July 1986.
71. Marshall to Walker, 15 July 1986.
72. PREM 19/1742. Cabinet Office, 'European Community draft directive on large combustion plants. E(A)(86) 37. Brief for the Prime Minister, 11 July 1986.
73. PREM 19/1742. Norgrove to Thatcher, 23 July 1986.
74. PREM 19/2142. Wybrew to Thatcher, 9 September 1986. Department of Environment press release, 'Government proposes new action to tackle problem of acid rain', 11 September 1986.
75. PREM 19/2142. 'Sulphur emissions, acid deposition and freshwater ecosystems: current scientific understanding', by chief scientist of Department of Energy, chief environment scientist of Department of Environment and director of technology, planning and research, CEGB, August 1986.

76. PREM 19/2142. Fairclough to Thatcher, 29 August 1986.
77. PREM 19/2142. Ingham to Thatcher, 9 September 1986.
78. *The Times*, editorial, 7 September 1984.
79. Jon Agar, *Science in the Twentieth Century and Beyond*. Cambridge: Polity, 2012, p.349.
80. ED 273/89. 'NERC. British Antarctic Survey. Note by Second Secretary', May 1980. In 1981 BAS employed 322 staff, half of whom worked in the Antarctic in the summer (82 over-wintering), at five stations (Rothera, Faraday, Fossil Bluff, Halley and Signy) in the Treaty zone and one outside (Grytviken), plus two research ships, RRS *John Biscoe* and RRS *Bransfield*.
81. ED 273/89. Minutes, NERC, 81/46 Supplement 2, September 1981. 'Grytviken' was the name used for this base between 1977 and 1982. Previously it was known as South Georgia or King Edward Point (the latter name used from 1982 onwards). M. A. Martin revised by J. Rae, 'A Brief History of the Research Stations andRefuges of the British Antarctic Survey and its Predecessors', 2016, https://www.bas.ac.uk/wp-content/uploads/2015/03/British-Antarctic-Stations-Refuges-v6.2–2016.pdf.
82. ED 273/89. Minutes, NERC, 81/59 Supplement A, 29 October 1981.
83. ED 273/89. Bowman to Heap, 7 December 1981.
84. Martin and Rae, *op cit*., p.10. After the conflict a return to South Georgia on 'political and strategic grounds' was mooted in 1985. FCO 7/6287. Eldon to Palmer, 9 January 1985.
85. PREM 19/1505. Howe to Pym, 25 June 1982. The Cabinet committee, Defence and Overseas Policy (OD), agreed that the activities of BAS should be increased at OD(82)13th.
86. ED 273/76. Thom to Tanner, 4 June 1982.
87. PREM 19/1505. 'Brief for the Prime Minister's meeting with Sir Alec Merrison and Sir Hermann Bondi on the British Antarctic Survey: Monday 12 July 1982', undated, July 1982. While the least likely option, the suggestion of cutting SSRC is interesting. 'Removing the SSRC even half of the additional amount needed for increased BAS activities … would virtually spell the end of the SSRC as its funding would then be almost too low to enable it to continue as a viable organisation. Ministers have previously rejected "death by starvation" as a dishonest way of abolishing the SSRC'.
88. PREM 19/1505. 'Note of a call on the Prime Minister by the Secretary of State for Education and Science, the Chairman of the Natural Environment Research Council and the Chairman of the Advisory Board for the Research Councils, at 18.30 hours on Monday 12 July, at No 10 Downing Street', 13 July 1982. ED 273/76. Thom to Tanner, 4 June 1982, shows that Thatcher had a long discussion with Dr Ray Adie, deputy director of BAS.
89. PREM 19/1505. Joseph to Thatcher, 6 August 1982. Thatcher to Merrison, 9 August 1982.
90. ED 273/89. Bondi, 'NERC. BAS. The context for economies', July 1981.
91. ED 273/76. 'British Antarctic Survey', undated (1982).
92. ED 273/76. 'NERC. Proposals for increased British research in the Antarctic', 2 December 1982 (as dated, although September is a more likely month).
93. Fred Pearce, *With Speed and Violence: Why Scientists Fear Tipping Points in Climate Change*. Boston: Beacon Press, 2007, p.218.
94. J. C. Farman, B. G. Gardiner and J. D. Shanklin, 'Large losses of total ozone in Antarctica reveal seasonal ClOx/NOx interaction', *Nature* 315 (16 May 1985): 207–10.
95. FCO 7/6303. Handwritten note on Jones to Palmer, 22 January 1985.
96. FCO 7/6303. 'Record of Mr Kenton's meeting with the Royal Society, 6 February 1985'.
97. ED 273/89. Shackleton to Thatcher, 2 September 1986. Crispin Tickell had sent a warning to the FCO about NERC's thoughts on cutting BAS in 1985. FCO 7/6287. Tickell to Braithwaite, 21 March 1985.
98. ED 273/89. Howe to Joseph, 15 May 1986.
99. ED 273/89. Powell to Smith, 9 September 1986.
100. ED 273/89. Powell to Smith, 5 March 1987.
101. ED 273/89. Baker to Thatcher, 11 March 1987.
102. DEFE 71/266. 'Protection of the stratosphere. Report by the working group of the official committee on environmental protection', undated (1977).
103. DEFE 71/266 Howell to Kaufman, 24 May 1977.
104. 'Environmental Affairs' creates a series of files from 1979 to 1990 and beyond. Part 5 (PREM 19/2142), which covers 1986–1988, has surprising gaps.
105. FO 949/434 contains the agreed text.
106. FO 949/592 contains a UK copy of the Montreal Protocol.

107. Margaret Thatcher, Speech to United Nations General Assembly (Global Environment), 8 November 1989, copy at https://www.margaretthatcher.org/document/107817 (accessed 1 May 2018).
108. PREM 19/2969. Patten to Thatcher, 17 July 1990.
109. Margaret Thatcher, Speech to Ozone Layer Conference, International Maritime Organisation, Albert Embankment, London, 27 June 1990. Copy at https://www.margaretthatcher.org/document/108133 (accessed 20 August 2018).
110. Margaret Thatcher, 'Speech to the Royal Society', 27 September 1988. Copy of the text (as prepared rather than as spoken) can be found in the online archives of the Margaret Thatcher Foundation. http://www.margaretthatcher.org/document/107346. Accessed 3 April 2017.
111. Jon Agar, ' "Future forecast – changeable and probably getting worse": the UK government's early response to anthropogenic climate change', *Twentieth Century British History* 26 (2015): 602–28.
112. CAB 184/567. Maude to Joseph, 27 July 1979. Agar, 'Future forecast', *op. cit.*, p.622.
113. John Campbell, *Margaret Thatcher. Vol 2: The Iron Lady*. London: Jonathan Cape, pp.642–3.
114. CAB 184/567. Courtney to Ashworth, 21 April 1980. The ambiguity here is that 'Climatic change' was the title of the report.
115. Interview with Sir Crispin Tickell by Malcolm McBain, 28 January 1999, pp.20–1. British Diplomatic Oral History Programme (BDOHP). https://www.chu.cam.ac.uk/media/uploads/files/Tickell.pdf. I am grateful to Alice Bell for pointing me towards this source.
116. PREM 19/2652. Tickell to Powell, 11 October 1988. 'Thank you for your nice letter about the Prime Minister's speech at the Royal Society. I am glad to have contributed to it'.
117. Agar, 'Future forecast', *op. cit.*, p.613. Interview with Tickell, *op. cit.*, p.20. 'I was giving her advice from 1984 onwards about environmental issues, in particular climate change'.
118. Thatcher's relevant personal papers after 1987 at Churchill College archives are currently closed. The documents relating to the preparation of, and thinking behind, the Royal Society speech, have not yet appeared in the PREM series at the National Archives.
119. Sonja Boehmer-Christiansen, 'Britain and the International Panel on Climate Change: the impacts of scientific advice on global warming. Part I: Integrated policy advice and the global dimension' and 'Part II: The domestic story of the British response to climate change', *Environmental Politics* 4 (1995): 1–18, 175–96. Uttam Kumar Sinha, 'Climate change and foreign policy: the UK case', *Strategic Analysis* 34 (2010): 397–408. Loren R. Cass, 'The indispensable awkward partner: the United Kingdom in European climate policy', in Paul G. Harris, ed., *Europe and Global Climate Change: Politics, Foreign Policy and Regional Cooperation*. London: Edward Elgar, 2007, pp.63–86.
120. PREM 19/2652. 'Climatic change. Note by the Cabinet Office', December 1988. Also included as an annex were further notes by the Department of the Environment, FCO/ODA, Energy, MAFF, the Cabinet Office's Science and Technology Secratariat and the Treasury.
121. PREM 19/2652. Wilson to Thatcher, 10 January 1989.
122. PREM 19/2652. Guise to Thatcher, 10 January 1989. This is a measured document, recognising the potential damage as well as the 'prisoner's dilemma' diplomatic bind of all countries recognising the need for action, but being unwilling to be punished for being the first to act. He calls for more research. See also PREM 19/2655/1. Guise to Thatcher, 17 April 1989, which starts with Lorenz's image of a butterfly causing a hurricane due to chaotic effects.
123. The pages came from UNEP, *The Greenhouse Gases*. UNEP: Nairobi, 1987, pp.31–2. Thanks to Michael Huber for helping to identify the UNEP document as the source.
124. PREM 19/2652. Guise to Thatcher, 20 January 1989.
125. PREM 19/3155. Guise to Thatcher, 9 March 1990.
126. Naomi Oreskes and Erik M. Conway, *Merchants of Doubt: How a Handful of Scientists Obscured the Truth on Issues from Tobacco Smoke to Global Warming*. New York: Bloomsbury Press, 2010. Oreskes and Conway detail the George C. Marshall's Institute's central role in the manufacture of doubt. See particularly chapter 6 for the Institute and climate change.
127. PREM 19/2955. Powell to Thatcher, 14 December 1989. See also PREM 19/2968. Institute of Energy press release, 'Climate forecasts are unreliable', 13 June 1990 – which entirely uncritically regurgitates George C. Marshall Insitute claims – was also almost certainly forwarded to Thatcher by Powell.
128. PREM 19/2654 has the draft and final programmes and lists of attendees. PREM 19/2656. 'Record of the prime minister's seminar on the global climate', 26 April 1989, is the re-

cord. Speakers were Tom Wigley (Climate Research Unit, UEA), Ken Currie (Head of Energy Technology Support Unit, Harwell), Tickell and Holdgate (now moved from DOE to be Director, International Union for the Conservation of Nature (IUCN)). Attendees included academics, representatives from industry, George Porter, Francis Tombs, James Goldsmith and James Lovelock, as well as the chief scientists. PREM 19/2656. Morris to Thatcher, 25 April 1989, contains the priceless advice: 'At Flag A is a note from Mr Parkinson about CEGB power station efficiency and methane form land filled sites (you are sitting next to Lord Marshall at lunch)'.

129. Note there was a connection between nuclear power, its public acceptability and climate change. Marshall, the leading proponent of PWR, said at Thatcher's climate change seminar: 'it was clear that nuclear electricity generation had a major contribution to make to mitigating the greenhouse effect. But first it would be necessary to restore public confidence following the Chernobyl disaster'. PREM 19/2656. 'Record of the Prime Minister's seminar on the global climate, 26 April 1989'.
130. PREM 19/2655/1. Handwritten comments on 'Prime Minister's seminar on global climate. Wednesday 26 April',
131. PREM 19/2655/1. 'Global climate. Note by the Cabinet Office', 14 April 1989. This was the cheapest of three no-exclusionary options Fairclough proposed: the climate modelling centre (£3–4m), participation in the proposed ERS-2 earth observation satellite (£60m) and participation of 10 per cent in the World Ocean Circulation Experiment (WOCE, £33m). Morris to Bright, 21 April 1989, contains the record of ministerial agreement to support the centre. The WOCE experiment was approved, both within the balancing constraints of the Public Expenditure Survey. The chief scientist would consult further on ERS-2.
132. AT 33/77. Royal Society, Minutes, 'British National Committee for the World Climate Programme', 7 November 1988.
133. AT 33/77. Houghton to Fisk, 13 January 1989.
134. AT 33/77. White, 'A national centre for climate modelling. A strategy document', March 1989.
135. Thatcher, speech opening Hadley Centre for Climate Prediction and Research, 25 May 1990. Copy available at the Margaret Thatcher Foundation https://www.margaretthatcher.org/document/108102.
136. PREM 19/2652. Thatcher's handwritten note on Tickell to Thatcher, 11 October 1988.
137. PREM 19/2654. Holdgate, 'Saving the Ozone Layer', undated (March 1989), is a report on the conference.
138. PREM 19/2654. Brundtland to Thatcher, 1 March 1989. The WWF, Greenpeace and Friends of the Earth launched its 'The Green Gauntlet' campaign, which listed policy proposals and promised to monitor performance on them closely.
139. PREM 19/2652. Parker to Powell, 6 January 1989.
140. PREM 19/2652. Thatcher's handwritten notes on Powell to Thatcher, 26 January 1989.
141. PREM 19/2653. Telegram, UKMIS New York to FCO, 18 February 1989.
142. PREM 19/2654. Tickell to Slater, 23 March 1989 (blind copied to Powell at Number 10).
143. PREM 19/2656. Handwritten comment on Peirce to Powell, 28 April 1989.
144. PREM 19/2656. Powell to Peirce, 3 May 1989.
145. PREM 19/2657/1. Bush to Powell, 23 August 1989. Tickell to Slater, 28 August 1989.
146. PREM 19/2657/1. Guise to Gray, 12 October 1989. Powell to Thatcher, 13 October 1989.
147. PREM 19/2657/1. Wadhams to Fisk, 23 October 1989.
148. Margaret Thatcher, Speech to United Nations General Assembly (Global Environment), 8 November 1989, copy at https://www.margaretthatcher.org/document/107817 (accessed 1 May 2018).
149. PREM 19/2657/1. Patten to Thatcher, 27 September 1989. Slocock to Thatcher, 3 October 1989.
150. David Pearce, Anil Markandya, Edward B. Barbier, *Blueprint for a Green Economy*, Report for the Department of the Environment. London Environmental Economics Centre. London: Earthscan, 1989.
151. PREM 19/2657/1. Morris to Thatcher, 22 August 1989. Morris was less impressed by the specifics: 'Pearce's conclusions are weak. He proposes simply looking further into fiscal incentives (eg a carbon tax) or transferable permits to pollute (which ... while they introduce a greater market discipline, are still a variant on the command and control means of enforcing environmental standards)'.
152. PREM 19/2657/1. Lawson to Patten, 6 October 1989.
153. PREM 19/2658. Wakeham to Thatcher, 23 November 1989.

154. PREM 19/2965. Wilson to Thatcher, 1 December 1989.
155. Department of the Environment, *This Common Inheritance: Britain's Environmental Strategy*. Cm. 1200. London: HMSO, 1990.
156. PREM 19/1217. Pascall to Thatcher, 25 July 1984.
157. PREM 19/1477. 'Agriculture and conservation. Report by the Department of the Environment of the findings of an inter-departmental working group', November 1984. Bodies represented were MAFF, Treasury, DAFS, SDD, the Welsh Office and the Forestry Commission.
158. PREM 19/1477. Jopling to Thatcher, 9 January 1985. 1.4 million hectares were designated SSSIs, 6 per cen of the country.
159. PREM 19/1477. Barclay to Ballard, 23 January 1985. The meeting did call for reviews of particular unresolved issues (costs of compensation, permanent Landscape and Nature Conservation Orders, planning controls in AONBs) and approved the Experimental Grazing Grant Scheme for the Broads.
160. FT 47/20. 'World Conservation Strategy', 8 February 1980. See also Martin Holdgate, *The Green Web: a Union for World Conservation*. London: Earthscan, 1999.
161. PREM 19/1744. Heseltine to Thatcher, 25 June 1980.
162. PREM 19/1744. Addison to Thatcher, 26 June 1980. Thatcher, handwritten comments.
163. FCO 76/2360. Holdgate to Gray, 14 October 1982.
164. FT 47/20. 'A paper prepared by the Programme Organizing Committee as part of a series on the UK response to the World Conservation Strategy', August 1981, records steps towards the objective.
165. PREM 19/1744. Thatcher, draft foreword, May 1986.
166. PREM 19/1165. Nicholson to Thatcher, 29 July 1982.
167. ACARD, *The Food Industry and Technology*, p.18.
168. PREM 19/1390. Thatcher, handwritten comments on Carlisle to Thatcher, 28 January 1981.
169. PREM 19/1390. Nicholson to Barclay, 16 December 1983.
170. Royal Commission on Environmental Pollution, *The Release of Genetically Engineered Organisms to the Environment*, 13th Report. London: HMSO, 1989, pp.114–16.
171. W. H. Williamson and K. C. Brown, 'The analysis and modelling of British invasions', *Philosophical Transactions of the Royal Society*, Series 8, 314 (1986): 506–22, cited in RCEP, *op. cit.*, p.28.
172. RCEP, *op. cit.*, p.42. EF 13/114/1 and 2 contains sample GENHAZ applications, for example of a GM brassica experiment at Newfound Farm, Norfolk.
173. PREM 19/2658. Patten to Howe, 1 November 1989.
174. *Environmental Protection Act 1990*. c.43. London: The Stationery Office, 1990, section 110.
175. She was particularly sceptical of the usefulness of the Official Committee on Biotechnology. See: PREM 19/3154. Nicholson to Thatcher, 5 January 1984.
176. Andrew Blowers, 'Transition or transformation? Environmental policy under Thatcher', *Public Administration* 65 (1987): 277–94, p.290.
177. Blowers, *op. cit.*, p.289.
178. John Lloyd, 'The green light', *Sunday Times Magazine* (26 February 1989), pp.49–50, p.49. John McCormick, *British Politics and the Environment*. London: Routledge, 1991. Neil Carter, *The Politics of the Environment: Ideas, Activism, Policy*. Cambridge: Cambridge University Press, 2007.
179. PREM 19/2654. Tickell to Howe, undated (1989).
180. PREM 19/2654. Morris to Thatcher, 1 March 1989. See also Michael McCarthy, 'Thatcher's "black record" on green issues', *The Times* (1 March 1989), p.2.
181. PREM 19/2965 contains Greenpeace, 'Margaret's favourite places', early 1990.
182. PREM 19/2654. Morris to Thatcher, 16 March 1989.
183. PREM 19/2747.1, Porritt to Thatcher, 29 September 1989. Slocock to Bright, 31 October 1989. Porritt had met Thatcher briefly at a *Good Housekeeping* awards ceremony and, feeling 'immensely honoured to be in such distinguished company' followed up with a request for half an hour 'to hear more of your thoughts on how best to address some of today's environmental issues'. They mostly discussed recycling, but the record does include a surprising comment on automobiles: 'The Prime Minister said that she was particularly concerned by the dependence of the Western economy on cars … The Prime Minister suggested that a "walk to work" campaign might help to reduce dependence'. PREM 19/2965. Slocock to Bright, 4 December 1989.
184. PREM 19/2967. Fairclough to Thatcher, 9 March 1990. PREM 19/2968. Fairclough to Thatcher, 3 May 1990. See also earlier advice on ERS-2 and other projects relevant to climate change.

8
Science policy under and after Thatcher

> [I]t was very complicated, the story of Thatcher and science, in my view. She, of course, greatly respected science; she thought that people who weren't scientifically literate were numbskulls and that's part of the reason she despised so many of my colleagues. But she also did fall into the hands, a bit, of the rather ideologically driven people, one or two of whom are still about, who said that the only Government funding of science should be for pure science and all the rest of it should be done by the private sector and look at Japan and so on and so forth.[1]

While 'very complicated' was how William Waldegrave, science minister under John Major, chose to summarise the story of Margaret Thatcher and science, it can be simplified helpfully by considering her actions and influence under four headings, of rising importance: 'science and image', 'science and power', 'science for policy' and 'policy for science'.

Science was an intermittent component of Thatcher's image. Having studied chemistry at the University of Oxford in the 1940s and worked as an industrial chemist, publicity for her first election campaigns featured her in a white coat surrounded by laboratory apparatus. When, three decades later, she moved into Number 10 Downing Street as Prime Minister, she installed a portrait of Isaac Newton and a bust of Michael Faraday. These icons of science were chosen as deliberate acts of self-fashioning. Nevertheless, science was a minor aspect of her public image. An analysis of caricature can provide a telling guide. I have found only a handful of cartoons featuring Margaret Thatcher in a white coat, and they were restricted to illustrations accompanying science policy articles. Far more frequent were the blue shoulder-padded skirt suits, pearls, exaggerated hair and handbag: the symbols of a powerful, female, Conservative politician.

Yet the fact that Thatcher, politician, was once Thatcher, scientist, adds an intriguing dimension to the question of how Thatcher wielded her authority. This book started because I had a fascination with how power operates in modern societies. Thatcher's power as Prime Minister was granted through the constitutional processes of a middle-sized, post-Imperial, industrialised democracy. The issues that she confronted were those of modern, industrial economy and society, and as such were saturated with science. Science is a modern form of authority, *par excellence*. I wanted to know what happened when constitutional, political power and the authority of organised, secular, scientific knowledge mixed at the top.

When Thatcher became Prime Minister in 1979 she inherited a vast array of ongoing and substantial scientific and science-based projects, including civil and military nuclear programmes, Cold War defence laboratories, private-sector research and commitments to international collaborations such as CERN, as well as a galaxy of institutes and universities. The decisions she took, and the advice she listened to, deeply affected all these bodies. She, unusually, reserved the right to answer science questions, as I discussed in Chapter 2. I have shown that sometimes she lent her authority to support projects, as was the case in Chapter 4, where I traced the mutual reinforcement of Thatcher's reputation and nuclear power. Yet it was also the case that her authority was primarily deployed to further other manifesto goals, notably the reduction of public spending and the pursuit of privatisation, that cut against these projects.

As Prime Minister, to put in bluntly, when it came to policy decisions she was the most important person in the room, and as an ex-scientist, sometimes this training mattered. Her heavyweight status emerges, in a way both trivial and profound, in an odd competition devised by William Waldegrave in 1993. With the United Kingdom committed to contributing funds to build the Large Hadron Collider at CERN, the Science Minister wanted an account of the Higgs boson that would be comprehensible to a lay-person. The prize was a bottle of vintage champagne. The winner, the physicist David J. Miller, offered the following explanation:

> Imagine a cocktail party of political party workers who are uniformly distributed across the floor, all talking to their nearest neighbours. The ex-Prime Minister enters and crosses the room. All of the workers in her neighbourhood are strongly attracted to her and cluster round her. As she moves she attracts the people she comes close to, while the ones she has left return to their even spacing.

> Because of the knot of people always clustered around her she acquires a greater mass than normal, that is, she has more momentum for the same speed of movement across the room. Once moving she is harder to stop, and once stopped she is harder to get moving again because the clustering process has to be restarted. In three dimensions, and with the complications of relativity, this is the Higgs mechanism.[2]

It is striking that a scientist, when asked to explain the fundamental processes of nature, albeit to a science minister, chose Thatcher's influence as a metaphor. Yet, as political scientist Andrew Gamble has argued, the agency of Thatcher has been over-estimated, both by her supporters and critics. 'The particular contexts in which Thatcher and her ministers were obliged to operate meant that their decisions were often ruled far more by particular circumstances and contingencies than they were by ideological goals and objectives,' Gamble wrote, adding that many of the 'policies that were adopted had consequences which were not foreseen; others did not achieve the results which were intended'.[3] The picture of Thatcher that Gamble finds in detailed accounts of her government reveal her to be 'cautious', 'aware of practical obstacles', 'adept at calculating the balance of forces confronting her' and 'particularly good at seizing opportunities … while presenting herself as always acting out of principle and conviction'. This description – which shows a politician responding flexibly to context and contingency, while only secondarily working out ideological goals – does not quite fit the detailed account I have given in previous chapters on how Thatcher conducted science policy.

For example, take Thatcher's engagement with 'science for policy' – the use of science to inform policy decisions. In previous chapters I have traced many such cases, including the atmospheric and industrial chemistry necessary for grasping the processes and consequences of acid rain (Chapter 7), molecular biology that underpinned developments such as monoclonal antibodies (Chapter 3) and the properties of radioactive materials essential to decision-making over nuclear projects and incidents (Chapters 4 and 5). In many of these cases the background science was prepared and delivered by her chief scientific advisers, and their submissions took account of, and sometimes appealed to, Thatcher's scientific knowledge.

Sometimes this provision of advice was reactive to events, as in the cases of Chernobyl or AIDS and these were certainly moments when contingencies came to the fore. At other times I have found caution, especially when deeply opposing views could be found in her own

Party – as in the case of embryological research, for which Thatcher unusually delegated the weighing up of evidence and the provision of advice to an independent inquiry under Mary Warnock, with primary legislation taking another six years to pass Parliament. I have argued that the issues of AIDS and bioethics were publicly prominent but atypical as science policy issues under Thatcher. Indeed in both cases I showed that Thatcher responded on moralistic grounds. In the case of AIDS she intervened to prevent government funding of a survey on sexual behaviour, which she thought was offensive and an invasion of individual privacy; in the case of embryological research she criticised her Chief Scientific Adviser's argument that a fertilised cell only had the potential to be a human individual as 'casuistry'. In these two cases, the Thatcher we see operating was closer to that identified by Florence Sutcliffe-Braithwaite in her historical analysis of Thatcherite social policy: an ideology that sprang from family-centred, moralistic individualism.[4]

In the central debates over 'policy for science', however, while Thatcher did emphasise the individual, it was not in a family-centred or moralistic manner. Instead Thatcher was hostile to a science policy that favoured the collective over the individual researcher – as in the case of the collective exploitation of patents generated by publicly funded research, which she saw as infringing and deincentivising the entrepreneurial individual researcher. Moreover, as a counterexample to Gamble's observation that 'decisions were often ruled far more by particular circumstances and contingencies than they were by ideological goals and objectives', the complex but crucial emergence of a new science policy in 1987 in the end was a case of ideological goals ultimately determining policy.

As I showed in Chapter 3, from 1979 to 1987 Thatcher, her ministers and advisers struggled, with rising frustration, with several key 'policy for science' issues, each of which impacted on each other. First, broad public-sector funding cuts, when applied to the specific case of universities, undermined the infrastructural support of academic research, causing the despair and anger vividly expressed by Save British Science. (This issue was indeed one that falls under Gamble's description of 'policies that were adopted had consequences which were not foreseen'.) Second, key ministries and advisers favoured the continuation of publicly funded research funds that supported emerging and strategic industries, notably information technology and biotechnology. Yet this industrial policy incensed some Thatcherites, especially within the Number 10 Policy Unit. Third, ministers clashed over defence funding – amid widespread concern, shared by Thatcher, that defence took a disproportionately high proportion of R&D spending. Fourth, international fundamental science projects, especially the 'grand but useless' CERN, were regarded as extravagant. Finally, stories of failure to exploit academic research for commercial gain enraged Thatcher, while mutterings about industry's underinvestment in research increased in volume.

1987 was the year in which Thatcher's 'policy for science' changed. Indeed, it should justly be called the first Thatcherite science policy. Three things happened. First, led by her Chief Scientific Adviser John Fairclough, the machinery of 'policy for science' was reformed and centralised, enabling stronger control. Second, the Advisory Board for the Research Councils published *A Strategy for the Science Base*. This called for a three-tiered structure with a few research-intensive universities distinguished from mere teaching centres, and a more mission-oriented approach. While the given justification for the *Strategy*, the efficient and restrained use of public funds, might seem to square with the aims of Thatcher's manifesto, the truly Thatcherite science policy was actually devised in opposition. Specifically, the third event of 1987 was the ascendancy of the science policy advice of George Guise, of the Number 10 Policy Unit, over that of the Chief Scientific Adviser (as well as the ABRC and ACARD).

Guise agreed with the Cambridge molecular biologist Max Perutz's furious attack on the *Strategy for the Science Base* when he said that it stifled innovation by seeking to micro-manage the independent researcher. Guise fed Thatcher with story after story – this was policy-making by anecdotal history of science – to argue that maximum economic benefit came, in the long run, from freely conducted, undirected pure science, while industry underinvested in research because public funding of 'near-market' research had crowded it out. Therefore, Guise argued, government should enthusiastically fund pure science (now branded 'curiosity-driven research') and cut 'near-market' support – essentially ending an active, interventionist, science-based, publicly funded industrial strategy.

Such a move deserves the epithet 'Thatcherite' because it was grounded on the values of championing the entrepreneurial individual researcher,[5] cutting public funding, encouraging privatisation (not least of the defence laboratories) and leaving private industry to judge its own investments better, confident that in the cut and thrust of the market it would invest in research. The 1987 shift in policy was not formally announced, but its language and values can be heard in Thatcher's flagship science speech at the Royal Society in September 1988, read in the Department of Trade and Industry's White Paper of January 1988, deduced indirectly from the sharply reduced hostility to CERN at the centre of government, but only uncovered in detail through painstaking historical research on the freshly released primary sources.

What is remarkable about the shift is that it happened, as I demonstrated, against the instincts, advice and evidence of the highest committees of science policy advice. It was also, I feel, a decision that could only have been taken by a politician for whom the experience of being a working scientist was now decades distant. Whereas the 1971 Rothschild decision, taken when she was Secretary of State for Education and Science under Heath, seemed to be one a scientist who had lived experience in the applied science of private industry might have naturally made, the 1987 near-market decision was that of an ideologue politician now far removed from a working knowledge of science in business. Did it matter, then, that Thatcher had been a scientist? In terms of 'science and image' and 'science for policy' the answer is yes. Ironically, however, in resolving the central debates in 'policy for science', in which she accepted a picture of science painted by Guise that was a parody of real, working science, it was her distance from scientific experience that told.

How do my findings square with the existing historiography of Thatcher and science policy? Edgerton and Hughes argued that 'what is distinctive about Mrs Thatcher is not that she is a scientist but rather than she is the first anti-technocratic prime minister Britain has had [in the twentieth] century'.[6] By 'anti-technocratic' they meant that Thatcher rejected the views that science and technological change were determinants of economic growth and development, and that an interventionist state was necessary to deliver them. Thatcher's overall political aim was instead to free private enterprise, reduce state intervention and cut public expenditure, 'and to shape what is left to serve industry directly'. So far, so good. This agenda, they argue, was, however, driven by an unstable mix of economic theory. On the one hand this consisted of neoclassical economics, which argued for the free market as more efficient (and which might require the state to intervene, say to hold the ring or to prevent monopolies) and on the other of an 'Austrian' tradition which celebrated the free market because it allowed the play of the unconstrained entrepreneur (and in which there was no role for the state). In practice, says Edgerton and Hughes, there was a contradiction. Thatcher's science policy in practice, they say, was 'highly centralising and dirigiste': industrial need primarily became expressed by 'an exclusive club', while the views of the Advisory Board for the Research Councils (ABRC), one of the chief avenues of scientific advice, were largely irrelevant, having failed to see that 'the technocratic and nationalistic policies' it advocated had been 'off the agenda since 1979'.[7]

The primary sources used by Edgerton and Hughes were understandably limited. Indeed they are focused on just three published documents: the DTI White Paper *DTI – the Department for Enterprise* (1988), the ABRC report *A Strategy for the Science Base* (1987) and the White Paper *Civil Research and Development* (1987). Of course these documents might reasonably, at the time, have been taken as accurate and representative statements of a *single* coherent science policy: but, crucially, they were not. 'As always in politics,' noted Gamble of the myths of Thatcher, 'policies were interpreted retrospectively and stories constructed which gave greater coherence than was intended at the time.'[8] Edgerton and Hughes were on the right lines when they noted that the ABRC views were off the agenda. The extent to which the older science policy was challenged (and by whom) and replaced has been one of the main findings of this volume.

But it was not the case that Thatcher simply rejected the view that scientific and technological change determined economic growth and development. Indeed, as I showed in Chapter 3 and summarised above, Thatcher was eventually persuaded that the long-term economic consequences of 'curiosity-driven research' were almost immeasurably immense. We can also ask: what evidence is there for economic theory driving policy? Edgerton and Hughes spot, for example, the influence of 'Austrian' theory in the language of Thatcher's 1988 Royal Society speech. What does further evidence show? This question can be quickly answered, although it has to be qualified by the fact that early 1980s policy was different from late. In the early radical years of the Thatcher administration, research and development, at least among the new, prominent band of economic advisers, was not seen as a major factor influencing national economic performance. It was a minor factor at best, and the reason is that if raised it was by the 'supply side' economists, who had been nearly completely pushed aside by the monetarist insurgents.

For example, at a luncheon seminar at Chequers on 13 July 1980 there was a gathering of the key economic advisers. They were led by Patrick Minford, the monetarist and author of the Liverpool Model that underpinned the Medium-Term Financial Strategy, the centrepiece of monetarist policy. Also in the room were Geoffrey Howe (Chancellor of the Exchequer), Terry Burns, Christopher Foster and the professors Robin Mathews (Cambridge), Brian Griffiths (City, later director of the Number 10 Policy Unit), Douglas Hague and James Ball (London Business School), all of whom 'could be said to be pretty much in sympathy with the Government strategy'.[9] The suggestion to invite 'one or two critics of the strategy', such as 'old Keynsians' (*sic*), in order 'to liven the discussion', was not taken up. In the briefing the Prime Minister was told that possible topics for discussion would fall into two categories: macro-economic issues 'associated with the Government's monetary and financial strategy' and micro-economic issues 'associated with public expenditure, technology, competition policy, and nationalised industries'.[10] Listed first among the micro-economic issues was 'research and development'. Yet in the seminar all the discussion focused on macro-economic, primarily monetary, matters; research and development was not raised.[11] This valuation of the topic was perhaps more generally reflective of the place of R&D policy in this period of contested economic high theory and its experimental application in the first years of Thatcher's government: the monetarists simply crowded it out.

Was it indeed the case, as Edgerton and Hughes say, that 'the main thrust of government R&D policy' concerned the 'restructuring of publicly funded civil R&D' and paid no attention to defence R&D? I demonstrated in Chapter 3 that attention was indeed paid to defence R&D. And, with respect to Edgerton and Hughes' central conclusion, was there really a fundamental desire to 'control the scientific community' in a way that aimed to 'let loose the industrial entrepreneurial spirit' by restricting other entrepreneurial activities, especially that of scientists? This seems to have been a conflation of two separate initiatives. Edgerton and Hughes read Fairclough's strengthening of the machinery of science policy as being motivated by an aim to control the scientific community – a framing that was not evident from primary sources even though such an external perception might have existed – while the separate aim to loosen the entrepreneurial spirit can be found in how Thatcher and her closest advisers talked about *both* industry and the individual academic researcher.

At this point it is useful to contrast Edgerton and Hughes's analysis with a second kind of historiography of 1980s science policy, exemplified by the journalist-historian Tom Wilkie.[12] His main point is that Thatcher undermined basic science. The reining in of public expenditure, a commitment to the marketplace and a perceived end to the exponential growth of science created the conditions for action, he argues, and then a 'curious compound of ideology and ignorance' led 'Mrs Thatcher's government [to bring to] an end the way basic, curiosity-driven scientific research' had been done in Britain since the First World War. Before 1988 this meant an attack on the dual-support system of University Grants Committee and the research councils, downgrading basic science and encouraging universities to do more applied research; after 1988 there was an 'ill-explained switch in policy', when the government 'decreed' that it would no longer fund 'near-market' research. Yet according to

Wilkie 'the evidence suggests' that the undermining of the dual support system 'was not intended'. It was rather an 'unhappy accident', while the switch in policy seemed to have originated in a DTI internal review of the Alvey programme (government-industry collaborative funding for new information technology research), referred to in passing in *The Department for Enterprise* White Paper that Edgerton and Hughes also commented upon.[13] Whereas the latter see evidence for deeper machinations, however, Wilkie sees carelessness; the former see conspiracy, the latter cock-up.

Wilkie's account can also be tested against the primary document record. There is some evidence for 'ideology and ignorance' driving the undermining of the dual-support system, although I would not go as far as 'by accident'. There was a late-1980s switch in science policy, although it took place in 1987, not 1988, and the origin of the switch was not a review of the Alvey programme. Other commentators have also noted the late 1980s switch in science policy. Stephen Wilks and Michelle Cini noticed that there was a 'redirection of science and technology policy', a 'radical change', a shift in 'norms' away from government funding for industrial research to a withdrawal from support for near-market research.[14] Ian Christie called it the result of a 'great debate' that involved 'politicians, senior administrators, industrialists and scientists', but one that started with a 'wave of criticism' following cuts.[15] Roger Williams, an academic research policy analyst who also advised the House of Lords Science and Technology Select Committee, also argues that the noticeable 'debate' of 1986–7 was primarily provoked by concern over cuts.[16] Nick von Tunzelmann, in an authoritative survey article, notes the existence of large-scale programmes of government support for industrial R&D in the early 1980s, such as LINK and Alvey; he also observes that they were 'effectively wound up' in the second half of the 1980s, but does not explain why.[17]

All of these commentators were informed observers, and some had more access than others. But only now, with the primary sources released at the National Archives, are historians able to trace in detail what changed in science policy, why, and who was responsible.

Science Policy under Major

I will now review the landscape of science policy in the UK as it developed after Thatcher. My main point is going to be that it displayed more continuity than change. John Major, the new Prime Minister, had a background in banking before becoming an MP for the first time in the epochal 1979 election that had brought Thatcher to power. His ministerial career, while late, was fast and fortunate. Major was promoted to Minister of State for Social Security in 1986 and appointed Chief Secretary at the Treasury in 1987; he joined the Cabinet as Foreign Secretary in July 1989 and, three months later, following Nigel Lawson's resignation, became Chancellor of the Exchequer. He had shown no particular interest in science and it was a while before he had his first prime ministerial meeting with his Chief Scientific Adviser.

I mapped the various people, committees and organisations that collectively shaped science policy in the 1980s in Chapter 2. Major inherited a Chief Scientific Adviser, William Stewart, who himself had not been long in the post. The Number 10 Policy Unit, by design a small, handpicked group working closely with the Prime Minister, inevitably underwent changes in the transition from Thatcher to Major. In particular George Guise, so influential on science policy in the later years of Thatcher's administration, immediately moved on. Other bodies continued. Major inherited the main central mechanisms for science policy as they had been settled in the late 1980s: a Cabinet committee for decision-making, the Advisory Council on Science and Technology (ACOST), the Advisory Board for the Research Councils (ABRC) for further advice, a Department of Education and Science, the research council system complemented by a business-oriented University Funding Council under the dual-funding mechanism for distributing funds to academic science, the Rothschild customer–contractor principle to frame government departmental commissioning of science and civil and defence research still largely considered separately.

While some initiatives started earlier, it was not until the general election of April 1992, in which the Major's Conservatives battled to a close and surprising victory over Neil Kinnock's Labour Party, that changes were made. First, Major gave the Chancellor of the Duchy of Lancaster a specific, Cabinet-level responsibility for science and technology policy. He appointed William Waldegrave to the task – a man with the reputation as a bright, intellectual politician who had worked in the CPRS in the 1970s. Waldegrave had, as we have seen in his brief appearances in earlier chapters, held junior posts under Thatcher at the Department for Education and Science and the Department of the Environment. Second, Major took the science policy elements of the Department of Education and Science and the Cabinet Office to form an Office of Science and Technology. In so doing he not only joined up these parts, but made them more visible too.

Green-lit by Major and led by Waldegrave, the government began a review of science policy and organisation. The output was a White Paper, *Realising Our Potential*, which was published in May 1993.[18] Describing itself as the result 'the first general review of policy and organisation since the early 1970s reports from Lord Rothschild and Lord Dainton',[19] and subtitled a 'Strategy for Science, Engineering and Technology', it was not quite as radical as it claimed. The switch in science policy in the late 1980s uncovered in Chapter 3 had been just as consequential (but the result of central, almost private, advice rather than formal review) and, as I will suggest, it could not or did not unpick the changes enough to be described as a full strategy. *Realising Our Potential*, however, did reorganise and add to the landscape of science policy-making. Industry, government and the scientific community were to work together to produce shared visions under a Technology Foresight Programme. Foresight, at least in aspiration, attempted to bring academic 'basic' researchers into conversation with industry, inspired by a Japanese model. Waldegrave recalled:

> we set up a so-called Foresight Programme, which was definitely not meant to be, and was easily ridiculed as being, an attempt to foresee the future: it was an attempt at what you might call iterative discussion about how different people saw the future. The model was the very best Japanese companies at that time who quite often had their salesmen talking to their basic research scientists, and the basic research scientists quite often would say well if you'd told me you wanted that, I could easily have seen a way of doing that for you and the salesman on the other hand, saying my goodness, this thing you're doing here is just absolutely fascinating, well there's real applications for it.[20]

Foresight's visions would inform a Council for Science and Technology (CST), developed out of the old ACOST. The research councils were chopped, changed and renamed: SERC was split into an Engineering and Physical Sciences Research Council (EPSRC) and a Particle Physics and Astronomy Research Council (PPARC), to try and resolve some of the tensions caused by Big Science. The Agricultural and Food Research Council (AFRC) became the more entrepreneurial sounding Biotechnology and Biological Sciences Research Council (BBSRC), while the ABRC was folded into the Office of Science and Technology.

One remarkable aspect of science policy-making in the 1980s is that the process spun off peripheral institutions and bodies that sometimes survived and grew to be more substantial in the 1990s. Heseltine's idea of a 'R&D Evaluation Unit' became the Science and Technology Assessment Office in the Cabinet Office, and ultimately the Government Office of Science (GO Science); ACARD's idea of exploitable areas of science became Tombs' extramural Centre for Exploitation of Science and Technology (CEST); Fairclough's plan for a renaissance in British industry led by picking national priorities gave us ACOST and then CST. Yet all of these bodies had initially been envisaged as being part of an industrial strategy – active, considered support for the innovation process from government – that ended when Guise persuaded Thatcher to end government-funded near-market research.

Nevertheless, *Realising Our Potential* did clarify policy, by responding to a perception of 'the absence of a clear statement of Government objectives, with the consequent transmission of mixed and sometimes contradictory signals to the scientific and engineering communities' by clarifying policy. First and foremost, *Realising Our Potential* emphasised the contribution of science to 'wealth creation'; by implication other reasons for supporting science, such as problem-solving, curiosity, informing policy, improved quality of life, while mentioned, were made secondary. The new research councils were rebranded partly to draw attention to new, explicit 'missions' that emphasised a 'commitment to wealth creation and quality of life' – a form of words that felt like a compromise, but were also read as placing more emphasis on the former than the latter.

Privatisation was confirmed as a major objective of Conservative policy, and I have summarised in Chapter 5 some of the consequences for nuclear power. By contrast, a novelty introduced by *Realising Our Potential* was greater forward scanning, if not planning. A regular Forward Look would be published, while 'technology foresight' would help coordination. 'Technology transfer' was proposed as something that would be 'developed to re-emphasise the importance of the interchange of ideas, skills, know-how and knowledge between the science and engineering base and industry'. In many ways 'technology transfer', as with the later emphasis on 'translation', can be seen as part of the slow crawl, post the 1987 reversal in policy, back to an industrial strategy.

But, overall, the main message of *Realising Our Potential*, for all its talk of wealth creation, was that 'Government cannot, and will not attempt to, remove from industry its responsibility for investing in innovation and bringing new products to market'. As Waldegrave recalled one 'thing about that White Paper that is not to be underestimated … is what was not in it':

> because when I said to the Cabinet that we needed to have a review of science policy, huge pressure came from my friend Michael Heseltine, who was then in charge of the Department of Trade and Industry as President of the Board of Trade, that all this blue sky stuff was a waste of time and what we should do is what the French allegedly did. I think the French had some years before put a huge amount of Government money into applied R & D, I mean development really, motor cars, *le car*. Anyway, this was what we should do and it was all a waste of time all this blue sky stuff. So the first battle that Bill [Stewart] and I had to fight was to fight off that utilitarian view of what the Science Budget was and we did so successfully.[21]

There was to be no immediate return to an industrial strategy of government-funded near-market applied research in priority areas identified by experts, although renewed interest was shown in the LINK scheme.[22] Rather faith was placed on an indistinct 'closer partnership and better diffusion of ideas'.[23] Diffusion is not the same as a strategy. In this way, Thatcher's science policy, especially in its 1987 transformed form, would cast a long shadow.

Europe became an increasingly important context and institutional forum for UK science. Each Framework Programme, from the First, which ran 1984–7 and had a budget of nearly 4 billion euro-equivalent, to the Fourth, which was negotiated during Major's administration, ran 1994–8 and redistributed over 13 billion euros, and since, was larger than the last. European science was viewed sceptically by Thatcher and some of her advisers, as I showed in the case of CERN in Chapter 3. Major had his own deeply divisive and prominent fights with Eurosceptics. But in terms of science policy, the attitude shown towards the European Framework programmes was warmer. This warmth was only partly a result of the net benefit that UK scientists felt – more money has come back from Europe in the form of grants than the UK has put in and under the Second Framework programme, between 1987 and 1991, British scientists secured 20 per cent of all the grants awarded, by value – since this was the case under the administrations of both Thatcher and Major.

The transition from Thatcher to Major coincided with the end of the Cold War. Also coincidental was the effect of the policy aim of privatisation on defence research. A list of the government laboratories (Table 8.1) that had been transferred to 'agency' status, a half-way house towards privatisation, proudly included in *Realising Our Potential*, shows a mix of both civil and defence entities affected, as well as the continuity of policy between the two administrations.[24]

Table 8.1 Government laboratories transferred to agency status, 1989–92.

Under Thatcher	
Warren Spring Laboratory, DTI	April 1989
Laboratory of the Government Chemist, DTI	October 1989
Central Veterinary Laboratory, MAFF	April 1990
Meteorological Office, MoD	April 1990
Building Research Establishment, Department of Environment	April 1990
Natural Resources Institute, ODA	April 1990
National Physical Laboratory, DTI	July 1990
National Engineering Laboratory, DTI	October 1990
Under Major	
Defence Research Agency, MoD	April 1991
Chemical and Biological Defence Establishment, MoD	April 1991
Forensic Science Service, Home Office	April 1991
Central Science Laboratory, MAFF	April 1992
Scottish Agricultural Science Agency, Scottish Office	April 1992
Transport Research Laboratory, Department of Transport	April 1992

However, in the 1990s and 2000s the proportion of defence R&D compared to civil R&D began to decline. This was a consequence, first, of the 1987 change in science policy that sought to remove government funding of research that, in the government's eyes, should be conducted by industry and, second, of the end of the Cold War. Thatcher and her ministers, as I showed in Chapter 3, had desired this rebalancing, but had been unable to achieve it.

In Chapter 7 I discussed how some environmental issues – acid rain, Antarctic science and the ozone hole, climate change – attracted considerable attention from the Prime Minister, while others – biodiversity conservation and the release of genetically modified organisms into the environment were my two examples – did not. Under Major, and under Tony Blair's New Labour government from 1997, intertwined environmental, science and food controversies – including GMOs – became matters of extraordinary national concern. In particular, some readers might be surprised that I have not discussed bovine spongiform encephalopathy (BSE) and its human version variant Creutzfeldt–Jakob Disease (vCJD), or salmonella, in this book so far. My reason is that while they had their origins in the 1980s – the first case of a mad cow was in 1984 – such controversies were only beginning to be prominent. Thatcher did appoint the Southwood committee in 1987, to investigate the implications of BSE for human health, which reported in 1989. But it was a slow-moving catastrophe. Even more so, along with salmonella, it became linked in 1990 as part of a pattern of failure in which there was a crisis in the role of expert advice. For example, in May 1990 *The Economist* commented on the death of Max, a Siamese cat from Bristol, from a brain condition similar to that found in cows with BSE:

> Although Max's death has failed to prove that mad-cow disease travels easily from species to species on a plate – a highly uncertain proposition – it has proved beyond scientific doubt that nobody trusts MAFF [the Ministry of Agriculture, Fisheries and Food] any more. Salmonella, botulism, listeria and now BSE (bovine spongiform encephalopathy) – Max belongs to a sequence of food scandals that the ministry seems powerless to prevent.[25]

The political history of these controversies was largely post-Thatcher, coming to a head in 1996.[26] While the first case of BSE was identified as such in November 1986, and by 1990 local authorities were banning beef from schools, while academics such as Richard Lacey were recommending at the time the slaughter of millions of cattle, a historical analysis based on reviewing the primary policy documents will have to wait until they are fully available.

Another topic that had its origins in the 1980s but was much more important for 1990s science policy was the public understanding of science. As I noted in Chapter 4, in the context of Chernobyl, anxieties over the public grasp of scientific knowledge sometimes reached discussions at the centre of government. Furthermore, in the context of the cuts affecting academic science, there was widespread belief within the academic scientific community that a falling respect for science was correlated to public ignorance.[27] In 1985 the chemist Walter Bodmer had chaired a Royal Society investigation into the subject, launched in 1983 and reporting in 1985.[28] The report was followed by a programme of action led by the Royal Society, British Association for the Advancement of Science and Royal Institution's Committee on Public Understanding of Science (COPUS). The ESRC also funded research on science and the public, which came to a diverse set of conclusions, including some which questioned the assumptions behind the anxiety surrounding 'public understanding'. However, the most publicised result of this research, a survey of 1,800 people that appeared in *Nature* in 1989, supported the

Royal Society's view that 'the solution to the public-understanding of science "problem" lay in the communication of science to non-scientists'.[29]

This knowledge 'deficit' model governed assumptions through much of the 1990s ('a new campaign to spread the understanding of science and technology in schools and amongst the public' was promised in *Realising Our Potential*).[30] After 2000 there was a turn towards two-way public dialogue as a better model of science–public interaction on policy-relevant issues. The BSE controversy which, as Gregory and Lock note, 'highlighted the failure of the communication of knowledge to defuse an issue of contested authority', was almost certainly one major factor in this turning point.[31] However, the public understanding of science, as a major programme, was never a priority science policy issue in Number 10 Downing Street.

In 2013 Denis Noble, one of the founders of the campaign against the public-sector cuts that affected academic science in the 1980s, could write in *Nature* that 'we are still saving British science from Margaret Thatcher'.[32] In 2017 Pallab Ghosh, the BBC's science correspondent, could ask, amid the resurgence of talk of an industrial strategy, 'why is another Conservative government reviving a policy of subsidising industrial research when an earlier one junked it in the 1980s?'[33] In some respects, therefore, the post-Thatcher world of the intersection of science and policy was a similar one of continuities – notably in movement towards privatisation, attitudes towards the orientation of science towards wealth creation and the restricted role, nevertheless, of government. Yet in other respects – notably in controversies over the food chain, public understanding of science and public engagement with science, differences would emerge. These will be the focus of further, future study.

Notes

1. 'Lord Waldegrave (Chancellor of the Duchy of Lancaster [i.e. Science Minister], 1992–4, and numerous other science-related Government posts). 'Edited transcript of a conversation between William Waldegrave and Peter Collins at the Royal Society on 13 July and 14 September 2011', Royal Society Library.
2. David J. Miller, 'A quasi-political Explanation of the Higgs Boson; for Mr Waldegrave, UK Science Minister 1993'. http://www.hep.ucl.ac.uk/~djm/higgsa.html (accessed 19 December 2018).
3. Andrew Gamble, 'The Thatcher myth', *British Politics* 10 (2015): 3–15, 5–6.
4. Florence Sutcliffe-Braithwaite, 'Neo-liberalism and morality in the making of Thatcherite social policy', *Historical Journal* 55 (2012): 497–520.
5. The following exchange recalled by Thatcher's first Chief Scientific Adviser, John Ashworth, is revealing about her expectations and assumptions regarding entrepreneurial capacities. Ashworth: 'Mrs Thatcher genuinely believed in the Schumpeterian principle that creative destruction was good of itself and that arising phoenix-like there would be a newer and better society.

Well I was at the receiving end of one of these homilies from her and she would spout them to everybody, whoever was within earshot at the time would get this. I can remember saying to her, Prime Minister I said, you have this image of entrepreneurs or businessmen or academics, whoever it is, of being like caged tigers, but what do you think will happen if when you fling open the doors you find that far from being caged tigers they are frightened little rabbits? Her response was that they should be eaten. And I think the universities for her fell into that sort of category.' Ashworth's interviewer then asked: 'And if some of them got eaten then they deserved it?'; Ashworth's reply: 'That's right, and the predatory ones would be the stronger thereby'. 'Sir John Ashworth. Edited transcript of a conversation between John Ashworth and Peter Collins at the Royal Society on 22 April 2009', Royal Society Library.

6. David Edgerton and Kirsty Hughes, 'The poverty of science: a critical analysis of scientific and industrial policy under Mrs Thatcher', *Public Administration* 67 (1989): 419–33.
7. The 'exclusive club' identified by Edgerton and Hughes were those associated with the Centre for Exploitation of Science and Technology, consisting of 16 large R&D spending manufacturing companies, and two financial institutions. This Centre was of marginal importance.
8. Gamble, *op. cit.*, p.6.
9. PREM 19/197. Lankester to Thatcher, 9 May 1980. Foster was at Coopers and Lybrand. He would later be one of the key architects of the poll tax. See: Michael Crick and Adrian Van Klaveren, 'Mrs Thatcher's greatest blunder', *Contemporary Record* 5 (1991): 397–416,
10. PREM 19/197. Burns, 'Chequers luncheon for academic economists', 11 July 1980.
11. PREM 19/197. Lankester, 'Note of a discussion at Chequers: Sunday 13 July', undated (15 July 1980).
12. Tom Wilkie, 'The Thatcher effect in science', in Dennis Kavanagh and Anthony Seldon, eds, *The Thatcher Effect*. Oxford: Clarendon Press, 1989. Wilkie, *British Science and Politics since 1945*. Oxford: Blackwell, 1991.
13. Wilkie, *op. cit.*, p.108. The decision about ending funding of near-market research 'was not formally announced at the time: the only policy statement … comes in the DTI's1988 White Paper "The Department for Enterprise"'.
14. Stephen Wilks and Michelle Cini, 'The redirection of science and technology policy under the Thatcher governments', *Policy Money & Management* 11 (1991) 11: 49–56.
15. Ian Christie, 'Research and development policy: the great debate', *Policy Studies* 8 (1988): 11–22.
16. Roger Williams, 'UK science and technology: policy, controversy and advice', *The Political Quarterly* 59 (1988): 132–44.
17. Nick von Tunzelmann, 'Technology in post-war Britain', in R. Floud, P. Johnson, eds, *The Cambridge Economic History of Modern Britain, Volume III: Structural Change and Growth, 1939–2000*. Cambridge: Cambridge University Press, 2004, pp.299–331, p.317.
18. *Realising Our Potential: a Strategy for Science, Engineering and Technology*. Cm. 2250. London: HMSO, 1993.
19. *Realising Our Potential, op. cit.*, pp.2–3.
20. 'Lord Waldegrave (Chancellor of the Duchy of Lancaster [i.e. Science Minister], 1992–4, and numerous other science-related Government posts). Edited transcript of a conversation between William Waldegrave and Peter Collins at the Royal Society on 13 July and 14 September 2011', Royal Society Library.
21. 'Lord Waldegrave (Chancellor of the Duchy of Lancaster [i.e. Science Minister], 1992–4, and numerous other science-related Government posts). Edited transcript of a conversation between William Waldegrave and Peter Collins at the Royal Society on 13 July and 14 September 2011', Royal Society Library.
22. *Realising Our Potential, op. cit.*, p.34.
23. *Realising Our Potential, op. cit.*, p.8.
24. *Realising Our Potential, op. cit.*, p.45.
25. 'Mad cows and ministry men', *The Economist* (19 May 1990), p.24.
26. Sheila Jasanoff, 'Civilization and madness: the great BSE scare of 1996', *Public Understanding of Science* 6 (1997): 221–32.
27. Jane Gregory and Steve Miller, *Science in Public: Communication, Culture and Credibility*. New York: Plenum, 1998.
28. Royal Society, *The Public Understanding of Science*. London: Royal Society, 1985.
29. Jane Gregory and Simon Jay Lock, 'The evolution of "Public Understanding of Science": public engagement as a tool of science policy in the UK', *Sociology Compass* 2 (1998): 1252–65,

p.1254. The survey was: John R. Durant, Geoffrey A. Evans and Geoffrey P. Thomas, 'The public understanding of science', *Nature* 340 (1989): 11–14.

30. *Realising Our Potential, op. cit.*, p.7.
31. Gregory and Lock, *op. cit.*, p.1257.
32. Denis Noble, 'We are still saving British science from Margaret Thatcher', *Nature* (17 April 2013). https://www.nature.com/news/we-are-still-saving-british-science-from-margaret-thatcher-1.12800 (accessed 20 August 2018).
33. Pallab Ghosh, 'Science funding: will "picking winners" work?', BBC News website (31 August 2017). https://www.bbc.co.uk/news/science-environment-41101892 (accessed 20 August 2018).

Bibliography

ABRC. *The Science Budget: a Forward Look 1982*. London: HMSO, 1982.

ABRC. *A Strategy for the Science Base*, Cm. 30. London: HMSO, 1987.

ABRC. *A Study of Commissioned Research* (Mason report). London: HMSO, 1983.

ABRC and ACARD. *First Joint Report by the Chairmen of the Advisory Council for Applied Research and Development (ACARD) and the Advisory Board for the Research Councils (ABRC)*, Cmnd 8957. London: HMSO, 1983.

ACARD. *Exploitable Areas of Science*. London: HMSO, 1985.

ACARD. *Exploiting Invention*. London: HMSO, 1980.

ACARD. *Industrial Innovation*. London: HMSO, 1980.

ACARD. *Technological Change: Threats and Opportunities for the United Kingdom*. London: HMSO, 1979.

ACARD. *The Applications of Semiconductor Technology*. London: HMSO, 1978.

ACARD. *The Food Industry and Technology*. London: HMSO, 1982.

ACARD, ABRC and Royal Society. *Biotechnology* (Spinks report). London: HMSO, 1980.

Agar, Jon. '"Future forecast – changeable and probably getting worse": the UK government's early response to anthropogenic climate change', *Twentieth Century British History* 26 (2015): 602–28.

Agar, Jon. 'Thatcher, scientist', *Notes and Records of the Royal Society* 65 (2011): 212–32.

Agar, Jon. 'The curious history of curiosity-driven research', *Notes and Records of the Royal Society*, published online 20 September 2017. DOI: 10.1098/rsnr.2017.0034.

Agar, Jon. *Science in the Twentieth Century and Beyond*. Cambridge: Polity, 2012, p.349.

Agar, Jon. *The Government Machine*. Cambridge, MA: MIT Press, 2003.

Agar, Jon and Brian Balmer. 'Defence research and genetic engineering: fears and dissociation in the 1970s', in Charlotte Sleigh and Don Leggett, eds, *Scientific Governance in Britain, 1914–1979*, 122–43. Manchester: Manchester University Press, 2016.

Alexievich, Svetlana. *Chernobyl Prayer: a Chronicle of the Future*. London: Penguin Modern Classics, 2016.

Allen, Les. 'Star wars: a paradox for our time. Whichever way you look at it, star wars doesn't make sense', *New Scientist* (1 May 1986): 53–5.

Anonymous. 'British R&D policy: a commonsense approach', *The Economist* (9 July 1983).

Anonymous. 'Crisis of science', *Times Higher Education Supplement* (28 January 1983).

Anonymous. Editorial, 'Death from the skies', *The Mirror* (9 September 1983)

Anonymous. Editorial, 'Government can help', *The Economist* (9 April 1983): 16.

Anonymous. 'Guidelines for helping industry', *Nature* 304 (14 July 1983): 101–2.

Anonymous. 'Her unique responsibility', *The Times* (19 February 1987): 13.

Anonymous. 'In search of missing links … ', *Times Higher Education Supplement* (8 July 1983).

Anonymous. 'Labour opposition to SDI spelt out by Kinnock', *Financial Times* (7 March 1985).

Anonymous. 'Mad cows and ministry men', *The Economist* (19 May 1990): 24.

Anonymous. 'No future for laser weapons in space', *New Scientist* (1 January 1981): 3.

Anonymous. 'Not just an academic question', *New Scientist* (7 July 1983): 2.

Anonymous. 'Publish the advice, says ABRC', *Times Higher Education Supplement* (17 September 1982).

Anonymous. 'Radio-active materials. New plant', *Financial Times* (8 April 1954).

Anonymous. 'Science based society', *The Times* (6 July 1983).

Anonymous. 'Science chiefs want more investment in new projects', *Financial Times* (28 October 1982).

Anonymous. 'Scientists in the market–place', *Financial Times* (19 November 1981).
Anonymous. 'Squeeze on science budget', *The Times* (12 February 1982).
Anonymous. 'Star Wars', *The Economist* (3 March 1984): 80–1.
Anonymous. '"Star Wars" condemned by Kinnock at NATO', *The Times* (7 March 1985).
Anonymous. 'US space shuttle to test laser weapons', *The Times* (7 January 1981).
Anonymous. 'US/UK officials hit snags on SDI co–operation', *Washington Post* (26 October 1985).
'Ariadne' column, *New Scientist* (25 September 1980): 976.
Baker, David. 'The making of Star Wars', *New Scientist* (9 July 1987): 36–40.
Baucom, Donald R. *The Origins of SDI, 1944–1983*. Lawrence: University of Kansas Press, 1992.
Bethe, Hans, Richard L. Garwin, Kurt Gottfried and Henry W. Kendall. 'Space–based ballistic–missile defense', *Scientific American* 251 (October 1984): 39–49.
Black, Douglas (chair). *Investigation of the Possible Increased Incidence of Cancer in West Cumbria. Report of the Independent Advisory Group*. London: HMSO, 1984.
Blowers, Andrew. 'Transition or transformation? Environmental policy under Thatcher', *Public Administration* 65 (1987): 277–94.
Blundell, John, ed. *Remembering Margaret Thatcher: Commemorations, Tributes and Assessments*. New York: Algora, 2013.
Boden, Rebecca, Deborah Cox and Maria Nedeva. 'The appliance of science? New public management and strategic change', *Technology Analysis and Strategic Management* 18 (2006): 125–41.
Boehmer–Christiansen, Sonja. 'Britain and the International Panel on Climate Change: the impacts of scientific advice on global warming. Part I: Integrated policy advice and the global dimension', *Environmental Politics* 4 (1995): 1–18.
Boehmer–Christiansen, Sonja. 'Britain and the International Panel on Climate Change: the impacts of scientific advice on global warming. Part II: The domestic story of the British response to climate change', *Environmental Politics* 4 (1995): 175–96.
British Library. National Life Story oral history of Lord Wakeham. British Library's Oral History of the Electricity Supply in the UK project. https://sounds.bl.uk/Oral–history/Industry–water–steel–and–energy/021M–C1495X0022XX–0001V0 .
Broad, William J. *Teller's War: the Top–secret Story behind the Star Wars Deception*. New York: Simon & Schuster, 1992.
Brooks, Harvey. 'The scientific advisor', in R. Gilpin and C. Wright, eds, *Scientists and National Policy–making*. New York: Columbia University Press, 1964, pp.73–96.
Brundtland, Gro Harlem. *Madam Prime Minister. A Life in Power and Politics*. New York: Farrar, Straus and Giroux, 2002.
Brzezinski, Zbigniew, Robert Jastrow, Max Kampelman. 'Search for security: the case for SDI', *International Herald Tribune* (28 January 1985).
Bundy, McGeorge, George F. Kennan, Robert S. McNamara and Gerard Smith. 'The President's choice: Star Wars or arms control', *Foreign Affairs* 63 (1984): 264–78.
Butler, Stuart A. 'National prestige and in(ter)dependence: British space research policy 1959–73', PhD thesis, University of Manchester, 2016.
Cabinet Office. *Biotechnology*, Cmnd 8177. London: HMSO, 1980.
Cabinet Office. *The Organisation and Management of Government R&D* (Rothschild Report), published alongside Council for Scientific Policy (CSP), *The Future of the Research Council System* (Dainton Report), in the Cabinet Office Green Paper, *A Framework for Government Research and Development*, Cmnd 4814. London: HMSO, 1971.
Cabinet Office. *Realising Our Potential: a Strategy for Science, Engineering and Technology*, Cm. 2250. London: HMSO, 1993.
Cabinet Office. *Review of the Framework for Government Research and Development (Cmnd 5046)*, Cmnd 7499. London: HMSO, 1979.
Cabinet Office. *Review of the Scientific Civil Service*, report of a working group of the Management Committee for the Science Group (CSD). London: HMSO, 1980.
Callahan, Michael and Kosta Tsipis. 'High Energy Laser Weapons: a Technical Assessment', Program in Science and Technology for International Security report no.6. Cambridge, MA: MIT Press, 1980.
Calver, Neil and Miles Parker. 'The logic of scientific unity? Medawar, the Royal Society and the Rothschild controversy 1971–72', *Notes and Records of the Royal Society* 70 (2016): 83–100.
Campbell, John. *Margaret Thatcher. Volume 1: The Grocer's Daughter*. London: Jonathan Cape, 2000.
Campbell, John. *Margaret Thatcher. Volume 2: The Iron Lady*. London: Jonathan Cape, 2003.
Campbell–Kelly, Martin. 'Sir Maurice Vincent Wilkes. 26 June 1913–29 November 2010', *Biographical Memoirs of Fellows of the Royal Society* 60 (2014): 433–54.

Carter, Neil. *The Politics of the Environment: Ideas, Activism, Policy*. Cambridge: Cambridge University Press, 2007.

Cass, Loren R. 'The indispensable awkward partner: the United Kingdom in European climate policy', in Paul G. Harris, ed., In *Europe and Global Climate Change: Politics, Foreign Policy and Regional Cooperation*, 63–86. London: Edward Elgar, 2007.

Chester, Peter F. Interviewed by Thomas Lean, *An Oral History of the Electricity Supply Industry in the UK, British Library*, C1495/15.

Chester, Peter F. 'Perspectives on acid rain. The Sir William Jackson Pope Memorial Lecture, Royal Society of Arts, 13 April 1983', *Journal of the Royal Society of Arts* 131 (1983): 587–603.

Christie, Ian. 'Research and development policy: the great debate', *Policy Studies* 8 (1988): 11–22.

Coghlan, Andy. 'Research councils face abolition threat', *New Scientist* (15 July 1989): 24–5.

Collins, Peter. *The Royal Society and the Promotion of Science since 1960*. Cambridge: Cambridge University Press, 2016.

Conservative Party. *The Conservative Manifesto 1979*, 1979.

Conservative Party. *The Conservative Manifesto 1983*, 1983.

Conservative Party. *The Next Moves Forward*, 1987.

Coopey, Richard. 'The white heat of scientific revolution', *Contemporary British History* 5 (1991): 115–27.

Cottrell, Alan. 'Research, risk and reward', the Sir Clavering Fison Lecture, on the 'anomaly': 'If industrial research is good for industry why have we, who have spent so much on research, done so poorly industrially', *Chemistry and Industry* (17 January 1981): 42–6.

Crick, Michael and Adrian Van Klaveren. 'Mrs Thatcher's greatest blunder', *Contemporary Record* 5 (1991): 397–416,

Dalyell, Tam. *A Science Policy for Britain*. London: Longman, 1983.

Danby, Grahame. 'Nuclear privatisation', Research Paper 96/3, House of Commons Library, 1996.

Davies, Hunter, ed. *Sellafield Stories: Life with Britain's First Nuclear Plant*. London: Constable, 2012.

Davies, Stephen. 'Rothschild reversed: Explaining the exceptionalism of biomedical research, 1971–1981', *British Journal for the History of Science*, published as a FirstView article online 28 August 2018. DOI: 10.1017/S0007087418000523.

de Chadarevian, Soraya. *Designs for Life: Molecular Biology after World War II*. Cambridge: Cambridge University Press, 2002.

Department for Education and Science. *The Science Budget: a Forward Look 1982*. London: DES, 1982.

Department of the Environment. *Acid Rain. The Government's Reply to the Fourth Report from the Environment Committee*. London: HMSO, 1984.

Department of the Environment. *This Common Inheritance: Britain's Environmental Strategy*. Cm. 1200. London: HMSO, 1990.

Department of Health and Social Services. *Human Fertilisation and Embryology: a Framework for Legislation*, Cm. 259. London: HMSO, 1987.

Department of Health and Social Services. *Report of the Committee of Inquiry into Human Fertilisation and Embryology* (Warnock Report), Cmnd 9314. London: HMSO, 1984.

Department of Industry. *Research and Development. Report 1981–82*. London: Department of Industry, 1982.

Department of Trade and Industry. *DTI – the Department for Enterprise*, Cm. 278. London: HMSO, 1988.

Department of Trade and Industry. *The Prospects for Nuclear Power in the UK*, Cm. 2860. London: HMSO, 1995.

Departments of the Environment and Transport, *Review of Research on Radioactive Waste Management and Radioactivity in the Environment*, Research Report 32, 1979.

DeYoung, Karen. 'British seek "Star Wars" answers', *Washington Post* (2 July 1985): 9.

Dickson, David. 'California set to cash in on British discovery', *Nature* 279 (1979): 663–4.

Durant, John R., Geoffrey A. Evans and Geoffrey P. Thomas. 'The public understanding of science', *Nature* 340 (1989): 11–14.

Dyson, Freeman J. *Disturbing the Universe*. New York: Harper & Row, 1979.

Edgerton, David. 'Liberal militarism and the British state', *New Left Review* (1991) 185: 138–69.

Edgerton, David. 'The "Haldane Principle" and other invented traditions in science policy', *History and Policy*. http://www.historyandpolicy.org/papers/policy-paper-88.html .

Edgerton, David. 'C. P. Snow as anti-historian of British science: revisiting the technocratic moment, 1959–1964', *History of Science* 43 (2005): 187–208.

Edgerton, David. *Warfare State: Britain, 1920–1970*. Cambridge: Cambridge University Press, 2006.

Edgerton, David and Kirsty Hughes. 'The poverty of science: a critical analysis of scientific and industrial policy under Mrs Thatcher', *Public Administration* 67 (1989): 419–33.

Edwards, Paul. *The Closed World: Computers and the Politics of Discourse in Cold War America*. Cambridge, MA: MIT Press, 1996.

Environmental Protection Act 1990. c.43. London: The Stationery Office, 1990.

Evans, Eric J. *Thatcher and Thatcherism*. London: Routledge, 1997.

Evans, H. John and Anne McLaren. 'Unborn Children (Protection) Bill', *Nature* 314 (14 March 1985): 127–8.

Fagan, Mary. 'ABRC's research plan meets united assault', *New Scientist* (5 November 1987): 19.

Fairclough, John. 'Setting priorities for science and technology', *Midland Bank Review* (Winter 1987): 18–23.

Farman, J. C., B. G. Gardiner and J. D. Shanklin. 'Large losses of total ozone in Antarctica reveal seasonal ClOx/NOx interaction', *Nature* 315 (16 May 1985): 207–10.

Ferry, Georgina. *Dorothy Hodgkin: a Life*. London: Granta Books, 1998.

Fishlock, David. 'Acid rain: UK's electric and coal industries find an impartial judge', *Energy Daily* (8 September 1983): 3.

Fishlock, David. 'Amersham International will come to the market shortly', *Financial Times* (11 January 1982).

Fishlock, Peter. 'Acid rain', *Financial Times* (6 September 1983).

FitzGerald, Frances. *Way Out There in the Blue: Reagan, Star Wars and the End of the Cold War*. New York: Simon & Schuster, 2001.

Flather, Paul. 'Scientists called to Downing St', *The Times* (4 May 1984).

Fontaine, Philippe. 'Blood, politics, and social science. Richard Titmuss and the Institute of Economic Affairs, 1957–1973', *Isis* 93 (2002): 401–34.

Francis, T. J. G. 'A personal view', *Underwater Technology* 10 (1984): 1.

Gamble, Andrew. 'The Thatcher myth', *British Politics* 10 (2015): 3–15.

Garfield, Eugene. 'Shame on you, Mrs Thatcher', *The Scientist* (9 March 1987): 9.

Ghosh, Pallab. 'Science funding: will "picking winners" work?', BBC News website (31 August 2017). https://www.bbc.co.uk/news/science-environment-41101892 .

Giddens, Anthony. *The Nation-State and Violence*. Cambridge: Polity Press, 1985.

Godin, Benoît. 'The linear model of innovation: the historical construction of an analytical framework', *Science, Technology, & Human Values* 31 (2006): 639–67.

Government Response to the Report 'Science and Technology' by the House of Lords Select Committee on Science and Technology, Cmnd 8591. London: HMSO, 1983.

Gowing, Margaret. *Britain and Atomic Energy, 1939–1945*. London: Macmillan, 1964.

Gowing, Margaret, assisted by Lorna Arnold. *Independence and Deterrence: Britain and Atomic Energy, 1945–1952. Volume 1: Policy Making*. London: Macmillan, 1974.

Gowing, Margaret, assisted by Lorna Arnold. *Independence and Deterrence: Britain and Atomic Energy, 1945–1952. Volume 2: Policy Execution*. London: Macmillan, 1974.

Grandin, Karl, Nina Wormbs and Sven Widmalm. *The Science–Industry Nexus: History, Policy, Implications*. New York: Science History Publications, 2004.

Greenhalgh, Geoffrey. 'The Soviet drive to nuclear power', *New Scientist* (8 May 1986): 18.

Greenpeace, 'Margaret's favourite places', 1990.

Gregory, Jane and Simon Jay Lock, 'The evolution of 'Public Understanding of Science': public engagement as a tool of science policy in the UK', *Sociology Compass* 2 (1998): 1252–65.

Gregory, Jane and Steve Miller. *Science in Public: Communication, Culture and Credibility*. New York: Plenum, 1998.

Guise, George. *Inside the Tank*. Epsom: Bretwalda Press, 2015.

Guise, George. 'Margaret Thatcher's influence on British science', *Notes and Records of the Royal Society* 68 (2014): 301–9.

Gummett, Philip. *Scientists in Whitehall*. Manchester: Manchester University Press, 1980.

Hansard. 'Human Fertilisation and Embryology (Warnock Report)' (23 November 1984): 526–90.

Hansard. 'Nuclear Waste (Sellafield)', HC Deb (21 November 1983) vol.49, cc19–22.

Hansard. 'Research and development. ACARD/ABRC Report' (10 February 1984): 1393.

Hansard. 'Unborn Children (Protection) Bill' (15 February 1985): 642–706.

Harford, Barbara and Sarah Hopkins, eds. *Greenham Common: Women at the Wire*. London: Women's Press, 1984.

Health Education Council. *AIDS. What Everybody Needs to Know*. London: Health Education Council, 1986.
Health Education Council. *Some Facts about AIDS. Acquired Immune Deficiency Syndrome*. London: Health Education Council, 1985.
Helm, Dieter. *Energy, the State, and the Market: British Energy Policy since 1979*. Oxford: Oxford University Press, revised edition, 2010.
Henderson, Mark. *The Geek Manifesto: Why Science Matters*. London: Bantam Press, 2012.
Hendry, John. *Innovating for Failure: Government Policy and the Early British Computer Industry*. Cambridge, MA: MIT Press, 1990.
Hennessy, Peter. *Whitehall*. London: Macmillan, 1989.
Herran, Néstor. 'Spreading nucleonics: the Isotope School at the Atomic Energy Research Establishment, 1951–67', *British Journal for the History of Science* 39 (2006): 569–86.
Hill, John. 'The politics of nuclear power', *Atom* 243 (1977): 1–9.
Holdgate, Martin. *The Green Web: a Union for World Conservation*. London: Earthscan, 1999.
House of Commons Select Committee on Environment, *Report on Acid Rain*, 2 vols. HMSO: London, 1984.
House of Lords Select Committee on Science and Technology, 'Science and government', 1981.
House of Lords Select Committee on Science and Technology. *Session 1988–89 3rd Report* (HL Paper 24). HMSO: London, 1989.
Hull, Andrew. 'War of words: the public science of the British scientific community and the origins of the Department of Scientific and Industrial Research, 1914–16', *British Journal for the History of Science* 42 (1999): 461–81.
Ince, Martin (and Jon Turney). *The Politics of British Science*. Brighton: Wheatsheaf Books, 1986.
Irwin, Alan. *Citizen Science: A Study of People, Expertise and Sustainable Development*. London: Routledge, 1995.
Jasanoff, Sheila. 'Civilization and madness: the great BSE scare of 1996', *Public Understanding of Science* 6 (1997): 221–32.
Judd, A. M. and K. F. Ainsworth, 'Fast reactors in the U.K. 1946–1996', *Energy* 23 (1998): 609–17.
Junor, Penny. *Margaret Thatcher: Wife, Mother, Politician*. New York: Henry Holt, 1983.
Kandiah, Michael and Gillian Staerck. *The British Response to SDI, Witness Seminar held in the Chancellor's Hall, Senate House, University of London, 9 July 2003*. London: Centre for Contemporary British History, 2005.
Krohn, P. L. 'Solly Zuckerman Baron Zuckerman, of Burnham Thorpe, O.M., K.C.B. 30 May 1904–1 April 1993', *Biographical Memoirs of Fellows of the Royal Society* 41 (1995): 576–98.
Labour Party. *Britain Will Win with Labour*, 1987.
Labour Party. *The Labour Way is the Better Way*, 1979.
Labour Party. *The New Hope for Britain*, 1983.
Large, Peter. 'Thatcher "should spend £30m to bridge college–industry gap"', *Guardian* (29 June 1983): 5.
Lean, Tom. *Electronic Dreams: How Britain Learned to Love the Computer*. London: Bloomsbury, 2016.
Lécuyer, Christophe. *Making Silicon Valley: Innovation and the Growth of High Tech, 1930–1970*. Cambridge, MA: MIT Press, 2007.
Lee, Kent D. 'The role of scientific advisers in the Strategic Defense Initiative', *Technology in Society* 8 (1986): 291–8.
Lehtonen, Markku and Jenny Lieu. 'The rise and fall of the fast breeder reactor technology in the UK: between engineering "dreams" and economic "realities"?', SPRU Sussex Energy Group report, 2011.
Leslie, Stuart W. *The Cold War and American Science: the Military–Industrial–Academic Complex at MIT and Stanford*. New York: Columbia University Press, 1994.
Lewis, Russell. *Margaret Thatcher: a Personal and Political Biography*. London: Routledge & Kegan Paul, 1975.
Liberal Party. *The Real Fight is for Britain*, 1979.
Liberal–SDP Alliance. *Britain United: the Time Has Come*, 1987.
Liberal–SDP Alliance. *Working Together for Britain*, 1983.
Ling, K. A., M. R. Ashmore and R. B. Macrory. 'The use of world-based models to describe the development of UK acid rain policy in the 1980s', *Environmental Science & Policy* 3 (2000): 249–62.
Lloyd, John. 'The green light', *Sunday Times Magazine* (26 February 1989): 49–50.

Lock, Simon J. 'Lost in translations: discourses, boundaries and legitimacy in the public understanding of science in the UK', PhD thesis, University College London, 2008.
Lukes, Steven, ed. *Power*. Oxford: Basil Blackwell, 1986.
Macgill, S. M. *The Politics of Anxiety: Sellafield's Cancer–link Controversy*. London: Pion, 1987.
Maddox, John. 'Mrs Thatcher asks for decisiveness', *Nature* 316 (18 July 1985): 178.
Marks, Lara V. *The Lock and Key of Medicine: Monoclonal Antibodies and the Transformation of Healthcare*. New Haven: Yale University Press, 2015.
Martin, Ben and Richard Whitley. 'The UK Research Assessment Exercise', in Richard Whitley, Jochen Gläser and Lars Engwall, eds, *Reconfiguring Knowledge Production,* 51–80. Oxford: Oxford University Press, 2010.
Martin, M. A., revised by J. Rae. 'A Brief History of the Research Stations and Refuges of the British Antarctic Survey and its Predecessors', 2016. https://www.bas.ac.uk/wp–content/uploads/2015/03/British–Antarctic–Stations–Refuges–v6.2–2016.pdf.
McCarthy, Michael. 'Thatcher's "black record" on green issues', *The Times* (1 March 1989): 2.
McCormick, John. *British Politics and the Environment*. London: Routledge, 1991.
McCray, W. Patrick. *The Visioneers: How a Group of Elite Scientists Pursued Space Colonies, Nanotechnologies, and a Limitless Future*. Princeton: Princeton University Press, 2013.
Miller, David J. 'A quasi–political Explanation of the Higgs Boson; for Mr Waldegrave, UK Science Minister 1993'. http://www.hep.ucl.ac.uk/~djm/higgsa.html.
Ministry of Supply. 'The Radiochemical Centre. Amersham', 1949.
Minnion, John and Philip Bolsover, eds. *The CND Story*. London: Allison & Busby, 1983.
Mitchell, Gordon R. *Strategic Deception: Rhetoric, Science, and Politics in Missile Defense Advocacy*. East Lansing: Michigan State University Press, 2000.
Moncur, Andrew. '£5m research on ways to curb pollution and acid rain', *Guardian* (6 September 1983).
Mooney, Graham. 'Public health versus private practice: the contested development of compulsory infectious disease notification in late–nineteenth–century Britain', *Bulletin of the History of Medicine* 73 (1999): 238–67.
Moore, Charles. *Margaret Thatcher. The Authorized Biography. Volume One: Not for Turning*. London: Allen Lane, 2013.
Moore, Charles. *Margaret Thatcher. The Authorized Biography. Volume Two: Everything She Wants*, London: Allen Lane, 2015.
Morris, Norma. 'Authority relations as condition for, and outcome of, shifts in governance: the limited impact of the UK Research Assessment Exercise on the biosciences', in Richard Whitley, Jochen Gläser and Lars Engwall, eds, *Reconfiguring Knowledge Production,* 239–65. Oxford: Oxford University Press, 2010.
Mulvey, John H. 'Can we save British science?', *Laboratory Practice* (February 1986), 5.
Myelnikov, Dmitriy. 'Cuts and the cutting edge: British science funding and the making of animal biotechnology in 1980s Edinburgh', *British Journal for the History of Science* 50 (2017): 701–28.
National Blood Transfusion Service. *AIDS. Important New Advice for Blood Donors*. London: National Blood Transfusion Service, 1985.
Nehring, Holger. 'The British response to SDI: introductory paper', in Michael Kandiah and Gillian Staerck, eds, *The British Response to SDI, Witness Seminar held in the Chancellor's Hall, Senate House, University of London, 9 July 2003*, 17–24. London: Centre for Contemporary British History, 2005.
Nehring, Holger. 'Cold War, apocalypse and peaceful atoms. Interpretations of nuclear energy in the British and West German anti–nuclear weapons movements, 1955–1964', *Historical Social Research* 29 (2004): 150–70.
Noble, Denis. 'We are still saving British science from Margaret Thatcher', *Nature* (17 April 2013). https://www.nature.com/news/we–are–still–saving–british–science–from–margaret–thatcher–1.12800.
Nuclear Power and the Environment. The Government Response to the Sixth Report of the Royal Commission on Environmental Protection, Cmnd 6820. London: HMSO, May 1977.
Oldham, Geoffrey. 'The future of university research in Britain', *Papers in Science, Technology and Public Policy* no.2. London: Imperial College/SPRU, 1982.
Oreskes, Naomi and Erik M. Conway. *Merchants of Doubt: How a Handful of Scientists Obscured the Truth on Issues from Tobacco Smoke to Global Warming*. New York: Bloomsbury Press, 2010.
Ortolano, Guy. *The Two Cultures Controversy: Science, Literature and Cultural Politics in Postwar Britain*. Cambridge: Cambridge University Press, 2009.

Owen, Geoffrey and Michael M. Hopkins. *Science, the State and the City: Britain's Struggle to Succeed in Biotechnology*. Oxford: Oxford University Press, 2016.

Owens, Susan. *Knowledge, Policy, and Expertise: the UK Royal Commission on Environmental Pollution 1970–2011*. Oxford: Oxford University Press, 2015.

Parker, David. *The Official History of Privatisation. Volume I: the Formative Years 1970–1987*. London: Routledge, 2009.

Parker, David. *The Official History of Privatisation. Volume II: Popular Capitalism, 1987–1997*. London: Routledge, 2012.

Parker, Mike and John Surrey. 'Contrasting British policies for coal and nuclear power, 1979–1992', *Energy Policy* 23 (1995): 821–50.

Parker, Miles. 'The Rothschild report (1971) and the purpose of government-funded R&D – a personal account', *Palgrave Communications* 2 (2016). Online only, paper reference 16053. DOI: 10.1057/palcomms.2016.53.

Parsons, Talcott. 'On the concept of political power', *Proceedings of the American Philosophical Society* 107 (1963): 232–62.

Patterson, Walt. 'Thatcher's failed romance with nuclear power', *Bulletin of the Atomic Scientists* 46 (1990): 9–12.

Patterson, Walter C. *Going Critical: an Unofficial History of British Nuclear Power*. London: Paladin Books, 1985.

Pearce, David, Anil Markandya and Edward B. Barbier. *Blueprint for a Green Economy, Report for the Department of the Environment*. London Environmental Economics Centre. London: Earthscan, 1989.

Pearce, Fred. 'Acid rain: new fears prompted cleanup', *New Scientist* (18 September 1986): 22–3.

Pearce, Fred. 'Warning cones hoisted as acid rainclouds gather', *New Scientist* (24 June 1982): 70.

Pearce, Fred. *With Speed and Violence: Why Scientists Fear Tipping Points in Climate Change*. Boston: Beacon Press, 2007.

Perutz, Max. 'How to stifle research', *New Scientist* (10 December 1987): 57.

Peterson, John. 'EUREKA: a historical perspective', in John Krige and Luca Guzzett, eds, *History of European Scientific and Technological Cooperation*, 323–45. Luxembourg: Office for Official Publications of the European Communities, 1997.

Phillips, Richard. 'The impact agenda and geographies of curiosity', *Transactions of the Institute of British Geographers* 35 (2010): 447–52.

Plokhy, Serhii. *Chernobyl: History of a Tragedy*. London: Allen Lane, 2018.

Prowse, Michael. 'Why bigger many not be better', *Financial Times* (21 December 1988).

Read, Nicholas. 'The "near market" concept applied to UK agricultural research', *Science and Public Policy* 16 (1989): 233–8.

Reagan, Ronald. 'Address to the Nation on Defense and National Security', 23 March 1983. Copy of text of speech at http://www.reagan.utexas.edu/archives/speeches/1983/32383d.htm.

Royal Commission on Environmental Pollution. 10th report, *Tackling Pollution – Experience and Prospects*, Cmnd 9149. London: HMSO, 1984.

Royal Commission on Environmental Pollution. 13th report, *The Release of Genetically Engineered Organisms to the Environment*. London: HMSO, 1989.

Royal Commission on Environmental Protection. 6th report, *Nuclear Power and the Environment*, Cmnd 6618. London: HMSO, 1976.

Royal Society, *The Public Understanding of Science*. London: Royal Society, 1985.

Rutherford, Malcolm. 'Who is Robin Nicholson?', *Financial Times* (5 March 1982).

Salam, Abdus. 'Particle physics: will Britain kill its own creation', *New Scientist* (3 January 1985): 16–18.

Schrafstetter, Susanna and Stephen Twigge. 'Spinning into Europe: Britain, West Germany and the Netherlands – Uranium Enrichment and the Development of the Gas Centrifuge 1964–1970', *Contemporary European History* 11 (2002): 253–72.

Scott, Peter. *National AIDS Manual*. London: NAM Publications Ltd, 1st edition November 1988, 1st update April 1989.

Sharp, Paul. *Thatcher's Diplomacy: the Revival of British Foreign Policy*. London: Macmillan, 1997.

Sheail, John. 'Torrey Canyon: the political dimension', *Journal of Contemporary History* 42 (2007): 485–504.

Sinha, Uttam Kumar. 'Climate change and foreign policy: the UK case', *Strategic Analysis* 34 (2010): 397–408.

Slayton, Rebecca. *Arguments that Count: Physics, Computing, and Missile Defense, 1949–2012*. Cambridge, MA: MIT Press, 2013.

Smallman, R. E. and J. F. Knott. 'Sir Alan Cottrell FREng. 17 July 1919–15 February 2012', *Biographical Memoirs of Fellows of the Royal Society* 59 (2013): 93–124.

Smith, David C., P. M. D. Collins, D. M. Hicks and S. Wyatt. 'National performance in basic research', *Nature* (23 October 1986): 681–4.

Smith, Robert Angus. *Air and Rain: the Beginnings of a Chemical Climatology*. London: Longmans, Green and Co., 1872.

Stocker, Jeremy. *Britain & Ballistic Missile Defence, 1942–2002*. London: Frank Cass, 2004.

Sutcliffe-Braithwaite, Florence. 'Neo-liberalism and morality in the making of Thatcherite social policy', *Historical Journal* 55 (2012): 497–520.

Taylor, Peter. 'The state as container: territoriality in the modern world-system', *Progress in Human Geography* 18 (1994): 151–62.

Thatcher, Margaret. *The Downing Street Years*. London: HarperCollins, 1993.

Thatcher, Margaret. *The Path to Power*. London: HarperCollins, 1995.

Thatcher, Margaret. Speech opening Hadley Centre for Climate Prediction and Research, 25 May 1990. https://www.margaretthatcher.org/document/108102.

Thatcher, Margaret. Speech to Ozone Layer Conference, International Maritime Organisation, Albert Embankment, London, 27 June 1990. Copy at https://www.margaretthatcher.org/document/108133.

Thatcher, Margaret. 'Speech to the Royal Society', 27 September 1988. Copy of the text (as prepared rather than as spoken) can be found in the online archives of the Margaret Thatcher Foundation. http://www.margaretthatcher.org/document/107346.

Thatcher, Margaret. Speech to United Nations General Assembly (Global Environment), 8 November 1989. Copy at https://www.margaretthatcher.org/document/107817.

Tickell, Crispin. Interview with Sir Crispin Tickell by Malcolm McBain, 28 January 1999, *British Diplomatic Oral History Programme (BDOHP)*. https://www.chu.cam.ac.uk/media/uploads/files/Tickell.pdf.

Titmuss, Richard. *The Gift Relationship: from Human Blood to Social Policy*. London: Allen & Unwin, 1970.

Tombs, Francis. *Power Politics: Political Encounters in Industry and Engineering*. London: I. B. Tauris, 2011.

Tsipis, Kosta. 'Laser weapons', *Scientific American* 245 (December 1981): 35–41.

Turner, Frank M. 'Public science in Britain, 1880–1919', *Isis* 71 (1980): 589–608.

Turney, Jon. 'Science research "in danger"', *Times Higher Education Supplement* (3 September 1982).

Twigge, Stephen. 'A baffling experience: technology transfer, Anglo-American nuclear relations, and the development of the gas centrifuge 1964–70', *History and Technology* 19 (2010): 151–63.

United Nations Environment Programme. *The Greenhouse Gases*. Nairobi: UNEP, 1987.

Vinen, Richard. *Thatcher's Britain: the Politics and Social Upheaval of the Thatcher Era*. London: Simon and Schuster, 2009.

von Tunzelmann, Nick. 'Technology in post-war Britain', in R. Floud and P. Johnson, eds, *The Cambridge Economic History of Modern Britain, Volume III: Structural Change and Growth, 1939–2000*, 299–331. Cambridge University Press, Cambridge, 2004.

Walker, John R. *Britain and Disarmament: the UK and Nuclear, Biological and Chemical Weapons Arms Control and Programmes 1956–1975*. Farnham: Ashgate, 2012.

Ward, Stephen. 'Charity set to fund sex survey vetoed by PM', *Independent* (13 October 1989).

Warnock, Mary. *A Memoir: People and Places*. London: Duckworth, 2000.

Werskey, Gary. *The Visible College*. London : Allen Lane, 1978.

Wilkie, Tom. 'The Thatcher effect in science', in Dennis Kavanagh and Anthony Seldon, eds, *The Thatcher Effect*. Oxford: Clarendon Press, 1989.

Wilkie, Tom. *British Science and Politics since 1945*. Oxford: Blackwell, 1991.

Wilkie, Tom and Roger Milne. 'Chernobyl: sorting fact from fiction', *New Scientist* (8 May 1986): 17.

Wilks, Stephen and Michelle Cini, 'The redirection of science and technology policy under the Thatcher governments', *Policy Money & Management* 11 (1991): 49–56.

Willetts, David. 'The role of the Prime Minister's Policy Unit', *Public Administration* 65 (1987): 443–54.

Williams, Roger. 'UK science and technology: policy, controversy and advice', *The Political Quarterly* 59 (1988): 132–44.

Williamson, W. H. and K. C. Brown, 'The analysis and modelling of British invasions', *Philosophical Transactions of the Royal Society*, Series 8 314 (1986): 506–22.

Wilson, Duncan. *The Making of British Bioethics*. Manchester: Manchester University Press, 2014.
Wilson, Harold. *The Relevance of British Socialism*. London: Weidenfeld and Nicolson, 1964.
Winskel, Mark. 'Autonomy's end: nuclear power and the privatization of the British electricity supply industry', *Social Studies of Science* 32 (2002): 439–67.
Wittner, Lawrence S. 'The forgotten years of the world nuclear disarmament movement, 1975–78', *Journal of Peace Research* 40 (2003): 435–56.
Wright, Pearce. 'Cabinet needs a chief scientific adviser', *The Times* (2 December 1975): 3.
Wright, Pearce. 'Sir John Fairclough', *Guardian* (11 June 2003): 23.
Wright, Susan. *Molecular Politics: Developing American and British Regulatory Policy for Genetic Engineering, 1972–1982*. Chicago: University of Chicago Press, 1994.
Wright, Susan. 'Recombinant DNA technology and its social transformation, 1972–1982', *Osiris*, 2nd Series, 2 (1986): 303–60.
Wynne, Brian. *Rationality and Ritual: the Windscale Inquiry and Nuclear Decisions in Britain*. Chalfont St Giles: British Society for the History of Science, 1982.
Yarchoan, Robert, et al, 'Administration of 3'–azido–3'–deoxythymidine, an inhibitor of HTLV–III/LAV replication, to patients with AIDS or AIDS–related complex', *The Lancet* 327 (15 March 1986): 575–80.
Young, Hugo. *One of Us: a Biography of Margaret Thatcher*. London: Macmillan, 1989.
Zuckerman, Solly. *From Apes to Warlords: the Autobiography (1904–1946) of Solly Zuckerman*. London: Hamilton, 1978.
Zuckerman, Solly. *Monkeys, Men and Missiles: an Autobiography, 1946–88*. London: Collins, 1988.

Index

www.ingramcontent.com/pod-product-compliance
Lightning Source LLC
LaVergne TN
LVHW010731020926
846742LV00029B/38